AF823638

DESIGN AND DEVELOPMENT OF

Fiber Optic Gyroscopes

DESIGN AND DEVELOPMENT OF

Fiber Optic Gyroscopes

Eric Udd and **Michel Digonnet**
Editors

SPIE PRESS
Bellingham, Washington USA

Library of Congress Cataloging-in-Publication Data

Names: Udd, Eric, editor. | Digonnet, Michel J. F., editor.
Title: Design and development of fiber optic gyroscopes / Eric Udd and Michel Digonnet, editors.
Description: Bellingham, Washington, USA : SPIE Press, [2019] | Includes bibliographical references and index.
Identifiers: LCCN 2018054142 (print) | LCCN 2018060316 (ebook) | ISBN 9781510626102 (pdf) | ISBN 9781510626119 (epub) | ISBN 9781510626126 (mobi) | ISBN 9781510626096 | ISBN 9781510626096 (softcover) | ISBN 1510626093 (softcover)
Subjects: LCSH: Optical gyroscopes. | Fiber optics.
Classification: Classification: LCC TL589.2.O6 (ebook) | LCC TL589.2.O6 D47 2019 (print) | DDC 681/.753–dc23
LC record available at https://lccn.loc.gov/2018054142

Published by

SPIE
P.O. Box 10
Bellingham, Washington 98227-0010 USA
Phone: +1 360.676.3290
Fax: +1 360.647.1445
Email: books@spie.org
Web: http://spie.org

Printed in the United States of America.
First Printing.
For updates to this book, visit http://spie.org and type "PM303" in the search field.

SPIE.

Table of Contents

Preface

In the early years of aviation, guidance was provided by mechanical gyros based on spinning wheels or disks. According to the conservation of angular momentum, the orientation of the spinning object axis is unaffected by tilting or rotation of the support on which the spinning object is mounted. The spinning top therefore defines a direction in space that is used as a reference. By the end of the 1930s, the performance of mechanical gyros had improved considerably, and their use was widespread in commercial and military aircraft. World War II resulted in the mass production of mechanical gyros on an unprecedented scale, with increased accuracy and resolution. In subsequent years, the boom in commercial air travel and military requirements for improved aviation significantly expanded the marketplace for inertial navigation systems based on these gyroscopes.

The implementation of navigation systems for aerospace platforms remained an important issue as mechanical gyros were responsible for nearly 50% of aircraft departure delays. Thus, the demonstration of the ring laser gyro shortly after the invention of the laser became an area of extreme interest for both military and commercial aviation. The US Department of Defense spent hundreds of millions of dollars to support research and development, followed by funds to support the establishment of manufacturing lines at US companies in the 1960s and 1970s. These efforts led to the introduction of ring laser gyro systems onto military and commercial aerospace platforms in the late 1970 and early 1980s.

In the 1970s, the fabrication of the first low-loss single-mode optical fiber occurred at Corning. Shortly thereafter, Dr. Victor Vali and Professor Richard Shorthill at the University of Utah constructed and operated the first open-loop fiber optic gyro. Their idea was simple: construct a Sagnac interferometer with a multi-turn fiber coil, which increases the total area subtended by the coil in proportion to the number of turns and enhances the Sagnac phase shift by the same ratio. This opened up the possibility of moving away from the severe requirement associated with manufacturing ring laser gyros in ultra-clean environments with ultra-pure gases, very-low-expansion-coefficient ceramics, and very-low-backscatter mirrors.

An immediate issue with the fiber optic gyro involved the need for eight orders of magnitude of dynamic range for the navigation of aircraft and extreme linearity. The open-loop fiber optic gyro at the time seemed capable of a dynamic range of three or four orders of magnitude with sufficient levels of linearity. The solution introduced by Cahill and Udd at McDonnell Douglas Astronautics Company used a closed-loop fiber gyro approach that solved in principle the dynamic range and linearity issue with performance and underlying equations similar to those of the ring laser gyro. Like in other closed-loop systems, the output signal of the gyro, which is proportional to the rotation rate, is fed back to the phase modulator in the Sagnac loop in order to cancel the output signal. The readout of the gyro is then the feedback voltage applied to the modulator, which is also proportional to the rotation rate. The benefits of the closed-loop gyro stem from the fact that the output always equals or is very near zero, no matter how large the rotation rate is, up to a very large value imposed mostly by the large voltage dynamic range of the feedback circuit. The dynamic range is therefore greatly increased, and its linearity is excellent because the signal never deviates far from zero. This solution offered the potential for an all-solid-state rotation sensor with a lower overall cost.

Realizing the potential of the fiber optic gyro, like the ring laser gyro, has been a long and expensive process. Many researchers have made important enabling contributions, and many more engineers have worked diligently for many years on solving the problems associated with realizing viable inertial navigation and guidance at affordable costs. This book contains contributions from key engineers and scientists who have worked from as early as 1977 to the present on manufacturing high-performance fiber gyros for many applications.

In this book, Eric Udd provides a chapter that overviews early work on developing open-loop and closed-loop fiber gyros at McDonnell Douglas. These efforts resulted in the first solid-state fiber optic gyros and were highly directed toward demonstrating feasibility for a range of aerospace and oil and gas applications. In parallel, Professor John Shaw at Stanford University obtained funding from Litton Guidance and Control that fueled many successful years of research to improve the performance of fiber optic gyros. In particular, his research group pioneered a series of novel all-fiber components in its early years—especially fiber couplers with extremely low loss and backscattering, and a fiber polarizer with an exceedingly high extinction ratio—that were implemented to eliminate the bulk components used in McDonnell Douglas early prototypes and produced gyros with record-breaking rotation sensitivities. Many of Professor Shaw's graduate students went on to make major contributions to fiber optic gyro technology, including Hervé Arditty and Hervé Lefèvre (at Thomson CSF, then Photonetics, and now IxBlue), George Pavlath (at Litton Guidance and Control, now

Northrop Grumman), Ralph Bergh (who has founded and operated a series of companies supporting fiber gyros), and Michel Digonnet, who succeeded Professor Shaw at the Edward L. Ginzton Laboratory at Stanford.

Several people from the Stanford group have contributed chapters to this book. Hervé Lefèvre provides a "potpourri of fortunate events" that serves as a broad overview of the history and fundamental physics of the fiber optic gyroscope, and the events that turned out just right for fiber optic gyros. With Hervé Arditty, Hervé Lefèvre promoted the "minimum configuration" fiber optic gyro, i.e., the configuration that comprises the minimum number of components required to enforce reciprocity, a key property that was ultimately instrumental in the remarkable overall performance of the fiber optic gyroscope. These components were eventually implemented in an integrated-optic chip fabricated in lithium niobate, a technology that was also critical to the gyroscope's success. These insights, as well as the early development of effective phase modulation techniques, were among the key contributions they both made to fiber optic gyro technology. George Pavlath of Northrop Grumman overviews the state of the art of closed-loop fiber optic gyros and their applications. In the early 1980s, Litton Guidance and Control selected him to lead their fiber optic gyro program, and over the decades he has guided that group to many important achievements, including the implementation of fiber optic gyros on major aerospace platforms. Most notably, Litton Guidance and Control provided the compact closed-loop fiber gyros that navigated all of the Mars rovers, including Spirit, Opportunity, and Curiosity. Pavlath's chapter outlines the achievements of Litton Guidance and Control and Northrop Grumman. Ralph Bergh's chapter outlines a recently improved signal-processing approach for optimizing the closed-loop fiber gyro operation. The work at the Ginzton Laboratory that Professor Shaw started continues under the direction of Professor Michel Digonnet. The chapter by Digonnet and his former graduate student Dr. Jacob Chamoun describes some of the latest efforts toward interrogating the fiber gyro with a coherent light source, instead of the conventional broadband light source, in order to produce the next generation of fiber gyros with improved scale-factor stability and reduced noise.

In the late 1970s, Professor Shaoul Ezekial at MIT demonstrated a different type of optical rotation sensor: the passive ring resonator. With James Davies, he later independently demonstrated a closed-loop fiber optic gyro similar to that of McDonnell Douglas. One of his students, Glen Sanders, joined Honeywell in Minneapolis in 1983. Honeywell was a leader in ring laser gyros but initiated research efforts in fiber optic gyros and resonant fiber optic gyros in the mid-1980s. This position increased in October 1986 when Honeywell acquired Sperry and their active fiber gyro program in Phoenix. Glen Sanders joined the Phoenix group in the late 1980s and became a leader of the fiber gyro program there. He was joined by key co-developer

Lee Strandjord and, later, by Steve Sanders in 1998. They continued to develop fiber optic gyros, particularly for high-performance applications, and they have demonstrated state-of-the-art approaches in RFOGs. They, and other Honeywell co-authors, summarize the history and status of this work in their chapter.

Also in the early 1980s, Richard Dyott of Andrew Corporation led his group in developing D-shaped optical fiber with an elliptical core. The D shape enabled the fabrication of fiber polarizers and polarization-preserving optical fibers. Andrew Corporation made satellite dishes, and their focus was on stabilizing these units. KVH Industries, Inc. acquired the fiber gyro capabilities of Andrew Corporation and improved the linearity and range of the open-loop fiber gyro. The result has been successful at producing units for the middle range of the rotation sensor market. Jay Napoli of KVH outlines the state of the art of these developments in his chapter.

Other companies continue to enter the fiber gyro marketplace as key patents have expired and new methods for enhanced performance are developed. The chapter by Al Cielo Inertial Solutions, Ltd provides an example of this type of company.

One of the keys to success of the fiber optic gyro are components and associated packaging that meet stringent requirements to reduce error sources. Examples of these components include polarization-maintaining optical fiber with thin coatings suitable for winding, polarizing optical fiber packaged for maximum and stable extinction ratios, and fiber couplers. Overall, the properties of polarization-maintaining fibers, fiber polarizers, and fiber couplers have enabled reductions in the fiber gyro bias drift by many orders of magnitude. Chris Emslie describes the specialty optical fibers and components that have played a significant role in fiber gyro development, and offers examples produced by the University of Southampton, Fibercore, and other key players.

In a fiber optic gyro, the configuration and packaging of the fiber coil is particularly important to reduce the errors induced by temperature variations, acoustic waves, and strains, as required to achieve high performance. Steve Yao at General Photonics offers a close look at quadrupole fiber-coil windings and the associated test procedures that are used to meet this goal.

The last chapter of the book is a personal history of the fiber gyro by Eric Udd. It provides a glimpse of some of the motivations, events, and people associated with the fiber gyro development and its introduction as an important product for many applications from 1977 to the present.

This book arose from efforts to form a special session to commemorate the 40th anniversary of the first hardware demonstration of the fiber gyro in 1976 by Vali and Shorthill. The invited expert papers published in the conference proceedings were extended and new material added in an effort to present both a historical perspective and a more in-depth representation of the existing state of the art. New chapters were prepared that extend the range of topics covered. We would like to thank the contributors to this book for their efforts over more than four decades to convert the dream of high-performance solid-state rotation sensors into reality.

Eric Udd
Columbia Gorge Research, LLC

Michel Digonnet
Stanford University

List of Contributors

Arnon Arbel
Al Cielo Inertial Solutions Ltd

Alejo Arrizon
Honeywell

Ralph A. Bergh
Fibernetics LLC

Jacob N. Chamoun
Stanford University

Michel J. F. Digonnet
Stanford University

Chris Emslie
Fibercore

Waymon Ho
Honeywell

Lisa Koenigsberg
Al Cielo Inertial Solutions Ltd

Neil A. Krueger
Honeywell

Clarence Laskoskie
Honeywell

Hervé C. Lefèvre
iXblue

Derek Mead
Honeywell

Shinji Mitani
Japan Aerospace Exploration Agency

Tadahito Mizutani
Japan Aerospace Exploration Agency

Sorin Mosor
Honeywell

Jay Napoli
KVH

Chellappan Narayanan
Honeywell

Noam Pasternak
Al Cielo Inertial Solutions Ltd

George A. Pavlath
Northrup Grumman

Tiequn Qiu
Honeywell

Meir Rosilio
Al Cielo Inertial Solutions Ltd

Shin-ichiro Sakai
Japan Aerospace Exploration Agency

Mary Salit
Honeywell

Glen A. Sanders
Honeywell

Steven J. Sanders
Honeywell

Marc Smiciklas
Honeywell

Lee K. Strandjord
Honeywell

Eric Udd
Columbia Gorge Research

Wes Williams
Honeywell

Jianfeng Wu
Honeywell

X. Steve Yao
General Photonics

Chapter 1
A Potpourri of Comments about the Fiber Optic Gyro for Its Fortieth Anniversary: How Fascinating It Was and Still Is!

Hervé C. Lefèvre
iXblue

1.1 Introduction

Forty years after its first experimental demonstration by Vali and Shorthill in 1976,[1] the fiber optic gyro (FOG) remains a fascinating device. Reciprocity and serendipity have allowed it to reach a unique performance of long-term bias stability: pure, unaided strapdown inertial navigation with a drift of less than one nautical mile in a month, which corresponds to a rate bias stability of 10 μdeg/h, i.e., 0.3 nanoradian in terms of interferometer phase stability.[2]

This chapter begins with the historical context of the Sagnac effect and proposes that it be renamed, as we shall see, as the Sagnac–*Laue* effect. It then recalls the various points that make the FOG potentially perfect and finishes with a potpourri of comments that should be outlined.

1.2 Historical Context of the Sagnac–Laue Effect

If Huygens proposed in the 17th century a wave theory of light, Newton imposed his views of a corpuscular theory in the early 18th century. It is only in the early 19th century that Young's double-slit experiment reopened the wave theory, knowing that it was not easily admitted: you did not contradict Newton! It required the exceptional quality of the theoretical and experimental work of Fresnel to convince the physics community. However, for the thinking of the time, a wave needed some kind of propagation medium, as for acoustic waves. It was called "luminiferous Aether," and light was seen as propagating at a constant velocity c with respect to this fixed aether.

Even when Maxwell showed in 1864 the electromagnetic nature of light waves, aether was not questioned. It required the famous Michelson and Morley experiment in 1887 to have a clear demonstration that the concept of aether should be revised, which yielded the special theory of relativity in 1905: based on earlier theoretical works of Lorentz, Poincaré, Planck, and Minkowski, Einstein abandoned the concept of aether and stated that light propagates at the same velocity c in any inertial frame of reference in linear translation, despite its own velocity. This revolutionary conceptual leap was very difficult to accept by many, and Sagnac's experiment (Fig. 1.1), the basis of present optical gyroscopes, was actually performed to demonstrate that aether did exist, as clearly stated in the title of the publication:[3] "The luminiferous Aether demonstrated by the effect of the relative wind of Aether in an interferometer in uniform rotation."

It is now understood that Sagnac's experiment, which takes place in vacuum (actually in air, but it can be considered as in vacuum), can be either explained by relativity or aether theory, and it does not allow one to demonstrate which theory is right or wrong since the explanation is evaluated from a "rest" frame.

It has not been well known in our community, but it was explained very clearly by von Laue (Fig. 1.2) as early as 1911,[4] two years before Sagnac's publication.[3] The Sagnac effect should be renamed the Sagnac–Laue effect!

Another important point of aether theory was the hypothesis made by Fresnel in 1818 regarding the drag of aether by matter, which was demonstrated experimentally by Fizeau in 1851.[5] In the "rest" frame, the

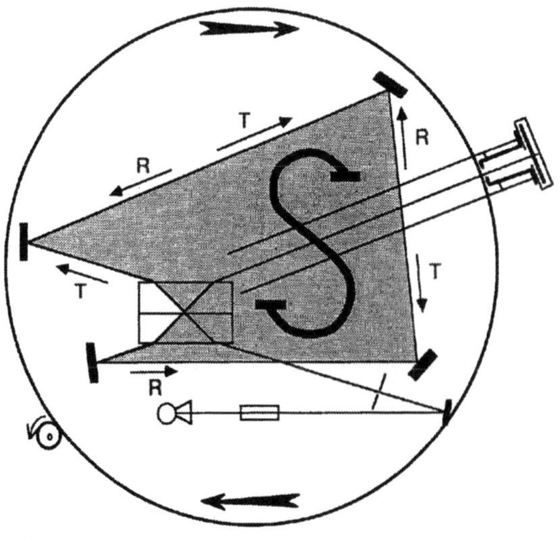

Figure 1.1 Original figure of Sagnac's experiment (S stands for *surface*, i.e., "area" in French).

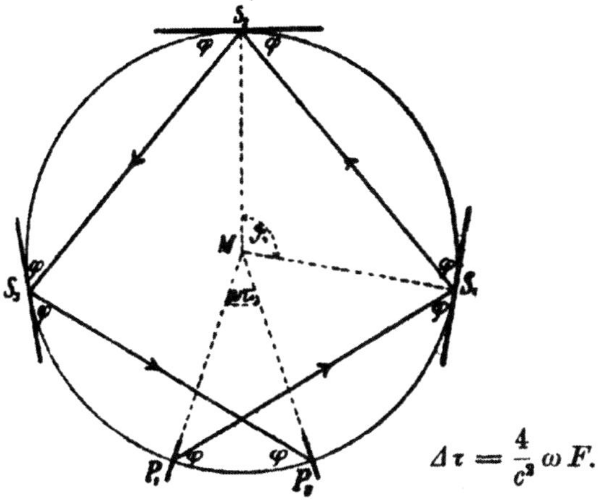

Figure 1.2 Original figure of von Laue's publication of 1911,[4] explaining the Sagnac–Laue effect and giving the formula of the effect (F stands for *Fläche*, i.e., "area" in German).

velocity v of a wave that propagates in a medium with an index of refraction n and that moves at a speed v_m is no longer c/n but

$$v = \frac{c}{n} + \left(1 - \frac{1}{n^2}\right) v_m. \tag{1.1}$$

This Fresnel–Fizeau drag effect was explained by von Laue in 1907[6] as related to the law of the addition of speeds of special relativity:

$$v = \frac{v_1 + v_2}{1 + \frac{v_1 v_2}{c^2}}, \tag{1.2}$$

where v_1 is the speed of an object in a frame moving at v_2 with respect to the "rest" frame, and v is the speed of this object in this "rest" frame. One sees that Eq. (1.2) yields Eq. (1.1) to first order, considering $v_1 = c/n$, the speed of light in the moving medium, $v_2 = v_m$, the speed of the medium with respect to the rest frame, and $v_m << c$. The Fresnel–Fizeau drag effect is actually a *relativistic* effect!

The Sagnac–Laue effect does not depend on the index of refraction n of the propagation medium because of this Fresnel-Fizeau drag effect,[7] as it is clearly experienced in the fiber optic gyro. The path difference looks longer in a medium (Fig. 3), but the velocities are different because of the Fresnel–Fizeau drag of the medium moving with respect to the rest laboratory frame. The speed of the co-rotating wave is increased, and the one of the counter-rotating wave is decreased. If the Sagnac–Laue effect in vacuum can be explained by aether or relativity theory, in a medium, it is related to the Fresnel–Fizeau drag effect and is, then, also a relativistic effect.

Note that Fizeau's experiment that demonstrated the drag effect is usually described in textbooks with a Mach–Zehnder interferometer, whereas it was actually performed with a *reciprocal* ring interferometer (Fig. 1.4). The

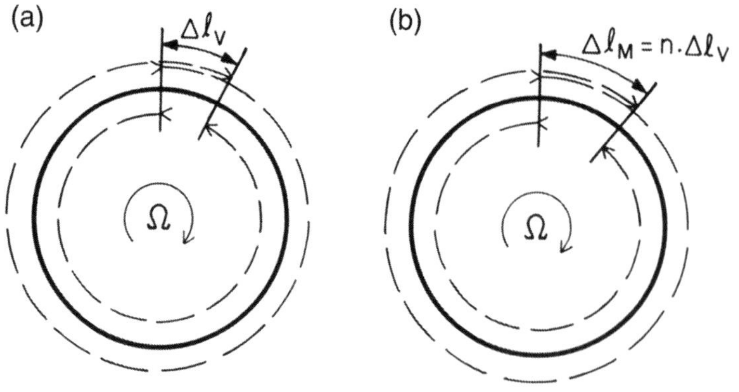

Figure 1.3 (a) Sagnac–Laue effect in vacuum, and (b) Sagnac–Laue effect in a medium: $\Delta\ell_M$ is longer than $\Delta\ell_V$, but this difference is compensated for by the difference of velocities induced by the Fresnel–Fizeau drag.

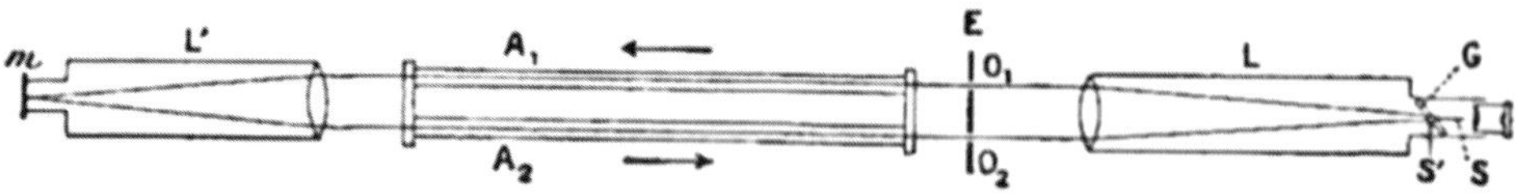

Figure 1.4 Reciprocal ring interferometer used by Fizeau to demonstrate the matter drag effect in 1851: a source S is placed at the focal point of a lens, and the collimated beam goes through two holes (O_1 and O_2) to yield two parallel beams that propagate in the moving water and that are swapped on a cat's eye (on the left) to return "reciprocally."

publication of Fizeau[5] clearly stated (page 352), more than 160 years ago, the importance of what we call today reciprocity, which is a key point for optical gyroscopes: "It is easy to see that, with this set-up, all the points that are on the path of one beam, are also on the one of the other beam, so that a change of density at any point of the path has the same effect on both beams, and then has no influence on the fringe shift." Beautiful!

1.3 Fascinating Serendipity of the Fiber Optic Gyro

1.3.1 The proper frequency

The first experimental demonstration of a fiber gyro[1] showed that a simple fiber ring interferometer is not perfectly reciprocal. Complementary interference fringes were observed at both ports of the interferometer, depending on the alignments of the fiber ends. All of these problems could have been a severe limitation to high performance, but they can be solved very simply with a so-called reciprocal configuration (Fig. 1.5), proposed independently by Ulrich[8] and by Arditty et al.[9] It is sufficient to feed light into the interferometer through a truly single-mode waveguide (single spatial mode and single polarization) and look at the returning interference wave, which is filtered through this same waveguide in the opposite direction. In this case, the alignments are needed solely to optimize the throughput power (and its related

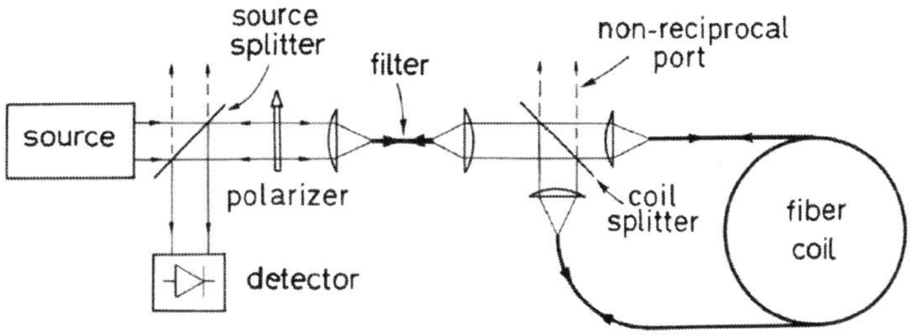

Figure 1.5 Reciprocal configuration with a single-mode filter on the common input–output port.[8,9]

signal-to-noise ratio), which requires difficult but nevertheless reasonable mechanical tolerances. It is now ensured that both returning waves have propagated along exactly the same path in the opposite direction and that they interfere perfectly in phase when the system is at rest. This simple modification has made both opposite paths perfectly identical, zero rotation giving zero phase difference.

The reciprocal configuration provides an interference signal of the Sagnac–Laue effect with perfect contrast since the phases as well as the amplitudes of both counter-propagating waves are perfectly equal at rest. The optical power response is then a raised cosine function of the rotation-induced phase difference $\Delta\phi_R$, and it is maximum at zero. To get high sensitivity, this signal must be biased about an operating point with a nonzero response slope. It is obtained with the use of a *reciprocal* phase modulator placed at one end of the coil. This coil then acts as a delay line (Fig. 1.6).[10] Because of reciprocity, both interfering waves carry exactly the same phase modulation ϕ_m but shifted in time. The delay equals the difference $\Delta\tau_g$ of the group transit time between the long and short paths that connect the modulator and the splitter. This yields a biasing modulation $\Delta\phi_m(t)$ of the phase difference, despite the reciprocity of the phase modulator:

$$\Delta\phi_m(t) = \phi_m(t) - \phi_m(t - \Delta\tau_g). \tag{1.3}$$

The ring interferometer behaves like a perfect delay line filter with a sinusoidal transfer function 2 sin $(\pi \cdot f_m \cdot \Delta\tau_g)$, that is maximum (in absolute

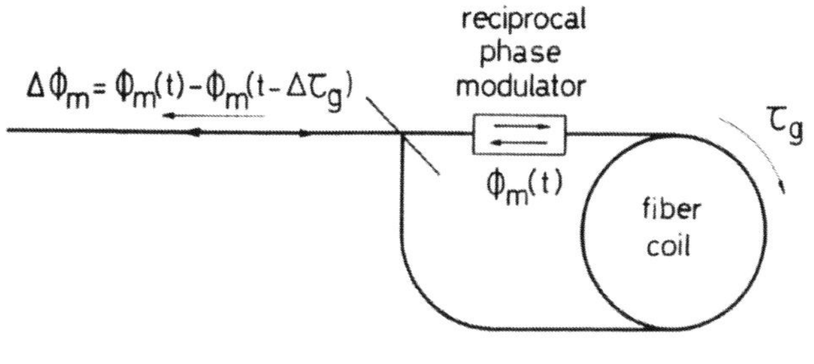

Figure 1.6 Generation of the biasing phase modulation $\Delta\Phi_m$ with a reciprocal modulator, using the delay difference $\Delta\tau_g$ through the fiber coil.

value) at the so-called proper (or *eigen*) frequency f_p, defined as $f_p = 1/(2\Delta\tau_g)$ and its odd harmonics, and that is null at DC and all of the even harmonics. However, a phase modulator is not perfect. It has some spurious harmonic content and parasitic intensity modulation, and the first example of serendipity in the fiber optic gyro is the fact that at the proper frequency f_p, all of these phase modulator defects are completely washed out.[11] The product proper frequency times the coil fiber length is about 100 kHz·km.

1.3.2 Perfection of the digital phase ramp

This biasing modulation–demodulation detection scheme provides very good bias stability since it preserves the reciprocity of the ring interferometer. However, if a high-performance gyroscope must have a stable and low-noise bias, it also requires good accuracy over the whole dynamic range, not only about zero. The measurement of interest is the integrated angle of rotation and not simply the rate. Any past error will affect the future information. This constraint implies the need for an accurate measurement at any rate (i.e., an accurate scale factor). Furthermore, the intrinsic response of an interferometer is sinusoidal, whereas the desired rate signal of a gyroscope should be linear.

This problem is solved with a closed-loop (or phase-nulling) signal-processing approach, as proposed independently by Davis and Ezekiel[12] and Cahill and Udd.[13] The demodulated biased signal (or open-loop signal) is used as an error signal that is fed back into the system to generate an additional feedback phase difference $\Delta\phi_{FB}$ that is opposite to the rotation-induced phase difference $\Delta\phi_R$. The total phase difference is servo-controlled on zero, which provides good sensitivity, since the system is always operated about a high-slope point. With such a closed-loop scheme, the new measurement signal is the feedback value $\Delta\phi_{FB}$. This yields a linear response with good stability since this feedback value $\Delta\phi_{FB}$ is independent of the returning optical power and of the gain of the detection chain. The current consensus holds that the best closed-loop method is the digital phase ramp proposed by Arditty et al. (Fig. 1.7),[14,15] on which serendipity also applies!

In addition, there is the possibility of a second servo loop to control the modulation efficiency (also called V_π, the voltage that induced a π radian phase shift) by controlling the resets on the 2π radian.[14,15] A better way is the so-called *four-state* modulation that allows one to get a permanent V_π control without waiting for a reset (Fig. 1.8).[16]

It is possible to directly view the reason why the digital ramp technique tolerates many defects (such as an imperfect 2π reset, quantization, nonlinearity of the electronic drive, nonlinearity of the phase modulator response, low frequency modulator drift, etc.) without degrading the scale factor performance. The actual phase ramp shift $\phi_{PR\ actual}(t)$ created by the modulator is the sum of an ideal phase ramp shift $\phi_{PR\ ideal}(t)$ and a defect $\phi_{PR\ defect}(t)$. The induced phase difference in the interferometer is

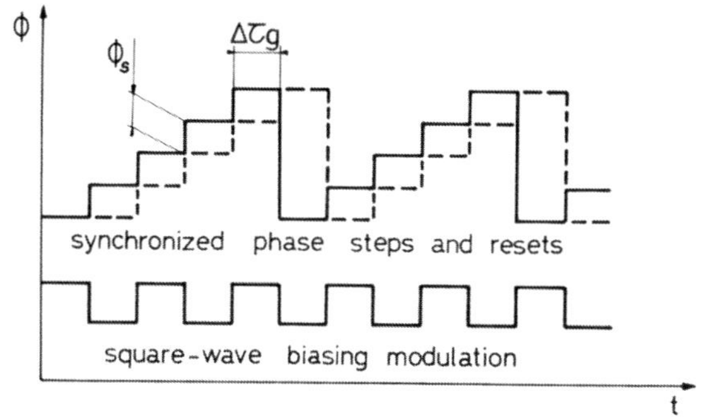

Figure 1.7 Digital phase ramp and resets synchronized with the square-wave biasing phase modulation.

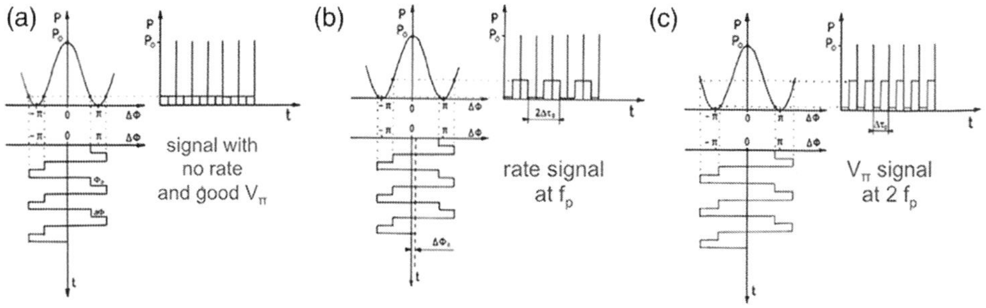

Figure 1.8 Principle of four-state modulation:[16] (a) Signal with no rate and good modulation efficiency (good V_π), (b) rate signal at the proper frequency (f_p), and (c) V_π error signal at twice the proper frequency ($2f_p$).

$$\Delta\phi_{PR\ actual}(t) = \Delta\phi_{PR\ ideal}(t) + \Delta\phi_{PR\ defect}(t). \tag{1.4}$$

Following the general equation (1.3) of any phase modulation, the defect $\Delta\phi_{PR\ defect}(t)$ of the phase difference is given by the delay difference $\Delta\tau_g$ through the coil with

$$\Delta\phi_{PR\ defect}(t) = \phi_{PR\ defect}(t) - \phi_{PR\ defect}(t - \Delta\tau_g). \tag{1.5}$$

Since the mean value of a difference is the difference of the mean values, the mean defect $<\Delta\phi_{PR\ defect}(t)>$ is

$$\begin{aligned} <\Delta\phi_{PR\ defect}(t)> &= <\phi_{PR\ defect}(t) - \phi_{PR\ defect}(t - \Delta\tau_g)> \\ &= <\phi_{PR\ defect}(t)> - <\phi_{PR\ defect}(t - \Delta\tau_g)>, \end{aligned} \tag{1.6}$$

and both mean values $<\phi_{PR\ defect}(t)>$ and $<\phi_{PR\ defect}(t - \Delta\tau_g)>$ are perfectly equal because of reciprocity, over a phase ramp period. Therefore, on average,

$$< \Delta\phi_{PR\ defect}(t) > = 0. \tag{1.7}$$

1.3.3 The optical Kerr effect

As analyzed by Ezekiel et al.,[17] an important case of truly nonreciprocal effect may arise due to the optical Kerr effect, which is nonlinear. Reciprocity is indeed based on the linearity of the propagation equation, but an imbalance in the power levels of the counter-propagating waves can produce a small nonreciprocal phase difference because of propagation nonlinearity induced by the high optical power density in the very small silica fiber core. Slow variations in the splitting ratio of the power divider feeding the sensing coil may therefore translate directly into bias drift. Experimentally, a power difference of 1 μW (e.g., arising from a 10^{-3} splitting imbalance of a 1-mW source) gives a nonreciprocal index difference as small as 10^{-15}, but when integrated along a kilometer of fiber, this produces a phase difference of a few 10^{-5} rad, at least three orders of magnitude above the theoretical sensitivity limit.

The Kerr-induced rotation-rate error results in fact from a complex four-wave mixing process[17] and not simply from an intensity self-dependence of the propagation constant of each counter-propagating wave. It has been shown[18,19] that if the mean value $<I>$ of the modulated intensity I equals its standard deviation, the Kerr-induced phase shift is balanced for both counter-propagating waves, which nulls out the effect on the phase difference—and this is actually the statistic of intensity fluctuation of a broadband spontaneous-emission source, clearly the ultimate serendipity!

1.3.4 Technological serendipity: erbium ASE fiber source and proton-exchanged $LiNbO_3$ integrated-optic circuit

As we just saw, serendipity is found in the mathematics of the signal-processing method and in the physics of the optical Kerr effect, but also in the technology:

- An amplified spontaneous emission (ASE) source,[20,21] derived from telecom erbium-doped fiber amplifier (EDFA) (Fig. 1.9), is a

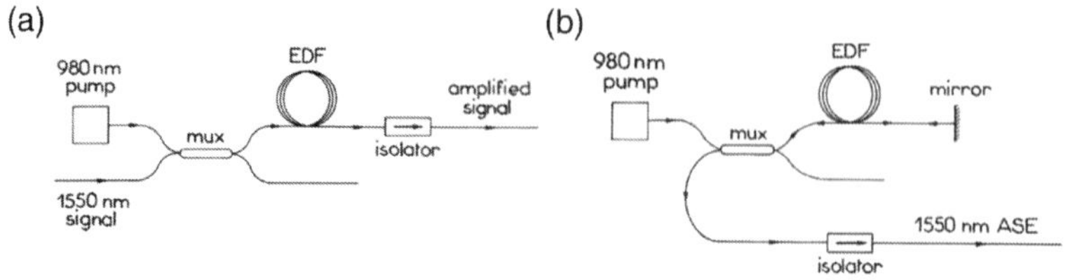

Figure 1.9 (a) Schematics of an erbium-doped fiber amplifier (EDFA) with a pump laser diode, wavelength multiplexer (mux), erbium-doped fiber (EDF), and isolator; and (b) an ASE source with the same components.

quasi-ideal source for the fiber gyro because it features a broadband and stable spectrum with unpolarized emission, which is very beneficial to reduce birefringence-induced non-reciprocity, as analyzed by Pavlath et al.[22]

- A proton-exchanged lithium niobate ($LiNbO_3$) integrated-optic circuit[23] provides excellent polarization rejection (Fig. 1.10), which is crucial since birefringence-induced non-reciprocity is related to the amplitude rejection ratio of the polarizer and not to the intensity, as analyzed by Kintner.[24]

1.4 Potpourri of Comments

1.4.1 OCDP using an OSA

The technique of optical coherence domain polarimetry (OCDP), also called distributed polarization extinction ratio (PER), has been a powerful tool to measure accurately the polarization and birefringence properties of the components of a fiber gyro since it measures the amplitude ratio.[25,26] It was first based on path-matched interferometry (Fig. 1.11) and has yielded several commercially available, dedicated instruments.[27]

However, a path-matched readout interferometer is actually a scanning interferometer, which is the base of Fourier-transform spectroscopy. One measures the full coherence function of the tested light, and through the Fourier transform the spectrum is retrieved. Conversely, a "classical" optical spectrum analyzer (OSA), grating based in particular, measures the spectrum, and through the (inverse) Fourier transform the coherence function can be retrieved. Since an OSA is now a very common multi-purpose test instrument following the development of wavelength domain multiplexing (WDM) telecoms, it is tempting to use an OSA for OCDP instead of a dedicated "single-purpose" test instrument.

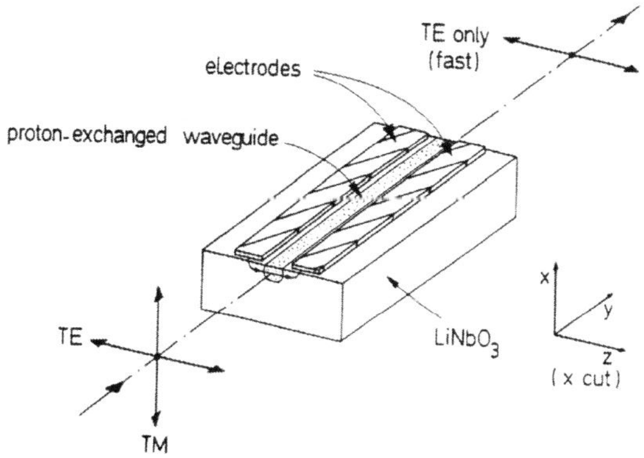

Figure 1.10 Proton-exchanged $LiNbO_3$ waveguide that guides a single-TE polarization.

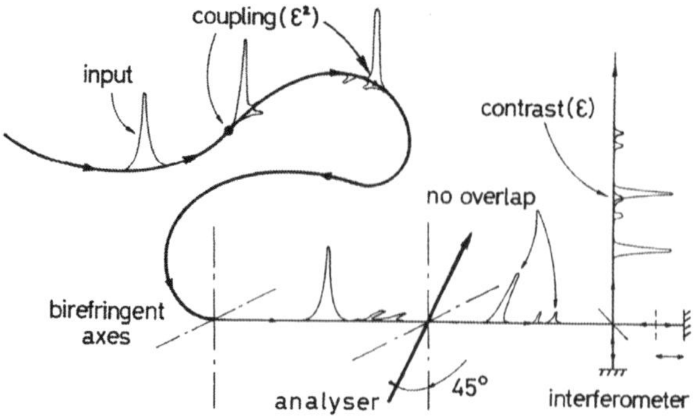

Figure 1.11 Principle of OCDP (or distributed PER) with a path-matched readout interferometer.

Both approaches are theoretically equivalent, but they have some practical differences. An OSA has a limited spectral resolution, typically $\Delta\lambda_r = 0.01$ nm, and can test a coupling only within an optical distance of $\lambda^2/\Delta\lambda_r$, i.e., 250 mm. For testing couplings in a PM fiber with a birefringence index difference $\Delta n_b = 5 \times 10^{-4}$, it corresponds to a length limited to $\lambda^2/(\Delta\lambda_r.\Delta n_b)$, i.e., 500 m. In terms of sensitivity, an OSA is limited by excess relative intensity noise (RIN) since it filters a very narrow "broad" spectrum. It requires usually an averaging of many scans to get a low noise, and since the interference pattern that is looked for varies in temperature, it is not stable enough to take full advantage of this averaging. In practice, a dedicated instrument allows one to measure contrast down to $1–3 \times 10^{-5}$, i.e., a coupling or rejection ratio of –90 to –100 dB, whereas an OSA is limited to $1–3 \times 10^{-4}$, i.e., –70 to –80 dB (Fig. 1.12).

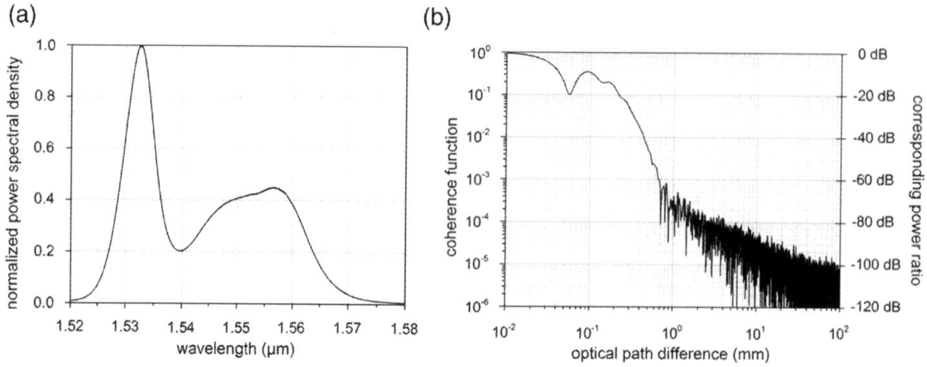

Figure 1.12 (a) Spectrum of an erbium-doped fiber ASE source, and (b) calculated coherence function showing a noise floor below 10^{-4}, i.e., a corresponding power ratio of –80 dB, above a 4-mm path difference.

1.4.2 Strain-induced "T dot" Shupe effect

Because of reciprocity, the two counter-propagating paths are equalized in a ring interferometer; but this is strictly valid only if the system is time-invariant. We have already seen (Section 1.3.1) that a biasing phase modulation can be generated with a reciprocal modulator placed at one end of the coil acting as a delay line, which provides a filtering behavior with a sinusoidal transfer function: $2\sin(\pi \cdot f_m \cdot \Delta\tau_g)$. This also applies to parasitic phase shifts generated by the environment, particularly with temperature change, as analyzed very early by Shupe.[28] The two interfering waves do not see the perturbation at exactly the same time, unless it is applied in the middle of the coil.

Considering that the phase perturbation has low frequencies compared to the proper frequency of the coil, the gyro behaves like a DC rejection filter with the usual 6 dB (electrical) per octave roll-off and differentiation properties: the phase error is proportional to the temporal derivative of the temperature, which may be very harmful, particularly during the warm-up. This temperature-induced phase shift can be also viewed as a spurious phase ramp, as the one used for closed-loop operation. This can be drastically reduced with the so-called quadrupolar winding, as proposed by Frigo.[29]

The so-called Shupe effect, as described in the original reference[28] and as it is still found in some recent publications, is incomplete: it considers that the thermo-optic coefficient $\alpha_T = dn/dT + n \cdot \alpha_{SiO_2}$ of accumulated phase is the same along the whole coil fiber, which is not at all the case because of non-uniform temperature-induced strains along this coil fiber due to the thermal expansion of the fiber coating as well as the one of the coil frame. These strains add phase changes to the ones due to the index dependence dn/dT and the thermal expansion α_{SiO_2}. This was described quite early by Cordova et al.[30] and analyzed in more detail recently by Mohr et al.[31]

Heat transfer in the fiber coil is related to a process of diffusion, and with the damping of the gyro housing, it can be shown that the pure Shupe effect is actually negligible and overcome by the strain-induced effect.[32] This strain-induced Shupe effect follows the temporal derivative (T dot) of the mean temperature of the coil and has a time signature that is very different than the pure Shupe effect.[30–32] It is often called the "T dot" Shupe effect, and this simple T-dot dependency eases its modeling.

The measurement of these temperature-induced strains can be performed very efficiently with a Brillouin optical time-domain analyzer (BOTDA).[33] Figure 1.13 displays the BOTDA signature of a self-standing quadrupolar coil. At +60°C, the curing temperature of the epoxy used to bond the fiber loops, there are only the strain fluctuations due to the variation of the tension of the winding machine (typically ±5 grams). At –60°C, there is a clear strain variation between the inner diameter (middle of the coil, in a quadrupolar

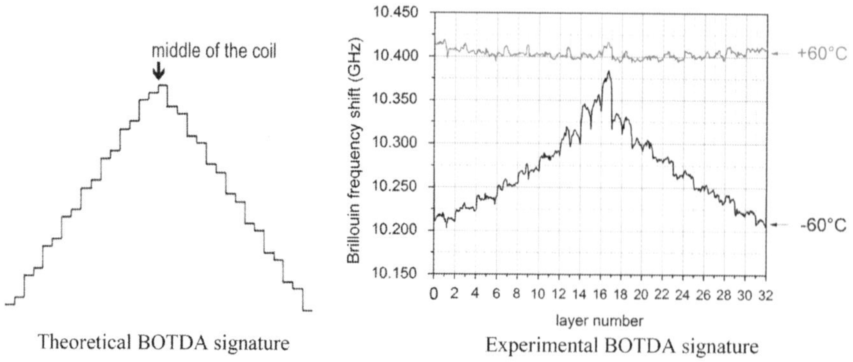

Figure 1.13 Experimental BOTDA signature of a quadrupolar coil at +60°C (grey curve) and –60°C (black curve) with the theoretical signature for a uniform thermal longitudinal strain variation difference between adjacent layers.

coil) and the outer diameter, and one can easily differentiate the strain of the various layers. The sensitivity of a BOTDA of 20 microstrains (με)/MHz shows a strain difference as large as 3,500 microstrains (0.35%) between the inner layer and the outer layer, which corresponds to about 1 με/°C for the mean variation of the difference between two adjacent layers. It is perfectly consistent with the observed strain-induced T-dot Shupe effect of 0.7 (°/h)/(°C/min) for this kind of 400-m coil, with 32 layers and a mean diameter of 45 mm.

Without any finite element analysis, but with a good physical understanding of the problem, one can show that, despite its stiffness, the silica fiber is actually unable to block the transverse thermal expansion/contraction of the softer coating, even in the radial direction. This 1 με/°C is perfectly consistent with a thermal expansion of 80 ppm/°C of the coating. This result is far from intuitive, but it is confirmed experimentally. It can be viewed as a coating expending/contracting thermally in the transverse direction *without* any radial resistance of the fiber loops. This changes the radius R of these loops and obviously their perimeter p since $p = 2\pi R$!

This result deserves an additional comment: this effect should be seen with *any* type of fiber, even with a hollow-core micro-structured fiber, since it is a matter of longitudinal strain of the coil fiber induced by the coating.

1.4.3 Transverse magneto-optic effect

As analyzed by Logozinski,[34] in addition to the well-known Faraday effect that depends on the magnetic field component parallel to the fiber, as analyzed by Hotate et al.,[35] there is also a smaller but significant magnetic dependence that is orthogonal to the fiber because the fundamental mode of a single-mode fiber has a small longitudinal component in addition to its main transverse component; it is actually a *hybrid* HE_{11} mode and not a purely transverse wave. The so-called LP_{01} (linearly polarized) is only an approximation.

The main transverse component of this hybrid mode follows a quasi-Gaussian function. The longitudinal component is proportional to the spatial derivative of this transverse component, and it is in phase quadrature, which means that, in the center of the mode w, the polarization is transverse but, on the sides, it becomes pseudo-elliptical in a plane parallel to the propagation z axis and then sensitive to a transverse magnetic B field (Fig. 1.14).

The effect of this small longitudinal component is generally negligible except in a few cases: it explains the circular birefringence created by elastically twisting a fiber, and it is also the reason of this transverse magnetic dependence. The derivation of Logozinski[34] concludes that this magnetic dependence, transverse to the fiber, is parallel to the coil axis, whereas the better-known Faraday effect dependence is perpendicular to the coil axis.[35] It yields a non-reciprocal phase difference $\Delta\phi_B$ proportional to the number N of turns of the sensing coil. At 1550 nm, it is

$$\Delta\phi_B / (\mathrm{B} \cdot N) \approx 0.4\,\mu\text{radian/gauss} \cdot \text{kiloturn}. \tag{1.8}$$

Considering a coil of 1 km over a diameter of 10 cm, i.e., 3200 turns, 3.2 kiloturns, it yields an axial dependence of about 1 μrad/gauss, whilst the Faraday effect yields an in-plane dependence more than ten times higher.

1.4.4 RIN compensation

A spontaneous emission source suffers from excess RIN, and an ASE source keeps the excess RIN of its seed spontaneous emission. Theory shows that the power spectral density (PSD_{RIN}) of the excess relative intensity noise is simply the inverse of the spectrum temporal frequency width Δf_{FWHM}:

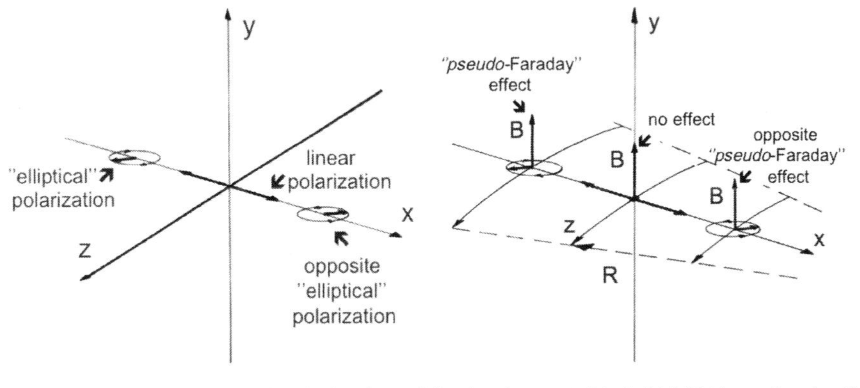

Figure 1.14 Pseudo-elliptical polarization of the fundamental hybrid HE11 mode of a fiber, lying in the horizontal plane that yields a transverse magnetic sensitivity to a vertical B field.

$$\mathrm{PSD}_{RIN} \approx 1/\Delta f_{FWHM}. \tag{1.9}$$

For a super-luminescent diode (SLD) with a mean wavelength $\lambda = 850$ nm, and $\Delta\lambda_{FWHM\text{-}SLD} = 20$ nm, the frequency width is $\Delta f_{FWHM\text{-}SLD} = c\,\Delta\lambda_{FWHM\text{-}SLD}/\lambda^2 = 8$ THz, and then:

$$\mathrm{PSD}_{RIN\text{-}SLD} = 1.25 \times 10^{-13}/\mathrm{Hz}\ \ (\text{i.e,}\ -129\,\mathrm{dB/Hz}). \tag{10}$$

This value of excess RIN corresponds to the theoretical photon noise of a power of 4 μW, which is in practice the returning power of a fiber gyro using such a SLD.

For an erbium-fiber ASE source, the spectrum is narrower, especially when filtering is used to stabilize the spectrum, and the width can be $\Delta\lambda_{FWHM\text{-}Er} = 7.5$ nm, yielding $\Delta f_{FWHM\text{-}Er} = 1$ THz, and then

$$\mathrm{PSD}_{RIN\text{-}Er} = 10^{-12}/\mathrm{Hz}\ \ (\text{i.e.,}\ -120\,\mathrm{dB/Hz}). \tag{1.11}$$

This is significantly higher than the one of a SLD, and in addition an ASE source provides much more power. Then the related photon noise is in the range of –140 dB/Hz, i.e., 100 times lower than the PSD of the RIN and 10 times in terms of relative rms values (the square root of the PSD). However, excess RIN is not a fundamental noise like photon noise: when a light beam is split, the photon noises of both outputs are uncorrelated, but their RINs are identical. It is possible to compensate for the RIN by measuring it with a reference and subtracting it from the signal, as proposed by Moeller et al.[36]

There is also a very simple method of RIN reduction by operating the gyro at a bias point close to a black fringe; that is, a phase bias close to π instead of $\pi/2$.[37] The sensitivity is proportional to the slope of the raised cosine response curve, i.e., $\sin\phi_b$ (where ϕ_b is the phase bias), whereas the excess RIN is proportional to the actual power on bias [i.e., the response $(1+\cos\phi_b)$]. Working, for example, at a 0.9π bias instead of $\pi/2$, the sensitivity is reduced by a factor of $\sin(0.9\times\pi)/\sin(\pi/2) = 0.3$, whereas the intensity noise experiences a reduction six times higher since $[1+\cos(0.9\times\pi)]/[1+\cos(\pi/2)] = 0.05$. Furthermore, this slightly improves the theoretical signal-to-photon-noise ratio.[37] These techniques to compensate for or reduce the excess RIN allows one in practice to get very close to the theoretical photon noise and to obtain a phase noise down to 10^{-7} rad/$\sqrt{\mathrm{Hz}}$ (Fig. 1.15).

1.4.5 Fundamental mode of an integrated-optic waveguide

An integrated-optic channel waveguide is quite similar to a fiber core but with a significant difference: the waveguide is buried just underneath the substrate top surface, yielding a very high index step in the direction perpendicular to

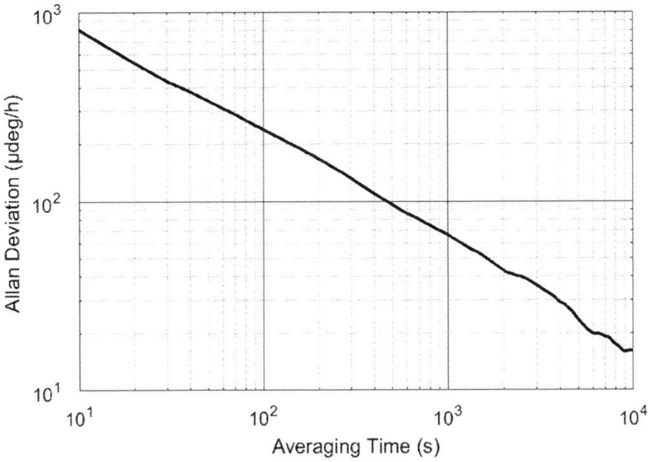

Figure 1.15 Allan deviation of an iXblue FOG prototype ($L = 5$ km, $D = 180$ mm, $A = 200$ m^2, $\lambda = 1530$ nm) using RIN reduction and showing an angular random walk of 32 μdeg/$\sqrt{}$h, which corresponds to a rotation rate white noise of 2 mdeg/h/$\sqrt{}$Hz, and an interferometer phase noise of 10^{-7} rad/$\sqrt{}$Hz.

this interface. Boundary conditions impose a null field on this interface. This interface can be considered to act as a perfect mirror working in total internal reflection (TIR) under grazing incidence. Such TIR yields a π-phase shift that leads to destructive interference between an incoming wave and its reflection. In the direction parallel to the interface, the mode profile is similar to the one of a single-mode fiber and follows the usual Gaussian curve. In the perpendicular direction, the mode profile is actually very close to the normalized derivative of a Gaussian function as with the second-order LP_{11} mode of a fiber but with only one lobe.

The similarity between the perpendicular distribution of the fundamental mode of a channel waveguide and the distribution of the second-order LP_{11} mode of a fiber may be pushed considering a virtual image with a π-phase shift. The fundamental mode of a channel waveguide actually has the same characteristics as the anti-symmetrical second-order LP_{11} mode of an equivalent fiber with a core that has the same width $2a_x$ in the parallel direction, and a width $2a_y$, which is the double of the height a_y of the waveguide in the perpendicular direction (Fig. 1.16).

This similarity is also found in the equivalent index that follows the law of this second-order LP_{11} fiber mode. In particular, if the fundamental mode of a fiber is unique for a normalized frequency V following $0 < V < 2.4$, and has theoretically no cut-off, the fundamental mode of a channel waveguide is unique for a normalized frequency V following $2.4 < V < 3.8$, as an LP_{11} fiber mode, and has a cut-off when $V < 2.4$, V being calculated considering that the core diameter of the equivalent fiber is twice the height a_y of the actual waveguide (Fig. 1.17).

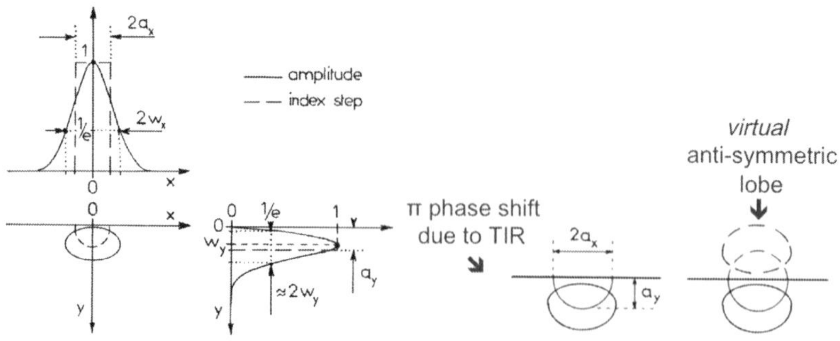

Figure 1.16 Equivalence between the fundamental mode of a channel waveguide with a size 2 $a_x \times a_y$, and the LP_{11} mode of a fiber with a core size $2a_x \times 2a_y$.

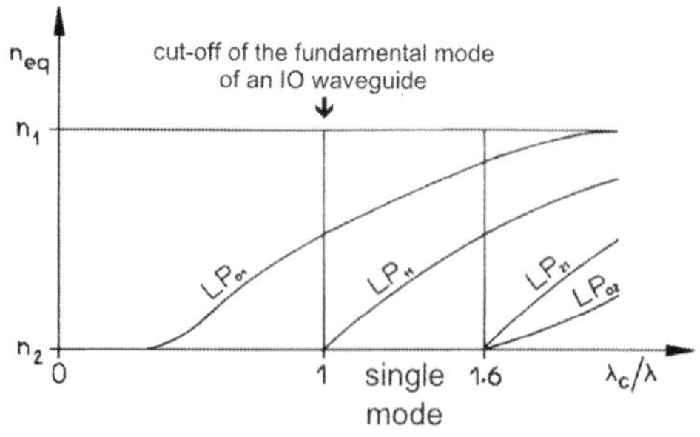

Figure 1.17 *Classical* curves of the equivalent index n_{eq} of the various modes of a fiber as a function of the spatial frequency $1/\lambda$, used to find the properties of the fundamental mode of an integrated-optic channel waveguide.

1.4.6 Limit of the rejection of stray light in a proton-exchanged $LiNbO_3$ circuit with absorbing grooves

The polarization rejection of a proton-exchanged $LiNbO_3$ circuit is not due to an absorption phenomenon: cross-polarized light (TM, for transverse magnetic, i.e., perpendicular to the top surface) is not guided and then is diffracted in the substrate. Therefore, some stray light may be partially coupled back into the output fiber. Very good polarization rejection is theoretically possible because the boundary conditions that nullify the light under the top face (also called the Lloyd mirror effect), as seen in Section 1.4.5, also apply to stray light. However, as it is well known, stray light reflected on the middle of the bottom surface limits this theoretical rejection (Fig. 1.18).

To increase the polarization rejection of a proton-exchanged $LiNbO_3$ circuit, one has to suppress, or at least reduce, this reflection on the bottom face of the substrate. Anti-reflection (AR) coating could be a solution, but it is very difficult technologically because of the high index of $LiNbO_3$ and the grazing incidence. Some reduction of the amount of reflected light may be also obtained by grinding the bottom face, but because of the grazing incidence, it is not very efficient, and the specular reflection remains important.

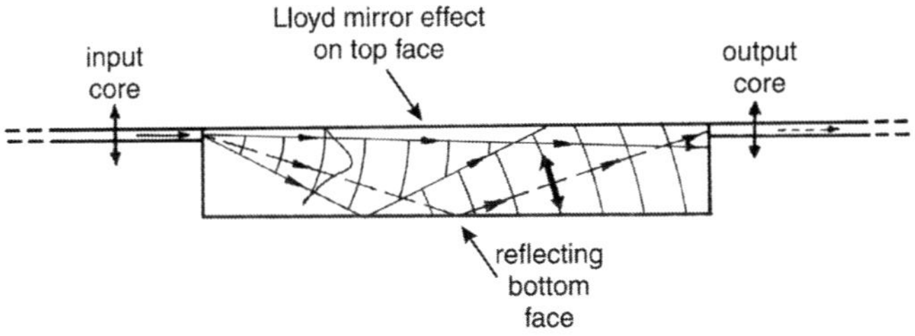

Figure 1.18 Propagation of the unguided TM polarization in a proton-exchanged waveguide.

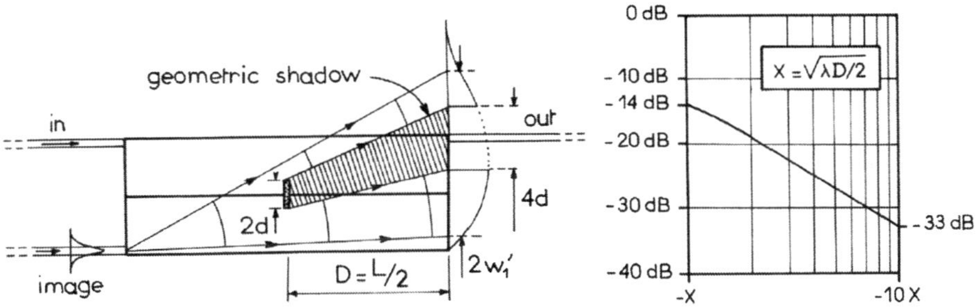

Figure 1.19 Power density attenuation in the geometric shadow region, due to knife-edge Fresnel diffraction.

The most common technique to improve the rejection is to fabricate absorbing grooves in the bottom face to block the light reflected in the middle, otherwise sent to the output fiber. That places the output core in the geometric shadow of the absorbing groove. The light power is not perfectly attenuated in this geometric shadow region because of knife-edge near-field Fresnel diffraction,[38] and the semi-darkness decay is not very fast. Figure 1.19 displays the actual attenuation due to Fresnel diffraction in the geometrical shadow, using log–log scales with respect to a normalized abscissa X, defined as $X = \sqrt{\lambda D/2}$, where λ is the wavelength in the medium (700 nm in $LiNbO_3$ for 1550 nm in vacuum), and D is the distance to the edge. Note that it is a linear function with a –2 slope, in logarithmic scales.

1.5 Conclusion

The design and the serendipity of a high performance fiber gyro was quite clear at the 20th Anniversary Conference in 1996,[39] but it required twenty additional years of very good engineering to achieve "ultimate" performance and, adding to the serendipity, be "fascinating." Today, it is almost reached, and the fiber optic gyro is clearly seen as a unique technology for the most-demanding strapdown inertial navigation applications. This is perfectly summarized by the words of Anthony Lawrence about the fiber optic gyro in his book *Modern Inertial Technology*:[40] "Nature is rarely that cooperative!"

Acknowledgment

Knowing that I have worked only for 39 years on the subject, this 40th-anniversary chapter is the opportunity to thank all of my colleagues who have shared this fascination for the fiber gyro (and the hard work) at Thomson-CSF (now Thales), Stanford University, Photonetics, iXsea, iXfiber, Photline, and today iXblue.

I want also to acknowledge the profound influence of the late Professor H. John Shaw during my post-doctoral scholarship at Stanford University in the early 1980s.

References

[1] V. Vali and R. W. Shorthill, "Fiber Ring Interferometer", *Applied Optics*, Vol. **15**, pp 1099–1100, (1976).

[2] Y. Paturel, J. Honthaas, H. Lefèvre, and F. Napolitano, "One Nautical Mile per Month FOG-based Strapdown Navigation System: a Dream already within Reach", Inertial Sensors and Systems 2012, DGON, Karlsruhe, paper 12, (2012).

[3] G. Sagnac, "L'éther lumineux démontré par l'effet du vent relatif d'éther dans un interféromètre en rotation uniforme," *Comptes rendus de l'Académie des Sciences*, Vol. **95**, pp. 708–710, (1913).

[4] M. von Laue, "Über einen Versuch zur Optik der bewegten Körper", *Münchener Sitzungsberichte*, pp. 405–411, (1911).

[5] H. Fizeau, "Sur les hypothèses relatives à l'éther lumineux, et sur une expérience qui parait démontrer que le mouvement des corps change la vitesse avec laquelle la lumière se propage dans leur intérieur", *Comptes rendus de l'Académie des Sciences*, Vol. **33**, pp. 349–355, (1851).

[6] M. von Laue, "Die Mitführung des Lichtes durch bewegte Körper nach dem Relativitätsprinzip", *Annalen der Physik*, Vol. **328** (10), pp. 989–990, (1907).

[7] H. J. Arditty and H. C. Lefèvre, "Sagnac Effect in Fiber Gyroscopes", *Optics Letters*, Vol. **6**, pp. 401–403, (1981).

[8] R. Ulrich, "Fiber optic Rotation Sensing with Low Drift", *Optics Letters*, Vol. **5**, pp. 173–175, (1980).

[9] H. Arditty, M. Papuchon, and C. Puech, "Ring interferometer device and its application to the detection of non-reciprocal effects", US Patent # 4, 480, 915, Foreign application priority of December 1979, (1984).

[10] J. M. Martin and J. T. Winkler, "Fiber optic Laser Gyro Signal Detection and Processing Technique", *SPIE Proceedings*, Vol. **139**, pp. 98–102, (1978).

[11] R. A. Bergh, H. C. Lefèvre, and H. J. Shaw "All-Single-Mode Fiber optic Gyroscope with Long-Term Stability," *Optics Letters*, Vol. **6**, pp. 502–504, (1981).

[12] J. L. Davis and S. Ezekiel, "Techniques for Shot-Noise-Limited Inertial Rotation Measurement Using a Multi-Turn Fiber Sagnac Interferometer," *SPIE Proceedings*, Vol. **157**, pp. 131–136, (1978).
[13] R. F. Cahill and E. Udd, "Phase-Nulling Fiber optic Laser Gyro," *Optics Letters*, Vol. **4**, pp. 93–95, (1979).
[14] H. C. Lefèvre, Ph. Graindorge, H. J. Arditty, S. Vatoux, and M. Papuchon, "Double Closed-Loop Hybrid Fiber Gyroscope Using Digital Phase Ramp," *Proceedings of OFS 3/'85, San Diego, OSA/IEEE*, Postdeadline Paper 7, (1985).
[15] H. J. Arditty, P. Graindorge, H. C. Lefèvre, P. Martin, J. Morisse, and P. Simonpiétri, "Fiber optic Gyroscope with All-Digital Processing," *Proceedings of OFS 6/'89, Paris, Springer-Verlag Proceedings in Physics*, Vol. **44**, pp. 131–136, (1989).
[16] H. C. Lefèvre and P. Martin, "Optical-Fiber Measuring Device, Gyrometer", US Patent # 5, **141**, 316, (1992).
[17] S. Ezekiel, J. L. Davis, and R. W. Hellwarth, "Intensity Dependent Nonreciprocal Phase Shift in a Fiberoptic Gyroscope", *Springer Series in Optical Sciences*, Vol. **32**, pp. 332–336, (1982).
[18] R. A. Bergh, B. Culshaw, C. C. Cutler, H. C. Lefèvre, and H. J. Shaw, "Source Statistics and the Kerr Effect in Fiber optic Gyroscopes", *Optics Letters*, Vol. **7**, pp. 563–565, (1982).
[19] K. Petermann, "Intensity-Dependent Nonreciprocal Phase Shift in Fiber optic Gyroscopes for Light Sources with Low Coherence", *Optics Letters*, Vol. **7**, pp. 623–625, (1982).
[20] K. A. Fesler, R. F. Kalman, M. J. F. Digonnet, B. Y. Kim, and H. J. Shaw, "Behavior of Broadband Fiber Sources in a Fiber Gyroscope," *SPIE Proceedings*, Vol. **1171**, pp. 346–352, (1989).
[21] P. F. Wysocki, K. Fesler, K. Liu, M. J. F. Digonnet, and B. Y. Kim, "Spectrum Thermal Stability of Nd- and Er-Doped Fiber Sources," *SPIE Proceedings*, Vol. **1373**, pp. 234–245, (1990).
[22] G. A. Pavlath and H. J. Shaw, "Birefringence and Polarization Effects in Fiber Gyroscopes", *Applied Optics*, Vol. **21**, pp. 1752–1757, (1982).
[23] T. Findakly, "Proton-exchanged Integrated Optical Components", Chapter 10, pp. 337–351, *Optical Fiber Rotation Sensing*, W. K. Burn Ed, Academic Press, (1993).
[24] E. C. Kintner, "Polarization Control in Optical-Fiber Gyroscopes", *Optics Letters*, Vol. **6**, pp. 154–156, (1981).
[25] K. Takada, J. Noda, and K. Okamoto, "Measurement of Spatial Distribution of Mode Coupling in Birefringent Polarization-Maintaining Fiber with New Detection Scheme," *Optics Letters*, Vol. **11**, pp. 680–682, (1986).
[26] H. C. Lefèvre, "Comments about the Fiber optic Gyroscope", *SPIE Proceedings*, Vol. **838**, pp. 86–97, (1987).

[27] P. Martin, G. Le Boudec, and H. C. Lefèvre, "Test Apparatus of Distributed Polarization Coupling in Fiber Gyro Coils Using White Light Interferometry," *SPIE Proceedings*, Vol. **1585**, pp. 173–179, (1991).

[28] D. M. Schupe, "Thermally Induced Nonreciprocity in the Fiber optic Interferometer", *Applied Optics*, Vol. **9**, pp. 654–655, (1980).

[29] N. J. Frigo, "Compensation of Linear Sources of Non-Reciprocity in Sagnac Interferometers", *SPIE Proceedings*, Vol. **412**, pp. 268–271, (1983).

[30] F. Cordova and G. M. Swabian, "Potted Fiber Optic Gyro Sensor Coil for Stringent Vibration and Thermal Environments", *US Patent # 5*, **546**, 482, (1996).

[31] F. Mohr and F. Schadt, "Error Signal Formation in FOGs through Thermal and Elastooptical Environment Influences on the Sensing Coil", *ISS Conference*, Karlsruhe, (2011).

[32] H. C. Lefèvre, *The Fiber optic Gyroscope*, Second Edition, pp. 99–105, Artech House, (2014).

[33] M. Niklès, L. Thévenaz, and P. A. Robert, "Simple Distributed Fiber Sensor Based on Brillouin Gain Spectrum Analysis", *Optics Letters*, Vol. **21**, pp. 758–760, (1996).

[34] V. N. Logozinski, "Magnetically Induced Non-Faraday Nonreciprocity in a Fiber optic Gyroscope", *Journal of Communications Technology and Electronics*, Vol. **51**, N° 7, pp. 836–840, (2006).

[35] K. Hotate and K. Tabe, "Drift of an Optical Fiber Gyroscope Caused by the Faraday Effect: Influence of the Earth's Magnetic Field", *Applied Optics*, Vol. **25**, pp. 1086–1092, (1986).

[36] R. P. Moeller and W. K. Burns, "Low noise fiber gyroscope system which includes excess noise subtraction", *US Patent #5*, **331**, 404, (1994).

[37] H. C Lefèvre, S. Vatoux, M. Papuchon, and C. Puech, "Integrated Optics: A Practical Solution for the Fiber optic Gyroscope," *SPIE Proceedings*, Vol. **719**, pp. 101–112, (1986).

[38] H. C. Lefèvre, *The Fiber optic Gyroscope*, Second Edition, pp. 347–351, Artech House, (2014).

[39] E. Udd, H. C. Lefèvre, and K. Hotate, Editors, "Fiber optic Gyros: 20th Anniversary Conference", *SPIE Proceedings*, Vol. **2837**, (1996).

[40] A. Lawrence, "Modern Inertial Technology: Navigation, Guidance, and Control", Second Edition, Springer-Verlag, (1998).

Hervé C. Lefèvre is Chief Science Officer of iXblue (www.ixblue.com) in France. He was born in 1954. He graduated from the *Ecole Normale Supérieure de Saint-Cloud* in Physics and was awarded a Doctorate in Optics-Photonics from the University of Paris-Orsay (France) in 1979. His doctorate research was performed at Thales (formerly Thomson-CSF) Central Research Laboratory and his thesis subject was pioneering work on the fiber optic gyroscope.

From 1980 to 1982, he was a post-doctoral research associate at Stanford University in California, continuing R&D on the fiber optic gyroscope. In 1982, he came back to Thales Central Research Lab and became head of the fiber optic sensor team. In 1987, he joined Photonetics, then a start-up, and became director of R&D. In addition to fiber optic sensors and gyroscopes, the company developed a very successful line of test instruments for optical fiber communications. As the company grew, he moved to the position of Chief Operations Officer in 1999. At the end of 2000, Photonetics was acquired by the Danish group GN-Nettest while its fiber-gyro activity was spun out to create iXsea. Dr. Lefèvre remained with GN-Nettest and managed its Photonics Division.

In 2004, he moved to iXcore, the parent company of iXSea, as vice president for R&D with a specific involvement in iXSea's activity, and became Chief Science Officer of iXBlue at its creation in 2010, as a merger of iXsea with several other subsidiaries of iXcore (iXfiber and Photline in particular).

He has authored and coauthored nearly 80 journal and conference publications about the fiber-gyroscope and fiber communication test instruments, and he has been granted more than 50 patents.

He is the author of a reference book on the fiber optic gyroscope, with a first edition published by Artech House in 1993 and a second one in 2014.

Hervé C. Lefèvre was also president of the French Society of Optics (SFO) for the 2005–2007 period, and president of the European Optical Society (EOS) for the 2010–2012 period. He is a member of the International Honorary Committee of the Optical Fiber Sensor conference (OFS).

Chapter 2
The Early History of the Closed-Loop Fiber Optic Gyro and Derivative Sensors at McDonnell Douglas, Blue Road Research, and Columbia Gorge Research

Eric Udd
Columbia Gorge Research LLC

2.1 Introduction

In 1976, Vali and Shorthill demonstrated for the first time an operational fiber optic gyro. McDonnell Douglas Astronautics Company in Huntington Beach, California (MDAC-HB) completed a project in that same year to redesign a new, lower-cost inertial measurement unit (IMU) for the Delta rocket based on dry-tuned mechanical gyros. NASA agreed to split the cost savings with McDonnell Douglas in exchange for the latter company funding the development costs of the new IMU. The project success benefited both NASA with savings and McDonnell Douglas with increased profit. This situation would continue to benefit both parties, although NASA in their contract included the option to replace the new McDonnell Douglas IMU at any time if a third-party vendor could provide a lower-cost option with similar performance. The principal concern involved the possibility of a new lower-cost, high-performance gyro becoming available. As a result, the guidance and control group at MDAC-HB funded a small project with their Electo-Optics Laboratory to investigate optical gyros. This chapter focuses on work associated with the Sagnac interferometer begun at McDonnell Douglas and, in the case of derivative inventions, continued at Blue Road Research and Columbia Gorge Research. Products and developments associated with the Sagnac interferometer continue worldwide, and collections of papers and books provide additional background.[1–6]

2.2 Invention and Demonstration of the Closed-Loop Fiber Gyro

When I started at MDAC-HB on September 6, 1977, Richard Cahill managed the Electro-Optics Lab. My first assignment involved devising an optical inspection tool for cryogenic foam used in the Delta rocket and for liquid-natural-gas tanks. After designing and demonstrating a breadboard, a technician built additional units, and Cahill assigned me to lead a new small optical gyro investigation. In late September, while discussing the relative merits of ring laser versus fiber optic gyros, Cahill remarked that "the fiber gyro would be an all solid-state device; too bad it has a sinusoidal output and is non-linear." Shortly thereafter, I came up with the "dispersive fiber gyro." The idea involved placing a frequency shifter in the Sagnac loop so that the wavelengths of the counter-propagating light beams would be balanced after a complete circuit but be sufficiently different so that the optical path difference would be controlled by the dispersion of the fiber and the frequency difference. This started a series of discussions between us and the quick realization that the dispersion effect was much smaller than the net phase shift induced by the frequency difference. On September 29, 1977, the first complete written and witnessed description of the closed-loop (or what was at that time called the phase-nulling) optical gyro occurred.

Realizing the importance of the invention, formal disclosure statements were filed with the McDonnell Douglas corporate patent department, and subsequent meetings were held between Cahill and me in the months to come extending and refining the initial ideas. At the time, McDonnell Douglas would not file a patent without a hardware demonstration. Components were ordered, including a Tropel He-Ne laser with a long coherence length, an acousto-optic modulator designed to operate at 50 MHz, and 100 m of Valtec optical fiber with a 2-micron core designed to be single mode at 633 nm. On July 28, 1978, the first signals were obtained from the demonstration unit shown in Fig. 2.1. The Delta rocket program monitored progress closely and demanded that a 10-cm-diameter unit be designed and built for the next phase of the project. The Electro-Optics Lab countered with 15 cm. After discussions, work began in mid-1978 on the design of a 12.5-cm unit.

Preparations began on a patent application to cover all aspects of the closed-loop fiber optic gyro known at the time, including the results from the breadboard demonstration of Fig. 2.1 and the design elements of the 12.5-cm unit. The patent application was filed on December 7, 1978, followed by publication of the first experimental results.

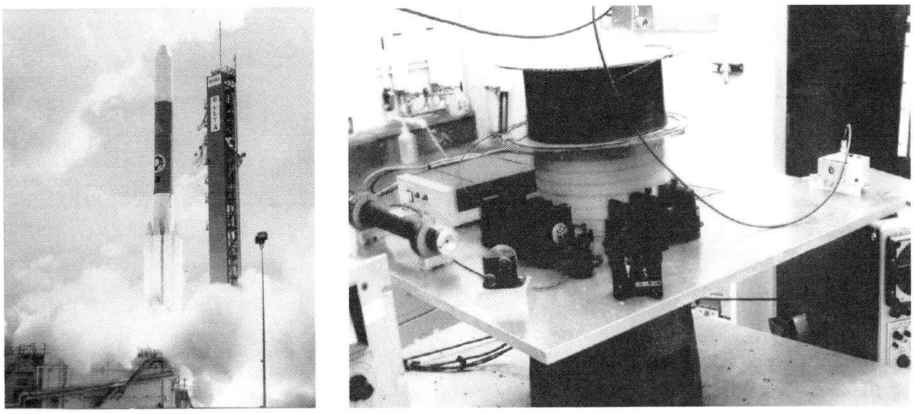

Figure 2.1 The Delta rocket program at MDAC-HB sponsored the first closed-loop fiber optic gyro demonstration on July 28, 1978.

The 12.5-cm fiber optic gyro represented the first solid state fiber optic gyro utilizing one of the first Hitachi single-mode laser diodes and some of the first low-loss, single-mode optical fibers produced by Corning. The original design included two balanced acousto-optic (AO) modulators, but one failed, and a single AO modulator configuration, shown in Fig. 2.2, was built and tested extensively.

Alignment of the 12.5-cm unit took about 4 weeks to complete. Bulk optic lenses were used with all housings epoxied in place. The ends of the 4-micron-core optical fiber coil were polished at an angle, and final alignment was enabled through the use of a set of three rotatable optical wedges placed in front of each fiber end. One surprise involved the 12.5-cm fiber gyro operating on the rotation stage for only 30 minutes before the Hitachi laser diode ceased to lase. The optical power dropped from about 1 milliwatt to few microwatts. Although the optical power dropped by about two orders of magnitude, so did

(a) (b)

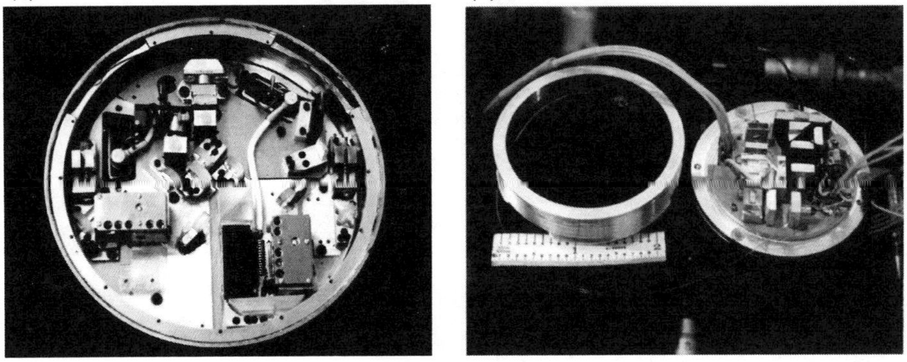

Figure 2.2 Early phase-nulling optical gyros: (a) 12.5-cm-diameter unit, built and demonstrated in 1979; (b) 6.3-cm diameter unit, built and demonstrated in 1980.

the noise floor with the signal-to-noise ratio remaining virtually the same. Realizing this was due to the change in coherence length, efforts were made to purchase short-coherence light sources thereafter.

Important work conducted by Ulrich regarding the need to use a second beamsplitter and polarizer led to design improvements that were included in the 6.3-cm design of 1980 (Fig. 2.2). The 6.3-cm unit responded to requests by MDAC sponsors for still-smaller demonstration models and improved alignment stations.

In an effort to make even-smaller packages, a 2.5-cm-wide, 7.5-cm-long open-loop fiber optic gyro was constructed shortly after the completion of the 6.3-cm circular design. This development piqued the interest of an oil and gas service company that sponsored work on a tool intended to navigate during the drilling process. MDAC used its expertise in molding electronics to support high vibrations and acoustic levels during launch to build orthogonal, oval open-loop fiber gyros capable of stable measurements of less than 1 degree per hour. Figure 2.3(a) shows a unit on display at an oil-drilling show in 1982. Improvements were also underway in packaging closed-loop fiber gyros. Figure 2.3(b) shows 9.5-cm-diameter units built for Eglin AFB. Packages such as those shown in Fig. 2.4 followed, which were 11 cm long, 4.5 cm high, and 5.5 cm wide. They could be stacked into a cube with accelerometers into a package about 11 cm on a side.

2.3 Looking for Error Sources, Finding New Sensors

In 1987, MDAC decided to license its closed-loop fiber gyro patents worldwide and concluded the first licenses during that year with US-based companies (extending licenses to Europe and Asia over the following years). I was not happy with the decision, as I had a plan in place to move toward production prototypes and felt that MDAC could establish itself in the inertial navigation field. MDAC management, in an effort to make the decision less painful, assured me that all internal funding going toward fiber gyros would

(a) (b)

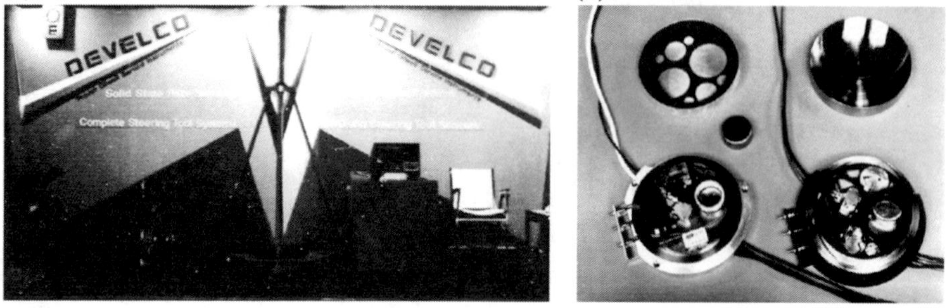

Figure 2.3 Improved rugged package for (a) a 1982 oil-drilling tool and (b) 1983 closed-loop fiber gyros.

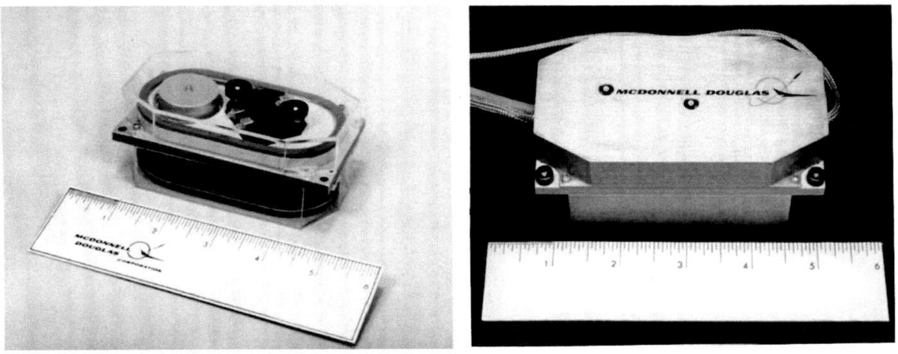

Figure 2.4 Closed-loop fiber gyros 11 cm long, 4.5 cm high, and 5.5 cm wide (1986).

now be applied to two other areas I was exploring: "fiber optic smart structures" and "secure fiber optic communication." This allocation allowed the fiber sensor group to remain intact. The technology supporting these efforts at the time was largely an outgrowth of fiber sensors derived from the Sagnac interferometer. The first sensor of this type was the Sagnac acoustic sensor,[7,8] which has a number of unique features, including low sensitivity in the center of the coil and higher sensitivity on fibers located near the central beamsplitter (see Fig. 2.5). Its response increases with frequency, and it can be used to create filters that are optimized for specific frequency ranges. This sensor has been used to support commercial site-security systems.

Another derivative sensor is the Sagnac strain sensor.[9,10] The early closed-loop fiber gyros that used acousto-optic modulators relied on a fixed-frequency offset that can be tuned to offset rotation-induced phase shifts. This offset results in a net phase shift between the counter-propagating light beams that is proportional to the length of the optical fiber, which causes the entire Sagnac loop to act as a strain sensor. For example, a unit with a 100-MHz offset and the ability to resolve frequency changes of 1 Hz can resolve changes in length on the order of 1 part in 10^8. This unit can be used as a very-long-gage-length strain sensor (Fig. 2.6). Initially, there was interest in using this functionality to monitor earthquake fault lines by integrating the strain

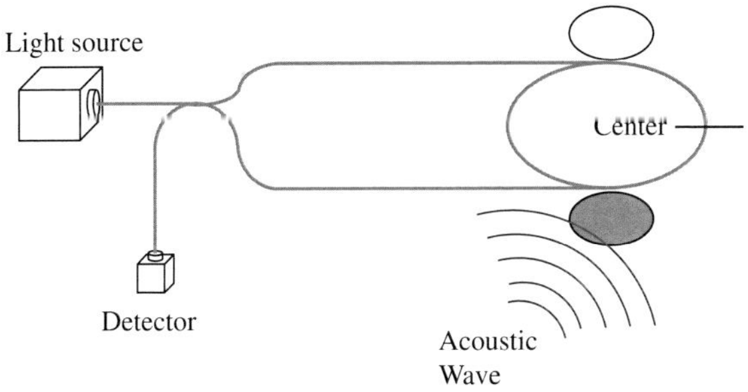

Figure 2.5 Sagnac acoustic sensor.

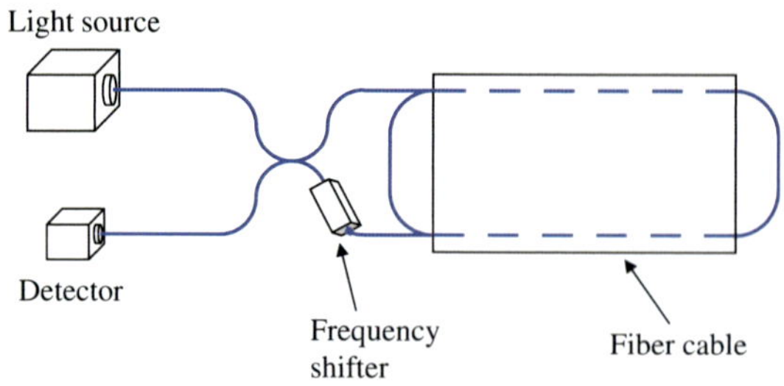

Figure 2.6 Long-gage-length Sagnac fiber optic strain sensor.

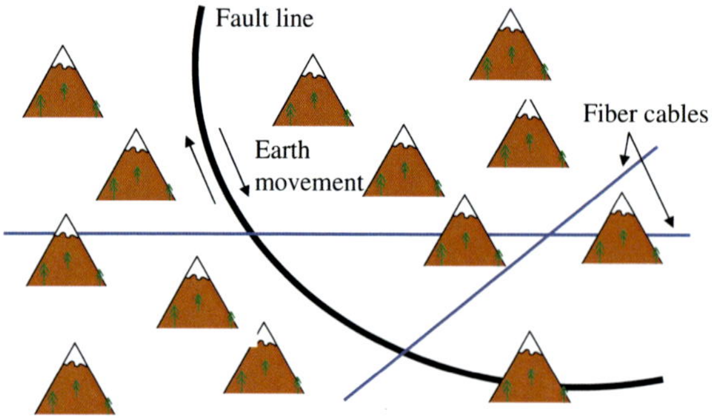

Figure 2.7 Measuring earth movement was an early application of the long-gage-length Sagnac fiber optic strain sensor.

sensors into cables (Fig. 2.7);[10] however, advances in cellphone and GPS technology reduced interest in this system.

By interlacing multiple Sagnac interferometers, distributed sensors intended to localize and measure time-varying events can be realized.[12] Figure 2.8 shows a configuration that was constructed using broadband light sources at 1300 and 1550 nm with standard 1300/1550-nm biconical taper WDMs. The sensitivity of each loop is zero in the center and increases as it moves away from the center position. By taking the ratio of a signature, a location can be identified, and by taking the sum, the amplitude may be measured.

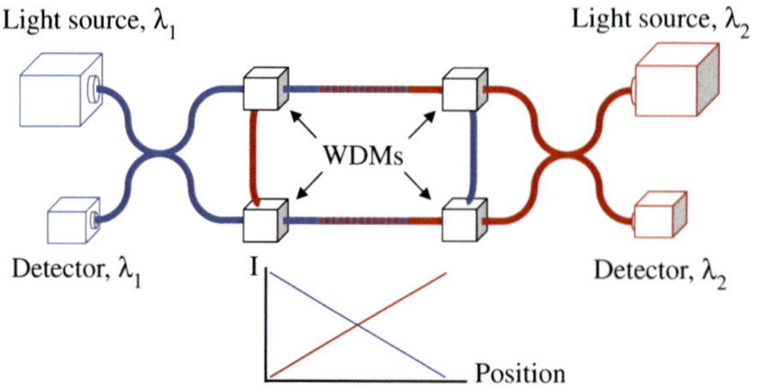

Figure 2.8 A Sagnac distributed sensor to locate and measure time-varying events formed by interleaving two Sagnac loops operating independently at 1300 and 1550 nm.

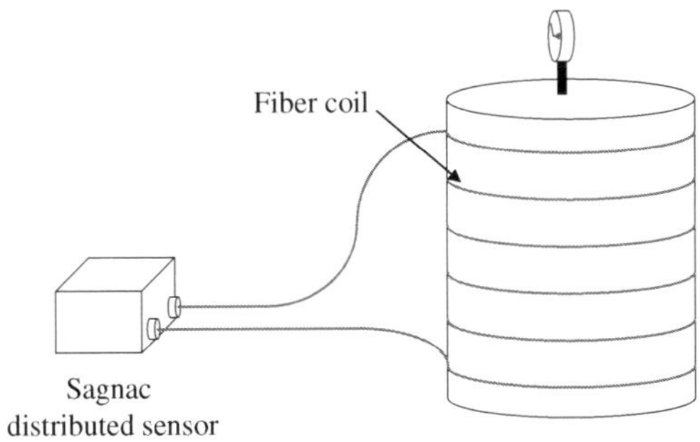

Figure 2.9 The Sagnac distributed sensor of Fig. 2.8 can be used to detect and localize insects in grain storage units and termites in wood.

One of the principal advantages of Sagnac acoustic and distributed sensors is that they can be supported by very-low-cost single-mode optical fiber, which enables a number of interesting applications, including the identification of leaks in pressurized pipes and containers, as well as the location of insects in grain-storage facilities or termites in wood. Figure 2.9 shows a potential application involving a pressurized tank or a grain silo.

A major Sagnac interferometer project that was undertaken as a derivative of the fiber gyro program involved secure fiber optic communication.[13,14] The system was intended to support 100 MBPS over 10 km without the use of encryption so that multiple sites could be rapidly and efficiently interconnected into networks. The prior art system used multimode optical fibers and was limited to 10 MBPS over 1 km. MDAC built and deployed this system, and then elected to license it to COMTEL, which in turn contacted Blue Road Research to support efforts to enhance the system. Link lengths operating at 1 GBPS over 50 km were achieved, and a series of physical improvements were devised.[15] Figure 2.10 shows the basic system and an implementation of inserting a datastream.

2.4 A Flow of Ideas

The motivation for optical gyros arose from issues associated with mechanical gyros that had major reliability issues for military and commercial aircraft and spacecraft. About half of the delays associated with mechanical issues delaying flights were associated with mechanical gyros. As demonstrated by Sagnac in 1913, rotation induces a change in the phase of counter-rotating light beams.[16,17] This was a difficult demonstration using bulk optics, and it was not until the invention of the laser that the possibility of a practical implementation occurred. This resulted in heavy investment by the US Air Force during the late 1960s and early 1970s to develop ring laser gyros.[18,19]

The ring laser gyro moved to production in 1976–1977 after hundreds of millions of dollars of government and industry investment. It began to replace

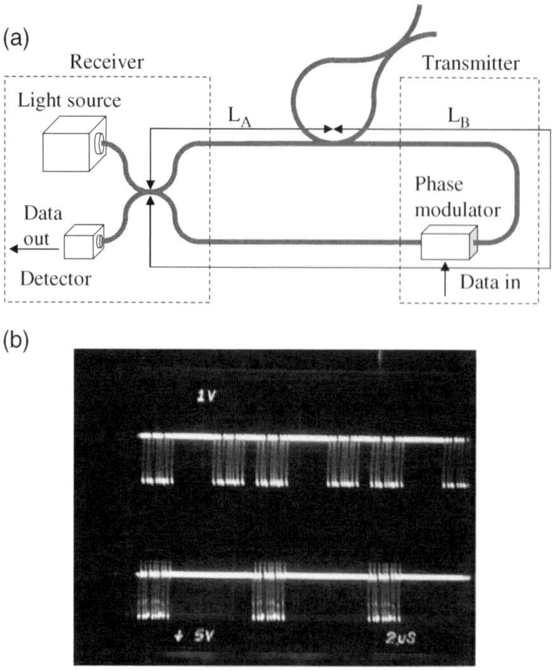

Figure 2.10 (a) Basic block diagram of the Sagnac secure fiber optic communication system and (b) input data blocks (lower trace) and output data blocks (upper trace).

mechanical gyros in major commercial and military aircraft by offering superior performance and improved reliability. Because a small amount of backscatter off mirrors and contaminating gases could cause a phenomenon called lock-in—where both counter-propagating light beams lased at the same frequency, resulting in no output from the ring laser gyro—ultra-clean manufacturing environments were required to produce high-quality ring laser gyros. Cost became a major issue, limiting the market for ring laser gyros to high-end aerospace platforms. Other major problems included difficulties in realizing smaller sizes and changing form factors. Smaller packages increased backscatter and lock-in issues, whereas changing form factors entailed major manufacturing costs. Furthermore, the ring laser gyro had the fundamental issue associated with a vacuum tube device.

Several inventors had suggested the possibility of using optical fibers to implement a rotation sensor. The lack of suitable components, notably single-mode optical fiber, precluded a demonstration until 1976, when Vali and Shorthill assembled the first bulk demonstrations.[20,21]

As described earlier, the fiber optic gyro of Vali and Shorthill was a nonlinear device, and the need for linearity and a wide dynamic range for launch-vehicle applications drove the invention of the closed-loop optical gyro. This was followed by a patent application that provides an overview of several methods of implementation,[22] including the use of acousto-optic, magneto-optic, and piezoelectric modulators to close the loop. The patent included a configuration using integrated optic components. Difficulties remained in producing suitable optic components, so the first closed-loop fiber optic gyro demonstrations were done using acousto-optic modulators and bulk optics supported by a gas laser[23] and an early semiconductor laser.[24] Many challenges remained, including polarization control. Ulrich and Johnson[25] first recognized the role that polarization plays in ensuring that counter-propagating light beams traverse the same path. Their work was followed by other papers by Ulrich and collaborators.[26,27] These efforts, as well as those by Shaw's group at Stanford,[28] found that a high-quality fiber gyro would require two beamsplitters and a polarizer in a configuration similar to that of Fig. 2.11.

The implementation of this configuration with the highest possible performance drove the development of high-quality components that could support polarization control. All optical versions were pioneered at Stanford,[29–31] followed by other groups in Germany and at the Naval Research Lab.[32–35] These developments showed that the fiber gyro could have the needed sensitivity and long-term stability.

Other key problems remained to be resolved. The cost and complexity of early closed-loop fiber gyros were prohibitive for an early introduction into the marketplace. It was realized that integrated optics could offer the prospect of a practical solution, and a sawtooth phase-modulated approach[36,37] made it possible to produce cost-effective optics with fiber gyros. Scale-factor accuracy and stability continued to be an issue until very stable, low-coherence light sources became available and improved methods were devised for controlled frequency shifts using integrated optical phase modulators.

As these issues were overcome, closed-loop fiber gyros increased their penetration into the marketplace.

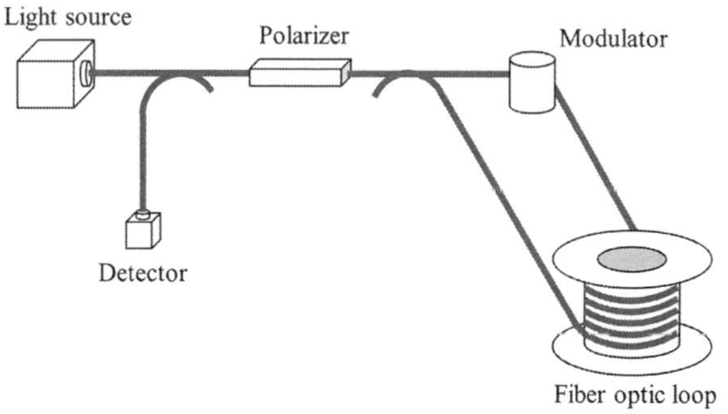

Figure 2.11 Minimum-configuration fiber optic gyro to ensure adequate polarization control.

2.5 Moving into Viable Products and Applications

In 1978, shortly after MDAC-HB started its fiber optic gyro program, Richard Cahill decided that he needed a better marketing tool than the bulk optics 1-m square closed-loop fiber gyro shown in Fig. 2.1, and so he made a few items to show internal and external customers what an all-solid-state fiber gyro could look like. A photo of these blocks is shown in Fig. 2.12. The fiber gyro envisioned had a diameter of 5 cm and a thickness of 1 cm. The three-axis fiber-gyro-based inertial system was 5 cm on a side (complete with accelerometers and all electronics). The long block used an elliptical-racetrack fiber gyro coil to demonstrate the flexible configurations of the fiber gyro. While the 6.3-cm fiber gyro built in 1980 approached the size of the wood disk, it was still supported by a rack of equipment. Since MDAC had advanced integrated circuit capabilities at the time, the availability of appropriate chips was accepted to be possible with sufficient funding.

Figure 2.13 shows a chart used for internal and external marketing in 1978 and reflects the perceived advantages of fiber optic gyro technology. Due to reliability and start-up issues, mechanical gyros were considered to be first-generation rotation sensors that would be replaced. The issue of cost would slow the initial introduction, but the performance advantages were so compelling that customers would be motivated. The issue with ring laser gyros versus fiber optic gyros revolved around the very large investment already made by the government and industry, and a ten-year development lead being difficult to overcome. The advantages of the fiber optic gyro with respect to ring laser gyros included an all-solid-state design that would be rugged and have long lifetimes, coupled with packaging

Figure 2.12 Wood models made in 1978 showing a 5 cm × 1 cm fiber gyro package, the geometry for a 2.5-inch three-axis fiber gyro inertial unit, and a single-axis package supported by an elliptical fiber gyro coil.

	Analog Fiber-Optic Gyro	Digital Fiber-Optic Gyro	Mechanical Gyro	Ring Laser Gyro
Digital Operation	No	Yes	No	Yes
Volume	Very Small	Small	Small	Large
Cost	Low	Moderate	Moderate	High
Fast Turn-on Time	Yes	Yes	No	Yes
Lifetime Limitation	Solid State	Solid State	Bearings	Gases
Packaging Flexibility	High	High	Moderate	Low

Figure 2.13 Marketing chart used in 1978 to support internal and external marketing for fiber gyros. The tradeoffs between mechanical, ring laser, and fiber gyros are shown.

flexibility that would be key for certain applications. The chart turned out to be quite accurate.

Figure 2.14 shows a second marketing chart from the late 1970s that was often used in combination with that of Fig. 2.13. It reflects the split between perceived markets for analog, open-loop fiber optic gyros (which would have

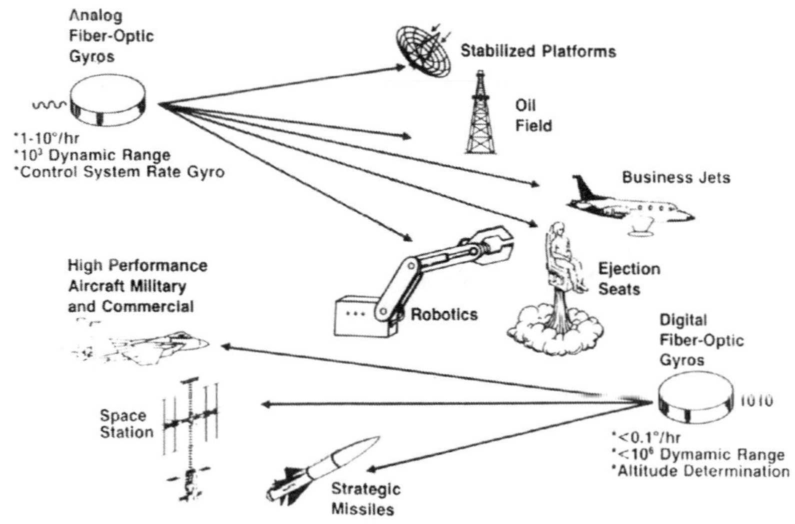

Figure 2.14 Marketing chart used in the late 1970s to support internal and external marketing for fiber gyros. Analog open-loop fiber gyros were considered a prime candidate for lower-accuracy applications where cost would be the key consideration, whereas closed-loop fiber gyros would be used for higher-performance projects with greater cost tolerance.

lower performance requirements and require the lowest possible cost) and digital, closed-loop fiber optic gyros (which would target the high-performance market and compete directly with ring laser gyros).

Just before MDAC licensed its fiber gyro patents worldwide, I estimated that the cost of moving toward a manufacturing base would be about ten million dollars. Privately, I have been told by people working for some of the major companies that the cost was closer to one hundred million. With advances in component development, these initial entry costs have dropped, and new entrants continue to move into the marketplace to compete with more-established companies.

So where are fiber gyros today in terms of fielded applications? They are used on launch vehicles, have supported missions to the moon, and are on all of the US rovers on Mars (Fig. 2.15). Space- and land-based antennas use them for stabilization, pointing, and tracking. They are used for the precise navigation of ships, commercial and private aircraft, and robotic and defense applications. Automobiles, delivery trucks, soccer-field grass cutters, and cleaning robots in malls have all used fiber optic gyros. There are competing technologies. MEMS-based devices have been introduced that are very low cost and compete directly with low-end fiber gyros. The combination of MEMS and GPS systems have replaced fiber gyros in the marketplace and enable navigation with cell phones. Low-end fiber gyros compete where higher bandwidth, stability, and sensitivity are required. High-end closed-loop fiber gyros continue to compete with ring laser gyros in the marketplace. They are competitive in applications where one or more factors are required: ultimate sensitivity, reliability, geometrical flexibility, and lifetime. Fiber-gyro and ring-laser-gyro manufacturers alike continue to improve their products to achieve greater penetration in each other's market.

When lunar and Mars colonies are established, fiber gyros can be expected to follow. They will be used to support functions similar to those being used

Figure 2.15 Fiber optic gyros are already on all Mars rovers and can be expected to play a bigger role in future lunar and Mars colonies, supporing surface- and space-based antenna stabilization, mining and tunneling, and robotic operations.

Figure 2.16 Launch vehicles will continue to use fiber optic gyros and with their long-term reliability and environmental hardness can be expected to play a major role in the next generation of affordable space access.

currently on earth, i.e., civil structures, mining, and tunneling. Figure 2.15 shows how they are and can be applied to the moon and Mars, but the utility of fiber gyros extends far beyond that. Their durability, passive nature, immunity to electromagnetic interference, high bandwidth, low maintenance, and performance capabilities will be necessary for colonies and for deep-space and interstellar missions. Many of these capabilities have begun to be realized in the past forty years, and as can be seen from Fig. 2.16, I have high hopes for the years to come and the next generation to lead the effort to implement the technology.

2.6 Summary and Conclusions

The closed-loop fiber optic gyro is one of a series of key developments that allowed the commercial realization of high-performance fiber optic gyros. As a result of its development at McDonnell Douglas, other fields associated with fiber optic sensors had early demonstrations, including "fiber optic smart structures," secure fiber optic communication, and a series of Sagnac acoustic and distributed sensors. The Sagnac interferometer continues to be a useful tool for a variety of sensing and communication applications.

The development of closed-fiber optic gyros into real products and viable applications resulted from the efforts of many organizations and companies. Hundreds of people have been actively involved in solving the many issues that limited its performance and allowed the realization of practical working devices that are now used widely.

Acknowledgments

I would like to thank my coworkers at McDonnell Douglas, Blue Road Research, and Columbia Gorge Research who have helped make many of my wild dreams associated with the Sagnac interferometer possible, especially my friend and mentor, Richard Cahill, during my early days at McDonnell Douglas; Bruce Turner and Paul Theriault, who patiently helped assemble the first solid state fiber gyros; Russ Johnson and Ed Hasagawa for electronics support; and Joe Kacin for wonderful mechanical designs. I would also like to acknowledge other members of the fiber sensor group at McDonnell Douglas: Jeff Eck, Stuart Higley, Ed Arriola, Ron Michal, Tim Clark, Steve Watanabe, Al Joseph, and Ray Wagoner. I would also like to single out Wil Otaguro and Roger Roberts, who offered encouragement and support for fiber optic gyro efforts. In addition to McDonnell Douglas sponsorship, I would like to thank the US Air Force for program support of the closed-loop fiber gyro from shortly after conception through 1989; in particular, I would like to thank Wright-Patterson AFB. Thanks to Wolfgang Schubel, our first technical monitor who patiently guided my efforts on producing clear and complete reports, and Dale Stevens, who continued to support our efforts after the untimely death of Wolfgang. I would also like to thank Captain John Ryan of Eglin AFB, who oversaw McDonnell Douglas efforts to produce their first tactical fiber gyros.

I would also like to thank the many people who have worked in other companies and organizations on the Sagnac interferometer, whose insights, inventions, and efforts have turned early dreams into reality.

References

[1] S. Ezekiel, and H. J. Arditty, Eds., *Fiber Optic Rotation Sensors*, Springer-Verlag, New York (1982).

[2] E. Udd, Ed., *Fiber Optic Gyros: 10th Anniversary Conference*, SPIE, Bellingham, WA (1986).

[3] R. B. Smith, Ed., *Selected Papers on Fiber Optic Gyros*, SPIE, Bellingham, WA (1989).

[4] S. Ezekiel, and E. Udd, *Fiber Optic Gyros: 15th Anniversary Conference*, SPIE, Bellingham, WA (1991).

[5] H. Lefevre, *The Fiber Optic Gyroscope*, Artech House, Boston (1993).

[6] W. K. Burns, Ed., *Optical Fiber Rotation Sensing*, Elsevier, St. Louis (1994).

[7] E. Udd, "Fiber optic acoustic sensors based on the sagnac interferometer," *Proc. SPIE*.**425**, 90 (1983).

[8] K. Krakenes, and K. Blotekjaer, "Sagnac interferometer for underwater sound detection: noise properties," *Opt. Lett.* **14**, 1152 (1989).

[9] R. J. Michal, E. Udd, and J. P. Theriault, "Derivative fiber optic sensors based on the phase nulling optical gyro," *Proc. SPIE* **719**, 150 (1986).
[10] E. Udd, R. Blom, D. Tralli, E. Saaski, and R. Dokka, "Application of the Sagnac interferometer based strain sensor to an earth movement detection system," *Proc. SPIE* **2191**, 126–136 (1994).
[11] E. Udd, "Usage of dispersive effects for scale factor correction in the fiber optic gyro," *Proc. SPIE* **1585**, 255 (1991).
[12] E. Udd, "Sagnac Distributed Sensor Concepts," *Proc. SPIE* **1586**, 46 (1991).
[13] E. Udd, "Secure Communication System," US Patent 5, 223, 967 (June 29, 1993).
[14] E. Udd, "Secure Communication System, US Patent 5, 274, 488 (December 28, 1993).
[15] E. Udd, M. M. Morrell, and P. F. Snawerdt III, "Single fiber Sagnac interferometer based secure communication system, US Patent 6, 690, 890 (February 10, 2004).
[16] G. Sagnac, "Sur la preuve de la realite de l'ether lumineux par l'experience de l'interferographe tournant," *Compt. Rend.* **157**, 1410–1413 (1913).
[17] E. J. Post, "Sagnac effect," *Rev. Modern Phys.* **39**, 475 (1967).
[18] J. Kilpatrick, "The laser gyro," *IEEE Spectrum* **4**, 44 (1967).
[19] F. Aronowitz, "The Laser Gyro," p. 131 in *Laser Applications*, M. Ross, Ed., Academic Press, New York (1971).
[20] V. Vali, and R. W. Shorthill, "Fiber laser gyroscope," *Proc. SPIE* **76**, 110 (1976).
[21] V. Vali, and R. W. Shorthill, "Ring interferometer 950 m long," *Appl. Opt.* **16**, 290 (1977).
[22] R. F. Cahill, and E. Udd, "Phase nulling optical gyro," US Patent 4, 299, 490 (1981).
[23] R. F. Cahill, and E. Udd, "Phase nulling fiber-optic laser gyro," *Opt. Lett.* **4**, 93 (1979).
[24] R. F. Cahill, and E. Udd, "Solid-state phase-nulling optical gyro," *Appl. Opt.* **19**, 3054 (1980).
[25] R. Ulrich, and M. Johnson, "Fiber ring interferometer: polarization analysis," *Opt. Lett.* **233** (1979).
[26] R. Ulrich, "Polarization and Depolarization in the Fiber Optic Gyroscope," p. 52 in *Fiber Optic Rotation Sensors*, Springer-Verlag, New York (1982).
[27] R. J. Fredricks, and R. Ulrich, "Phase error bounds of fibre gyro with imperfect polarizer/depolarizer," *Electron. Lett.* **20**, 330 (1984).
[28] G. A. Pavlath, and H. J. Shaw, "Birefringence and polarization effects in fiber gyroscopes," *Appl. Opt.* **21**, 1752 (1982).
[29] R. A. Berg, H. C. Lefevre, and H. J. Shaw, "All-single-mode fiber optic gyroscope," *Opt. Lett.* **6**, 198 (1981).

[30] R. A. Berg, H. C. Lefevre, and H. J. Shaw, "All-single-mode fiber optic gyroscope with long term stability," *Opt. Lett.* **6**, 502 (1981).
[31] H. C. Lefrevre, R. A. Bergh, and H. J. Shaw, "All-fiber gyroscope with intertial navigation short term sensitivity," *Op. Lett.* **7**, 454 (1982).
[32] K. Bohm, P. Martin, K. Peterman, E. Weidel, and R. Ulrich, "Low drift fibre gyro using a superluminescent diode," *Electron. Lett.* **17**, 352 (1981).
[33] K. Bohm, P. Russer, and E. Weidel, "Low noise fiber optic rotation sensing," *Opt. Lett.* **6**, 64 (1981).
[34] W. K. Burns, R. P. Moeller, and C. A. Villarruel, "Observation of low noise in a passive fibre gyroscope," *Electron. Lett.* **18**, 648 (1982).
[35] W. K. Burns, C. Chen, and R. P. Moeller, "Fiber optic gyroscope with broad band sources," *J. Lightwave Technol. LT-1*, 98 (1983).
[36] A. Ebberg, and G. Schiffner, "Closed loop fiber optic gyroscope with sawtooth phase modulated feedback," *Opt. Lett.* **10**, 300 (1985).
[37] H. C. Lefevre, S. Vatoux, M. Papuchon, and C. Puech, "Integrated optics: a practical solution for the fiber optic gyroscope," *Proc. SPIE* **719**, 101–112 (1986).

Eric Udd, President of Columbia Gorge Research LLC, has been deeply involved with fiber optic sensors since 1977 and helped pioneer early work on fiber optic gyros, fiber optic smart structures for health monitoring, multi-axis strain sensors, high-speed fiber optic sensors systems, and fiber optic pressure and temperature sensors. He worked for McDonnell Douglas from 1977 to 1993, where he managed over 25 DOD, NASA, and internally funded fiber optic sensor programs. He held a series of positions, moving from Engineer/Scientist to Fiber Optics Manager, and in 1987 was appointed as one of 40 McDonnell Douglas Fellows. He invented, designed, and built the first closed-loop fiber gyros. Udd's work on fiber optic smart structures resulted in the instrumentation of the composite hydrogen tank of the Delta Clipper to measure strain and temperature to assess its condition after each flight. The success of the Delta Clipper inspired Spacex and Blue Origin to further develop vertical landings and reusable launch vehicles. In 1993, he started Blue Road Research, Inc. (BRR) and directed the growth of the company through its acquisition by Standard MEMS in January 2000. In January 2006, he left BRR to work full time at his new company, Columbia Gorge Research LLC (CGR). His efforts at BRR and CGR have involved applications in aerospace, defense, civil structures, composite manufacturing, robotic surgery, oil and gas field services, electrical utilities, and energetic material.

Udd has 54 issued US patents with many issued internationally, written over 200 papers, and chaired and helped organize 39 international conferences on fiber optic sensor technology, including the 10th, 15th and 20th Fiber Optic Gyro Anniversary Conferences. His books include *Fiber Optic Sensors: An Introduction for Engineers and Scientists, Second Edition* (Wiley, 2011), *Fiber Optic Smart Structures* (Wiley, 1995), and *Field Guide to Fiber Optic Sensors* (SPIE Press, 2014). He is a Fellow of SPIE and OSA, and a member of IEEE. He was awarded the Richardson Medal in 2009 by the Optical Society for his work on fiber optic sensors and the field of fiber optic smart structures.

Chapter 3
20 Years of KVH Fiber Optic Gyro Technology: The Evolution from Large, Low-Performance FOGs to Compact, Precise FOGs and FOG–Based Inertial Systems

Jay Napoli
KVH Industries, Inc.

3.1 Introduction

Precision gyroscopes have become critical components within an expanding array of platforms and applications ranging from guided weapons to unmanned submersibles, and from camera and radar assemblies to automobiles, gun turrets, and virtual-reality simulators. KVH open-loop fiber optic gyros and integrated inertial systems, developed over the course of 20 years, offer tactical-grade performance and excellent reliability within a compact package for a reasonable cost. The KVH family of FOGs was made possible through innovations in several key fields, including:

- proprietary single-mode, polarization-maintaining, D-shaped optical fiber, with an elliptical core;
- revolutionary ThinFiber;
- innovative optical circuit design, component fabrication, and system integration; and
- patented digital signal processing (DSP) for improved performance.

This combination provides innovative technology and an outstanding cost-to-performance ratio that compare well against other competing tactical-grade technology currently available on the market.

All of the fiber components needed to assemble the FOGs are made at the KVH manufacturing facility in Tinley Park, Illinois. This allows KVH to produce its FOGs in a highly vertically integrated environment, with complete control over the quality, cost, and supply of the components. The gyros are either sold as single-axis sensors or assembled into higher-level, multi-axis products that can be integrated with accelerometers to produce inertial measurement units (IMUs). These IMUs are often coupled with GPS receivers to produce an inertial navigation system (INS).

In 1997, KVH announced that it had acquired the assets of Andrew Corporation's Sensor Products group, including a unique fiber optic gyroscope, vehicle land-navigation systems, and various fiber optic products and technologies. The intent of the acquisition was to integrate these technologies more fully into the wide range of KVH communications and navigation systems that it designed and manufactured for military, commercial, and marine applications, as well as to develop new products suitable for a number of new markets.

In 2011, the production of KVH FOG axes surpassed 50,000 units. By the end of 2016, that number had climbed to more than 100,000 gyros, both as standalone systems and as fully integrated components within high-performance KVH IMUs and INSs.

3.2 Superior Performance through End-to-End Manufacturing

KVH is one of the only FOG manufacturers in the world to fabricate 100% of the optical fiber and fiber components used in its FOG products. The company undertakes all aspects of FOG manufacture, from drawing the fiber to manufacturing optical components out of the fiber, to assembly of the optical circuit, and then final integration and calibration with the FOG electronics. This production chain gives the company complete control over the various processes involved and ensures the highest level of quality. KVH utilizes a group technology approach to every FOG design. There are four main optical components, all of which are assembled from proprietary KVH optical fiber. These optical components are then spliced together to make a total of three optical circuit variants that are then housed and mated to printed circuit boards to create three basic types of gyros. This versatile design concept means that once software is loaded, KVH can produce about 60 different types of single-axis products. The single-axis FOGs can then be assembled into multi-axis FOGs as well as products such as IMUs and INSs.

3.2.1 At the heart of the FOG: creating the fiber

More than 35 years of research into fiber design has resulted in the KVH E•Core fiber, a proprietary class of D-shaped-cross-section, non-stress-induced fiber with an elliptical core that acts as an optical waveguide for the FOG, resulting in stable performance over a wide temperature range. These attributes of the E•Core fiber make KVH FOGs extremely well-suited for stabilization, pointing, and navigation applications.

The E•Core fiber minimizes thermal effects and has a high tolerance for stress due to bending and other factors. The fiber provides distinctive sensor capabilities that facilitate the manufacture of couplers and polarizers. The fiber's D-shape permits easy access to the light-guiding region without grinding or polishing. Additionally, the major axis of the elliptical core is parallel to the flat of the D, allowing accurate alignment of the birefringent axis.

Most FOG manufacturers start instead with inexpensive telecommunication optical fiber that requires numerous additional processes related to coil winding and signal processing in order to perform rate measurement. KVH, on the other hand, draws its own polarization-maintaining (PM) fiber from pure silica tubes.

The process of creating the fiber begins by manufacturing a preform from a silica tube using chemical vapor deposition (CVD) techniques. The CVD process takes about eight hours to complete and involves numerous parameters that must be properly adjusted. The tube of silica glass is placed into one of several preform towers at the KVH manufacturing facility (Fig. 3.1). Digitally controlled mass-flow controllers meter out specific chemical gases into the interior of the 1-m-long silica tube. The tube travels past a heating element at 1700°C to sinter the gases to the inner surface of the tube.

Figure 3.1 Creating the glass preform: a chemical vapor deposition preform tower.

Ultra-pure chemicals, including various rare-earth dopants, are deposited into the interior diameter (ID) of the tube while slowly passing it through a heating element. This sinters those chemicals within the ID of the tube and begins the process of deposition of index layers—which ultimately becomes the cladding and core of the fiber. Another pass with the tube through a vacuum collapses the tube into a solid rod, which is essentially a short, inflexible macro optical fiber fully capable of guiding light.

Once the solid rod has been produced from the silica tube, it undergoes a number of processes, the end result of which is that the round, 1-m rod is transformed into a D-shape, with the flat side of the rod parallel to the long axis of the elliptical core.

The D-shaped preform rod is then placed into a tall fiber-drawing tower (Fig. 3.2), where it is precisely heated and drawn into a strand of PM fiber. The fiber can then be wound onto spools for later testing and use within the FOGs. A single 1-m preform is capable of producing up to 5 km of KVH E•Core fiber, which is no thicker than a human hair. The fiber itself, with an external coating (jacket) that is applied during the draw process, is either 170 or 240 microns in diameter; the light-guiding elliptical-shaped fiber core is 1×3 microns.

The optical components of a FOG are produced in batches and then spliced together to produce an optical circuit. The optics are housed and

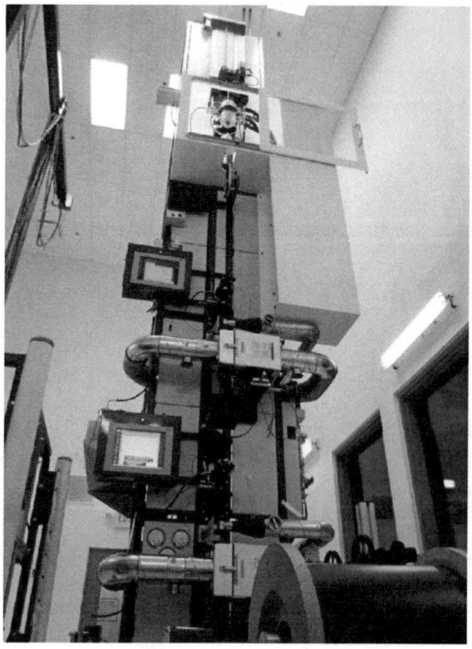

Figure 3.2 The fiber-drawing tower, which can produce up to 5 km of E•Core fiber from a single 1-m preform tube.

circuit boards are attached to make the final product. Before shipping, the gyros undergo extensive environmental screening and calibration over military temperature extremes. Complete documented acceptance-test reports are generated for both single-axis and multi-axis products.

Each component and assembly within every FOG is 100% tested, and each process and test machine is connected to a central database. All of the production and test data are recorded, which provides an enormous resource of real-time data to track quality yields and productivity with standardized charts. It also gives KVH the capability to correlate final-product performance data to the original fiber data for continual improvement and increased yields of each component. As an added bonus, customers appreciate the complete traceability from top-level assemblies back to actual workstations and build dates to facilitate root-cause analysis, if required.

Since each FOG component is serialized and tracked through assembly and test, it is possible to take any top-level serial number and trace any single component back to the rig on which it was made, on what day, and by which operator. This system extends back to raw tubes used to produce the optical fiber preforms. KVH can data mine the system and answer such questions as, "What lower limit on the coil-fiber cutoff wavelength is acceptable to produce a specific level of end-product performance?" The ability to answer this type of question allowed KVH to open up the tolerance band on one fiber parameter without affecting the end-product performance. This change improved fiber yield and produced a lower-attenuation fiber, further improving yield in the upper assembly areas.

FOG coil lengths typically vary from 80 to 200 m, which is usually wound on a bobbin using a standard thread- or level-winding technique (Fig. 3.3).

Figure 3.3 A quadrupole winder, which begins winding fiber onto the bobbin from the center point of the fiber rather than from end-to-end, permitting far more precise fiber alignment at each layer. This results in several performance benefits over the FOG's handling of temperature variations, shock, and vibration.

In a FOG with a level-wound coil, the fiber is wound onto the bobbin by starting at one end of the fiber and precisely rotating the bobbin, with the fiber under monitored tension, until the appropriate amount of fiber has been utilized.

When higher FOG accuracy is necessary, KVH utilizes more-advanced techniques such as the quadrupole winding method, where optical fiber is wound onto the bobbin starting from the center point of the fiber length. The fiber is wound precisely to ensure perfect alignment, which results in significant performance improvements in terms of reduction of the well-known Shupe effect compared to level-wound FOGs. This quadrupole winding process is used in the KVH TG-6000, as well as the KVH 1725, 1750, and 1775 IMUs, and the DSP-1750 and DSP-1760 FOGs.

3.2.2 The core design of KVH open-loop FOGs

KVH FOGs are based on an open-loop design consisting of a broadband solid state optical source and PM fiber, and fiber components fabricated from E•Core and ThinFiber (see Fig. 3.4). The light energy passes first through a proprietary 3-dB directional coupler that isolates the detector and then through a polarizer to ensure that a single polarization state enters the fiber coil. The polarized light is then split into two paths by the second directional coupler and directed to opposite ends of a coil of E•Core or ThinFiber. The fiber coil, with counter-propagating light waves, then serves as the rotational sensing element.

The scientific principle behind the FOG is the Sagnac effect, which is also the basis of the ring laser gyro. When the fiber coil is not rotating, the optical

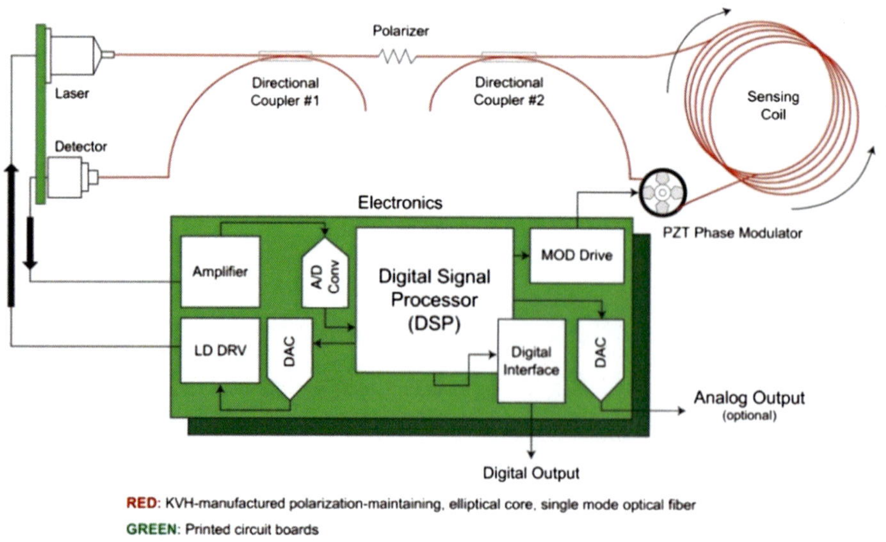

Figure 3.4 A basic functional diagram of a KVH open-loop FOG.

path length in both directions is the same, so the two counter-rotating optical signals are in phase upon their return to the detector. Rotating the coil introduces an optical phase difference in the counter-rotating light paths known as the Sagnac phase shift. The phase difference in the two paths results in a change in amplitude of the recombined signals proportional to the rotation rate. As such, the FOG is often referred to as an interferometer.

The piezoelectric (PZT) modulator is driven with a sinusoidal signal to bias the Sagnac interferometer at the most-sensitive operating point. The open-loop gyroscope has a first-order sinusoidal response to rotation, and the SF is dependent on the optical intensity and modulation depth. The interferometer converts this modulation into an output signal comprising harmonics correlating to the Bessel functions. All of the required information to determine the rotation rate and stabilize the SF is extracted from the fundamental signal up to its fourth harmonic.

This light amplitude is measured at the detector and, after processing, results in an output that is proportional to the rotation rate about the main axis of the sensing coil. This design permits KVH to create exceptionally high-performance open-loop gyros that use minimal lengths of fiber and number of couplers, as well as low-cost light sources.

The FOG has no moving parts, resulting in enhanced reliability. There are no cross-axis sensitivities to vibration, acceleration, or shock. In addition, the FOG is stable with temperature and time, making it useful in a wide variety of applications including land navigation, stabilization and positioning systems, robotics, and instrumentation.

3.2.3 Design advantages

Topping the list of design advantages held by KVH FOGs is its E•Core and ThinFiber D-shaped PM optical fibers. The elliptical core of the fiber acts as a light-guiding path that ensures that the desired polarization of light is maintained as it travels through the optical circuit, resulting in a strong optical signal with low attenuation. The geometry-induced birefringence of the elliptical core exhibits a high birefringence level to prevent polarization coupling along the coil, with minimal temperature-induced birefringence changes compared to typical stress-induced PM-fiber performances. As a result, signals returning to the coil coupler are well aligned with the fiber polarizer. Temperature-induced polarization drifts are minimized, which is crucial to reduce the bias drift in the output of the FOG.

For several decades, KVH produced its E•Core optical fiber with a diameter of ~240 microns. This fiber is still utilized today in some KVH FOG products. However, in an effort to reduce the physical size of its FOGs without sacrificing (and instead improving) performance, KVH developed its E•Core ThinFiber in 2011. ThinFiber has a diameter of ~170 microns, including the jacket. This thinner fiber means that a longer fiber length can be wound on a smaller bobbin, which translates into several improved performance attributes. In addition to the overall performance, the D-shaped E•Core and ThinFiber permits easy access to the light-guiding core, which is important when producing couplers and polarizers as it provides an excellent physical surface for mating fibers.

Another essential design element in high-accuracy KVH FOGs is the use of the company's patented DSP technology. Once the optical output from the FOG is picked up at the detector and transformed into a voltage, this voltage is passed through an analog-to-digital converter. The DSP approach allows KVH engineers to devise effective methods for analyzing and correcting this digitized data on the fly, a capability not possible with analog signal processing. A simple example is the way KVH FOGs monitor their internal temperature and then compensate for temperature errors by using a lookup table stored in memory. The DSP approach and other improvements in electronics and software have resulted in an increase in the accuracy of KVH FOGs by orders of magnitude compared to earlier KVH FOG variants, as will be shown next.

3.2.4 Key gyro performance factors

There are five critical areas where industrial- and tactical-grade gyros must outperform consumer-grade gyros while maintaining a small form factor and low cost.

- **Angle random walk (ARW)**: The output of a gyro includes a broadband, random (white) noise element. The ARW describes the average deviation or error that will occur as a result of this noise element. It can be obtained from the Allan variance value at the 1-s crossing time (IEEE-std-952, Specification Format Guide and Test Procedure for Single-Axis Interferometric Fiber Optic Gyros). At short averaging times (horizontal axis of the Allan variance plot), sensor noise dominates the Allan variance and is given as a −1/2 slope.

  Major contributors to random noise are the active elements of the gyro, such as the laser diode and photodiode in a FOG, and the silicon or quartz vibrating beam and detection electronics in micro-electro-mechanical systems (MEMS) gyros. There are large noise performance differences among the various FOG and MEMS gyro manufacturers, so designers typically pay particular attention to this parameter. An excessive amount of gyro noise results in loss of sensor precision and accuracy in the rate or position measurement.
- **Bias offset error**: A stationary gyro can incorrectly register some rotation; this is called "bias offset" error. Its deviation from zero is typically given at 25°C for an ideal environment (i.e., no temperature

change, vibration, shock, or magnetic field applied). The bias offset is typically measured by an end-user in their system or application and calibrated out of actual rate measurements.

- **Bias instability** (at constant temperature) is the stability of the bias offset at any constant temperature and in an ideal environment. The bias instability is best measured using the Allan variance (IEEE-std-952) measurement technique. It is a fundamental method of deriving performance specifications for all major gyro types (FOG, MEMS, and ring laser gyros (RLGs)). The standard definition of bias instability used by inertial sensor manufacturers is the minimum point on the Allan variance curve—the lower this value is, the better the bias performance. The Allan variance is a measurement of all noise processes, both electrical and optical, including the thermal noise fluctuations of the sensing coil's fiber waveguide effective index, shock, or vibration effects. While a gyro's constant bias offset is usually calibrated out, the bias instability introduces an error that may not be easy to calibrate. Due to bias instability, the longer a gyro operates, the greater its accumulated rate or position error—a lower bias instability is critical for applications that need excellent accuracy over long periods.
- **Temperature sensitivity** is a vital performance factor for gyros because this performance can be impacted by changes in temperature. The parameters sensitive to temperature include noise, bias offset, and SF stability. System integrators characterize these parameters over temperature, including temperature rate of change, to confirm that the performance meets their system targets. KVH FOGs have internal temperature-calibration tables for bias and SF correction versus temperature (see Section 3.2.3).
- **Shock and vibration sensitivity** are essential factors because the gyro performance can degrade under certain vibration and shock input beyond the specification limits of the FOG. Shock- and vibration-induced errors are not easily calibrated. Vibration performance is important in many industrial and military applications as gyros must perform accurately in the presence of a random vibration and/or shock inputs. To help minimize vibration issues, designers typically locate the gyro where vibrations are dampened and utilize anti-alias and decimation filtering. The more resistant a gyro is to vibration, the more reliable the performance and the more freedom the designer has for gyro location within the system.

Fiber optic gyros are known for high performance in the five key performance parameters vital for navigation, control, and stabilization, including a low ARW, small bias offset, excellent bias instability (or low drift), and reduced temperature, vibration, and shock sensitivity.

3.3 Evolution of the Technology

3.3.1 The creation of D-shaped, elliptical-core fiber

KVH D-shaped E•Core fiber (Fig. 3.5) offers a number of advantages, including high PM ability (greater than –40 dB-m) and low sensitivity to stress and temperature. These features make E•Core fiber ideal for fiber splicing and pigtailing to other optical components since preserving the state of polarization is critical to the operation of the sensor.

The elliptically shaped core region within KVH optical fiber has a significantly higher index than the surrounding cladding, creating the geometrical birefringence necessary to maintain the polarization of the light within the fiber. Because of its design, it is able to maintain its polarization and low-loss characteristics when exposed to physical stresses and temperature-induced changes far better than PM fiber created using stress-induced birefringence. Both E•Core and ThinFiber are manufactured using high-grade silica materials and various high-purity dopants. A dual UV-cured acrylate coating is applied during the drawing process, thereby preventing moisture from penetrating the fiber, which preserves its mechanical strength and improves its handling characteristics.

The D-shaped fiber maintains excellent optical qualities while adding a special external geometry that provides easy access to the light-guiding region without grinding or polishing. Also, the major axis of the elliptical core is parallel to the flat of the "D," allowing accurate alignment of the birefringent axes. The 1- × 3-micron elliptical core is also well matched to the 1- × 5-micron output facet of the light source, as shown in Fig. 3.6, and optimizing the core-to-facet alignment maximizes the power coupled into the FOG's optical circuit.

3.3.2 The first generation of KVH FOGs

The initial commercial FOGs produced by KVH in 1997 using E•Core fiber were offered as part of the E•Core series. Available in both analog and digital

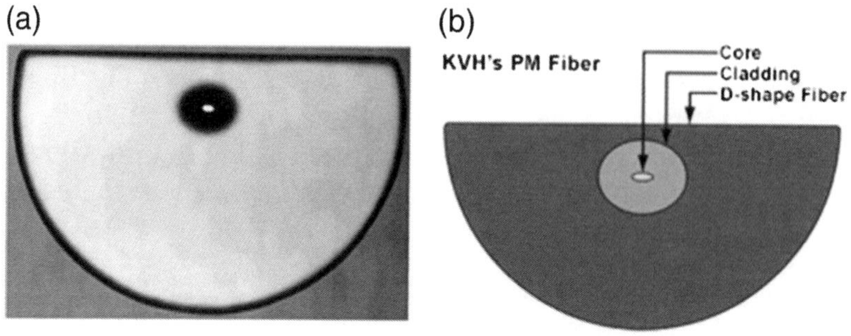

Figure 3.5 (a) An electron micrograph and (b) illustration of the cross-section of KVH D-shaped elliptical-core fiber.

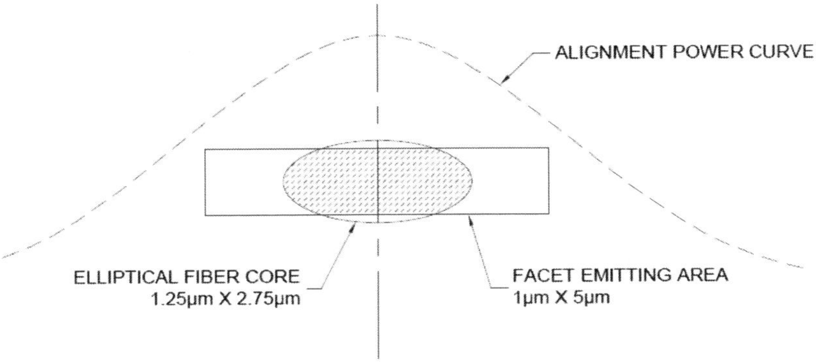

Figure 3.6 An illustration of the alignment between the elliptical core of KVH fiber and the output facet of the light source.

variants, the E•Core 1000 and E•Core 2000 were compact for their day (Figs. 3.7 and 3.8), with moderate but solid performance (see Table 3.1 for specifications).

3.3.3 The shift to digital signal processing

In 2000, KVH engineers refined all aspects of the FOG design, with an eye toward size and weight reduction and a substantial increase in performance. One of these improvements included a shift to digital signal processing (DSP). The new signal processing involves synchronously detecting the fundamental component of the piezoelectric modulation frequency, whose magnitude is proportional to the rotation rate. The system digitally samples the gyro detector output signal at a frequency that is an integer multiple of the modulator drive frequency. A fast Fourier transform (FFT) of the sampled

Figure 3.7 The original KVH E•Core 1000 open-loop fiber optic gyro.

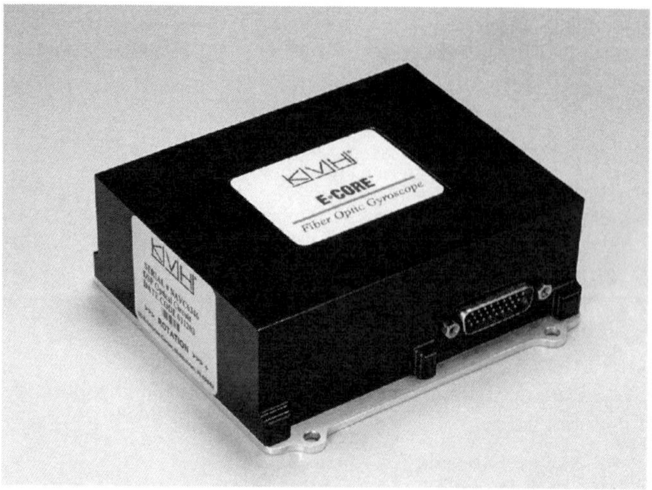

Figure 3.8 The original KVH E•Core 2000 open-loop fiber optic gyro.

Table 3.1 E•Core 1000 and E•Core 2000 basic specifications.

Attribute	Specification	Specification
Performance	**E•Core 1000**	**E•Core 2000**
Input rotation rate	± 100°/s	± 30, 60, or 100°/s
Angle random walk	$0.3°/\sqrt{h}$	$0.08°/\sqrt{h}$
Bias instability	≤ 20°/h	< 4°/h
Scale factor linearity	5,000 ppm	2,000–4,000 ppm
Physical		
Dimensions	4.3" × 3.3" × 1.6" (109 × 84 × 41 mm)	4.41" × 4.27" × 1.69" (112 × 108 × 43 mm)
Weight	0.55 lb. (0.25 kg)	0.75 lb. (0.34 kg)

Table 3.2 DSP-3000 basic specifications.

Attribute	Specification
Performance	
Input rotation rate	± 375°/s
Angle random walk	$\leq 0.067°/\sqrt{h}$
Bias instability	≤ 1°/h
Scale factor linearity	< 500 ppm
Physical	
Dimensions	3.5" × 2.3" × 1.3" (88.9 × 58.4 × 33 mm)
Weight	0.6 lb. (0.26 kg)

data generates the coefficients corresponding to the fundamental modulator drive frequency and at the second, third, and fourth harmonics. As a result, the system can maintain a constant modulation depth and provide sensor output data over a wide operating range (see Table 3.2).

Figure 3.9 The KVH DSP-3000 FOG family, a series of DSP–based open-loop gyros.

The result of KVH's migration to DSP was a new series of FOGs called the DSP-3000 that incorporated proprietary DSP technology (introduced in Section 3.2.3) in a package slightly larger than a deck of cards and dramatically more capable than the predecessor E•Core systems (Fig. 3.9).

3.3.4 Changing the game: the invention of ThinFiber

In 2011, KVH's engineering efforts led to a breakthrough in size and performance: E•Core ThinFiber, which has about 30% less volume than the standard KVH E•Core fiber. This development enabled KVH to utilize twice as much optical fiber on a significantly smaller bobbin and to dramatically increase performance. This breakthrough in fiber technology has set a new standard in fiber optic gyros for size, performance, and cost.

The first of the new FOGs to be launched with this new fiber was the DSP-1750 in 2009, the world's smallest high-performance FOG (see Table 3.3 for specifications). This super-compact commercial FOG offers revolutionary performance levels for a unit its size, including input rates five to ten times faster than the original KVH E•Core FOGs and an ARW five times better than the KVH DSP-3000 series. In addition, its bias stability is eight times better than the DSP-3000 series, achieving a level of performance previously only available in larger, more expensive closed-loop FOGs and RLGs.

The design of the DSP-1750 FOG provides system integrators with maximum versatility thanks to its single- or dual-axis configurations (see Figs. 3.10 and 3.11), a small size that enables various mounting and interfacing possibilities, and optional magnetic shielding. Due to its design, in which the

Table 3.3 DSP-1750 basic specifications.

Attribute	Specification
Performance	
Input rotation rate	± 490°/s (standard) or ± 1000°/s
Angle random walk	≤ 0.013°/√h
Bias instability	≤ 0.1°/h
Scale factor linearity	< 200 ppm
Physical	
Dimensions	1.7" diameter × 0.9" height (45.7-mm diameter × 22.9-mm height)
Weight	Single axis: 0.24 lb. (0.11 kg) Dual axis: 0.30 lb. (0.14 kg)

Figure 3.10 The KVH DSP-1750 FOG single-axis FOG, the first to be built using KVH ThinFiber technology.

Figure 3.11 The KVH DSP-1750 dual-axis FOG with tethered sensing coils for flexible installation.

optical sensor is tethered to the control electronics, the DSP-1750 is extremely easy to integrate into a broad range of systems and applications.

Just over a year later, KVH further enhanced ThinFiber-based FOGs with the introduction of the DSP-1760 (see Table 3.4 for specifications). Available as a housed or unhoused OEM variant in one-, two-, or three-axis configurations (Fig. 3.12), the DSP-1760 series is the logical progression in KVH's ongoing improvements in technology. It is built using the higher-performance quadrupole winding (discussed previously in Section 3.2.1) and improved software and algorithms, while taking advantage of higher processing power and improved electronics. The result is a versatile, single- or multi-axis FOG with improved performance but no increase in cost.

3.3.5 Expanding capabilities with high-performance fully integrated systems

Beginning in 2009, KVH offered IMUs and INS based on the DSP-3000, but the major breakthrough in size, price, and performance came in 2012 with the

Table 3.4 DSP-1760 basic specifications.

Attribute	Specification
Performance	
Input rotation rate	± 490°/s
Angle random walk	≤ 0.012°/$\sqrt{}$h
Bias instability	≤ 0.05°/h
Sale factor linearity	< 50 ppm
Physical	
Dimensions	3.5" diameter × 2.9" height (88.9-mm diameter × 73.7 mm height)
Weight	Single axis: 1.1 lbs. (0.5 kg)
	Dual axis: 1.2 lbs. (0.54 kg)
	Three axis: 1.3 lbs. (0.60 kg)

Figure 3.12 The KVH DSP-1760 FOG, which offers one-, two-, and three-axis configurations in both housed and unhoused variants for increased versatility.

introduction of the 1750 IMU, which combined the E•Core ThinFiber-based DSP-1760 with very-low-noise, solid-state MEMS accelerometers to create a commercial off-the-shelf IMU (Fig. 3.13). Since then, the high-performance DSP-1760 FOG has served as the foundation for all new KVH FOGs and integrated inertial system platforms. Considering the many design, performance, and cost advantages that KVH DSP-1760 and related ThinFiber FOGs offer, it follows that these versatile units present ideal solutions for integrated IMUs and INSs.

The KVH 1750 IMU provides

- highly accurate six-degree-of-freedom angular rate and acceleration data,
- excellent bias instability (<0.05 deg/h),
- a cylindrical package that is only 89 mm in diameter and 74 mm in height (3.5" diameter × 2.9" height), and
- weight less than 0.6 kg (1.4 lbs).

Recognizing that different applications may have dramatically different performance versus cost needs, the 1750 IMU served as the platform for an expanded family of IMUs, including the ultra-high-performance 1775 IMU and the 1725 IMU, which offers FOG performance for the cost of a MEMS system.

The 1775 IMU is designed to deliver the highest level of performance of the IMU product line offered by KVH. The 1775 IMU is an advanced inertial sensor system that utilizes the proven technology of the DSP-1760 fiber optic

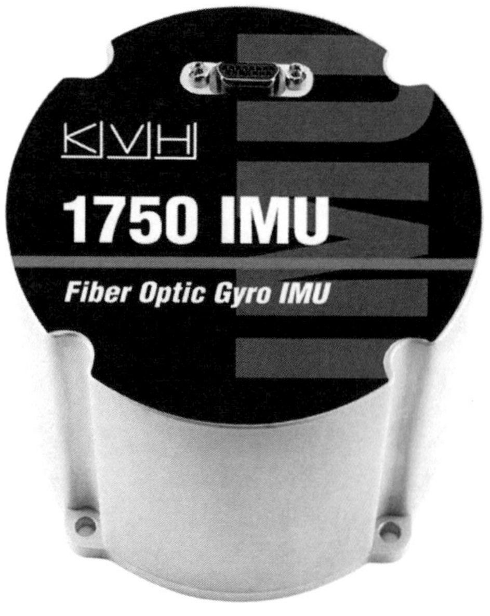

Figure 3.13 The KVH IMU 1750, the first high-performance KVH inertial system based on the E•Core ThinFiber technology.

gyro, integrated with three low-noise accelerometers, and three axes of magnetometers for automatic gyro-bias compensation even in the presence of strong magnetic fields. The 1775 IMU was developed for systems and applications where very high bandwidth and low latency, noise, and drift are critical to system performance.

The 1775 IMU (Fig. 3.14) offers easy integration for designers of higher-level inertial navigation, guidance, and stabilization systems by offering user-programmable features such as an adjustable baud rate, as well as an adjustable data rate as high as 5,000 Hz so that communication latency can be adjusted to receive accurate, timely data. These features are particularly important for such challenging applications as drilling, mining, pipeline inspection and maintenance, mobile mapping systems, and stabilization systems for radar, LIDAR, and very-high-speed gimbals.

Developed as a highly accurate alternative to MEMS–based inertial systems, the 1725 IMU offers the performance of a FOG–based inertial system at prices comparable to MEMS systems. The 1725 IMU includes three DSP-1760 FOGs and three low-noise accelerometers, and was designed for applications in which affordability, high performance, and high bandwidth are critical (Fig. 3.15). This cost-effective FOG–based IMU highlights flexibility with multiple interfaces, programmable message outputs, and baud and data rates. The 1725 IMU is scalable with high bandwidth and low noise, so it can provide the data output required for challenging applications and still

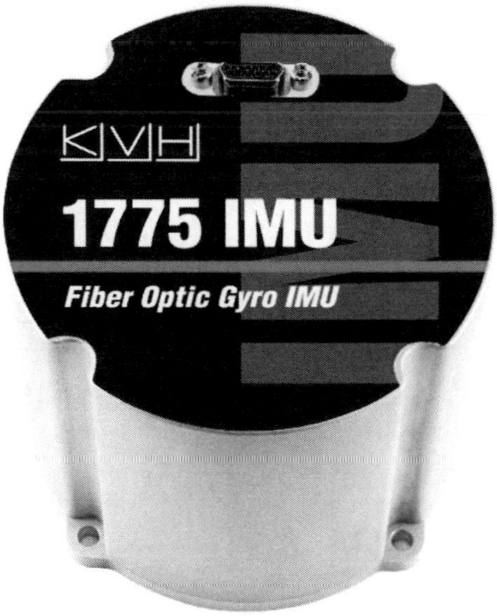

Figure 3.14 The KVH 1775 IMU is the highest-performing KVH FOG–based inertial system based on the E•Core ThinFiber technology. The 1775 IMU is available with ±10-g or ±25-g accelerometers.

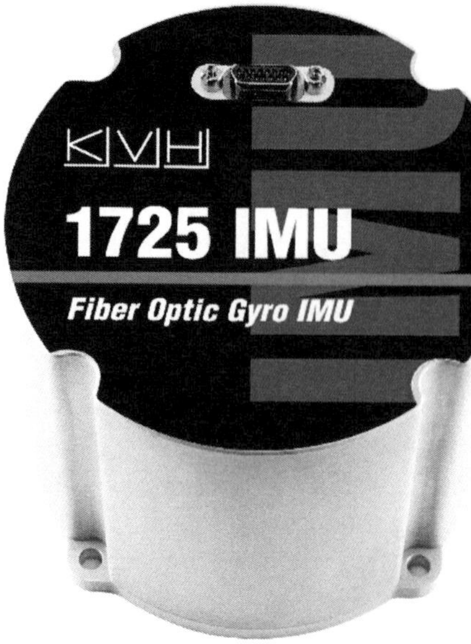

Figure 3.15 The KVH 1725 IMU brings FOG–based performance and accuracy to an affordable inertial system; it was designed to offer a high-performance alternative to MEMS–based inertial systems.

help maintain program costs. The goal of the 1725 IMU is to make the engineer's traditional choice between performance and price obsolete.

Most recently, the 1750 IMU became the foundation for an entirely new inertial navigation system, the GEO-FOG 3D and GEO-FOG 3D Dual. These new products offer full positioning and navigation capabilities for demanding applications in unmanned, autonomous, and manned aerial, ground, marine, and subsurface platforms, such as sub-sea remotely operated vehicles (ROVs) or mining systems. To achieve this performance level, the 1750 IMU is fully integrated with a three-axis magnetometer, a barometric pressure sensor, and a dual- or triple-frequency real-time kinematic (RTK) Global Navigation Satellite System (GNSS) receiver to deliver reliable, real-time, centimeter-level positioning and orientation measurements.

The system's breakthrough sensor-fusion algorithms automatically switch from loosely- to tightly-coupled filtering for improved performance under poor GNSS signal conditions. The system also offers high-speed update rates and rapid north-seeking gyrocompass capabilities for high-accuracy heading in environments when magnetometers and GNSS-aided heading cannot be used. The GEO-FOG 3D Dual INS is also an attitude and heading reference system (AHRS). This product features two GNSS antennas on a fixed RTK baseline that offers the same reliability and performance levels as the GEO-FOG 3D, with increased heading, pitch, and roll accuracy for static and dynamic applications where single-antenna systems can be problematic (see Fig. 3.16).

Figure 3.16 The KVH GEO-FOG 3D and GEO-FOG 3D Dual inertial navigation systems, which use three axes of DSP-1760 FOGs as the core of the embedded 1750 IMU.

3.4 Setting the Course for the Future of FOG Technology and Expanded Applications

From the start of FOG research, development, and commercial production efforts, KVH engineers have sought to improve continuously upon the most important attributes of the company's FOGs and FOG–based inertial systems. These efforts have resulted in

- smaller, more-accurate FOGs and inertial systems at a lower cost;
- faster, improved performance over shock and vibration; and
- better yields in manufacturing through the automation of splicing stations, as well as coupler and polarizer manufacturing stations, among others.

The future for FOG development is leading to several critical paths:

- the creation and implementation of a continuous-flow manufacturing process, which will eliminate the need for splicing optical components together, thereby improving performance and simultaneously reducing the manufacturing and calibration time required;
- continued research into smaller, more-accurate FOGs and FOG–based systems; and
- significant cost reductions via FOG design.

The increasing performance levels, reduced size and weight, versatility, and affordability achieved through the evolution of KVH FOGs and FOG–based inertial systems to date have made the use of these high-performance systems practical in an increasing number of applications.

3.4.1 Navigation and control

KVH high-performance FOGs and IMUs play a vital role in closing the gap between the needs of the manned, unmanned, or autonomous system user and the capabilities of the navigation and control solution. KVH FOG–based inertial systems capture highly accurate angular-rate and acceleration data for navigation solutions used in aerial, ground, marine, and subsurface unmanned systems. KVH IMUs and INS are integrated into unmanned, autonomous and manned navigation systems to both enhance and augment GPS/GNSS, and to provide accurate, reliable navigation when GPS is blocked or unavailable (Fig. 3.17).

Users of unmanned and autonomous systems rely on the data from KVH sensors for navigation and control functions and to produce accurate end products such as survey maps, imagery, 3D representations of different environments, search grids for infrastructure inspection, and pipeline mapping and inspection.

3.4.2 Positioning and imaging

Precise positioning and pointing enables the collection of geo-referenced data by scanners and sensors such as cameras and LIDAR, radar, sonar, and more. KVH customers rely on our inertial sensors and fiber optic gyros for precise data used in their unmanned mapping and surveying systems, as well as augmenting GNSS for unmanned or autonomous positioning and pointing

Figure 3.17 Navigation in GNSS–denied environments is a challenge for small unmanned vehicles such as those used for research, mapping and surveying, exploration, infrastructure monitoring, security monitoring, and more. In the Saab Sea Wasp, a groundbreaking waterborne anti-IED security platform, the KVH FOG–based inertial system provides accurate navigation data when the GNSS is jammed, blocked, or simply unavailable, e.g., underwater, in tunnels, and in other such environments.

applications. Integrating inertial and GNSS technology enables a system to seamlessly determine the precise position of the sensor within the mapping frame, even in the absence of a GNSS signal.

KVH inertial measurement units and inertial navigation systems provide fast, accurate data to the company's customers' unmanned aerial, subsurface, and land pointing, positioning, and orientation systems.

3.4.3 Stabilization and orientation

Stabilized platforms such as gimbals and a variety of manned and unmanned platforms prevent the motion of the vehicle (Fig. 3.18) from interfering with the gathering of data by a device such as LIDAR, radar, cameras, recording and measuring sensors, etc. Likewise, stabilization and positioning are critical for a wide array of other applications, such as humanoid robots (Fig. 3.19), and even robust and demanding systems such as vehicle-mounted weapons (Fig. 3.20). In order to obtain the best results, the movement of the platform

Figure 3.18 Ultra high-definition (UHD) cameras and high-speed gimbals require extremely precise stabilization and pointing in demanding environments. KVH UHD camera and gimbal customers require gyros with low noise, low drift, high bandwidth, and superior accuracy, and they have therefore selected KVH DSP-1750 or DSP-1760 FOGs to provide two or three axes of stabilization for their systems.

Figure 3.19 Eleven of the 23 robot finalists in the live hardware portion of the 2015 DARPA Robotics Challenge stayed on track thanks to the integration of a KVH FOG–based 1750 IMU within their systems. KVH IMUs were integrated into several of the entries, including the ATLAS robot platform by Boston Dynamics (shown here), as well as the eventual winner of the challenge.

Figure 3.20 Compact, robust KVH FOGs offer a combination of exceptionally low noise and drift, as well as excellent performance in environments with high shock and vibration, making them ideal for weapon-platform stabilization and other demanding applications, such as precision payload pointing and stabilization.

must be isolated from the data-gathering device. Otherwise, the movement results in errors in the line of sight of the device, creating blurred imagery or even making the device inoperative.

Single- and multi-axis KVH gyros and IMUs provide orientation and/or pitch, roll, and yaw data to enable seamless stabilization of payloads on manned, autonomous, and unmanned systems used in ground, air, marine, and subsurface applications.

3.4.4 Looking ahead

As KVH engineers look ahead, the next challenge to be met and overcome is the evolution of smaller, mass-produced, low-cost, high-performance FOGs that can truly support mass markets, such as cars, with a level of performance unmatched by alternative gyro technologies.

Jay Napoli has more than 35 years' experience managing international sales, business development, and marketing of a wide range of technical products and systems including fiber optic gyros (FOGs) and FOG-based systems, enterprise application software, aerospace systems and instrumentation, as well as industrial equipment and machinery, and special alloys and metals. He has served as KVH's Vice President of FOG and OEM sales since November 2008. He joined KVH in 1999 as the Director of FOG and OEM sales. During his tenure at KVH, Mr. Napoli has been instrumental in helping guide KVH's fiber optic gyro product line from the early product stage to the full commercialization of a versatile family of product offerings supporting defense and commercial applications worldwide. Prior to joining KVH, Mr. Napoli worked for Allegheny Teledyne, Inc., first as the Managing Director for Teledyne Italia and later as Vice President of European business development. He also worked for Management Science America (MSA) and Dun & Bradstreet Software as the Managing Director of MSA Italia. Prior to MSA, Mr. Napoli served in the US Army in various assignments both in the US and Italy as a field artillery officer. He is a graduate of the Defense Language Institute and the United States Military Academy at West Point.

Chapter 4
Fiber Optic Gyro Development at Honeywell

Glen A. Sanders, Steven J. Sanders, Lee K. Strandjord, Tiequn Qiu, Jianfeng Wu, Marc Smiciklas, Derek Mead, Sorin Mosor, Alejo Arrizon, Waymon Ho, Mary Salit, Neil A. Krueger, Clarence Laskoskie, Chellappan Narayanan, and Wes Williams
Honeywell International

4.1 Introduction

Guidance and navigation (G&N) represents one of the core business elements within Honeywell Aerospace. Figure 4.1(a) shows the breadth of platforms on which Honeywell G&N systems are deployed for a multitude of land, sea, air, and space applications. Inertial sensors are necessarily at the heart of these G&N systems. For this reason, Honeywell has been developing, designing, manufacturing, and selling gyroscopes for nearly five decades. Microelectromechanical systems (MEMS), ring laser gyros (RLGs), and fiber optic gyros (FOGs) all form key pieces of Honeywell's inertial sensor portfolio. Figure 4.1(b) shows the mapping of these various gyro technologies to applications according to their performance classes: MEMS gyros primarily target the tactical market needs, whereas RLGs satisfy a wide range of high-volume applications, spanning the spectrum from high-end tactical through navigation-grade performance. As can be seen in the figure, Honeywell FOGs service the niche of the most-demanding long-life applications in space, submarine, and strategic applications. Although this high-performance market segment is much lower in volume than mainstream navigation applications, the rigorous performance and life requirements (nav-plus through strategic grade) maintain an active demand for IFOG products.

To address the navigation-grade-plus through strategic market and beyond, over the last 25 years Honeywell has developed a number of interferometric FOG (IFOG) products utilizing both polarization-maintaining and depolarized closed-loop architectures as the high-performance complements to

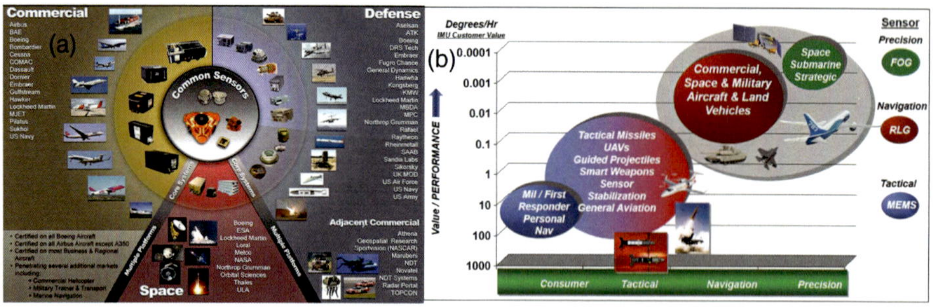

Figure 4.1 Honeywell's navigation and inertial sensing portfolio.

its navigation-grade RLG product lines. These IFOG designs provide precision guidance and pointing in the 0.001 deg/h to 0.0001 deg/h range of bias instability and represent mature and deployable products. The IFOG measures the interference of two counter-propagating light waves in the fiber optic loop of a Sagnac interferometer to measure the inertial rotation rate. Rotation sensitivity is tunable by scaling the length–diameter (LD) product of the Sagnac coil, and excellent signal-to-noise ratios (SNRs) can be obtained: minimum detectable signals can be below a single rotation per century. Excellent bias stability is obtained with a broadband light source and the well-known minimum reciprocal architecture,[1] while scale factor (SF) instability and SF nonlinearity in the range of single-digit parts per million (PPM) can be achieved using closed-feedback-loop electronics to null and linearize the rotation signal.[2] Section 4.2 gives a survey of active Honeywell IFOG products and developments.

The resonator fiber optic gyro (RFOG) measures the rotation rate by using the frequency shift between clockwise (cw) and counter-clockwise (ccw) resonances in a multi-turn fiber ring resonator.[3,4] Despite its less-proven performance, the development of the RFOG has been motivated by its potential to achieve the same rotation-sensing performance in a more-attractive form factor than the RLG or IFOG.[5] This improvement arises from the fact that a RFOG combines the SNR sensitivity enhancements of both the RLG and the IFOG by recirculating light around the loop many times (signified by the resonator finesse F), and multi-turn coils can be deployed to increase the path length well beyond that of a RLG. The same shot-noise-limited signal-to-noise sensitivity can be obtained in a RFOG of $F/2$ with less fiber length than that of an IFOG of comparable diameter and optical power. Thus, for a finesse of 100, the shot-noise-limited angle random walk (ARW) sensitivity of the RFOG should be obtainable with fifty times less fiber than the IFOG counterpart. Hence, RFOG development is motivated by the promise of an inexpensive, small optical gyro for commercial navigation applications. However, the technical barriers to performance become significantly more challenging due to the requirements of a resonant design and, in particular, the need for a highly stable, truly coherent narrow-linewidth light source and at least one high-quality frequency shifter. While the use of a broadband light source solved many issues in the IFOG case, a RFOG requires different solutions to mitigate problems such as optical backscatter, the Kerr effect, and stimulated Brillouin scattering. These solutions, and progress toward the mitigation of optical backscatter effects using a novel RFOG architecture, are discussed in Section 4.3.

4.2 IFOG Status

4.2.1 Navigation-plus-grade IFOGs

Honeywell IFOG efforts began in the mid-1980s, with the first products being tactical-grade gyros for attitude, heading, and reference systems (AHRS) on commercial aircraft.[6] Flight production started in the mid-1990s, and over 5000 open-loop gyros were built and sold for both commercial and business commuter jet use. For commercial aircraft, the AHRS was used on the Boeing 777 as a secondary or backup reference system. For business commuter jets, the AHRS was used in many applications, but the highest volume was with Embraer. Many of these IFOGs are still in service twenty years later.

More recent IFOG products at Honeywell are more focused on higher-performance,[7,8] radiation-hardened,[9] and long-life[10] niches inaccessible to other gyro technologies such as MEMS and RLGs. One such niche is the "nav-plus" performance range:

$$0.001\ \text{deg/h} \leq \textit{bias instability} \leq 0.01\ \text{deg/h},$$
$$0.0005\ \text{deg/sqrt(h)} \leq \textit{ARW} \leq 0.005\ \text{deg/sqrt(h)}.$$

Two product offerings in this class are mature and in active production today. The first is the SPIRIT inertial measurement unit (IMU), designed for long-life space applications to low-noise platform stabilization. Figure 4.2(a) shows a photo of the IMU, which boasts four IFOGs, two optional banks of four accelerometers, redundant low-voltage power supplies (28–50-V and 70–100-V inputs optional), and Mil-Std-1553 and RS-422 interfaces. This fault-tolerant unit completed space qualification in 2014 and is in production for commercial and military applications today. Several units have been delivered. These gyros utilize a depolarized sensing loop,[11] as shown conceptually in Fig. 4.2(b), rather than a simpler polarization-maintaining coil (Fig. 4.2(c)), to leverage the inherent radiation tolerance and cost benefits of non-polarization-maintaining, single-mode optical fiber.

With the depolarized architecture, SPIRIT is radiation-hardened and qualified for mid-earth orbit (MEO) environments, as well as artificial

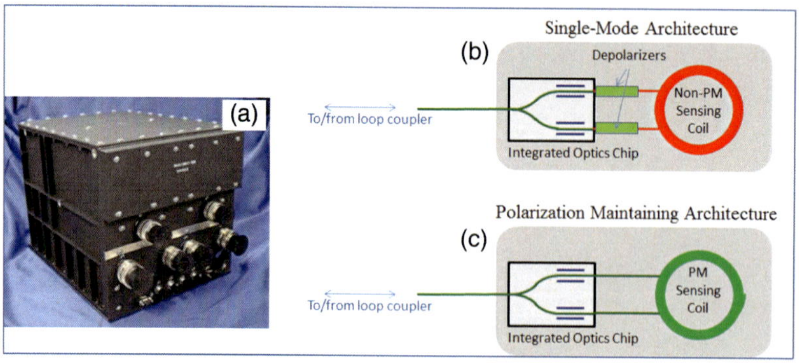

Figure 4.2 (a) Photo of the SPIRIT IMU. (b) Basic depolarized and (c) polarization-maintaining IFOG architectures.

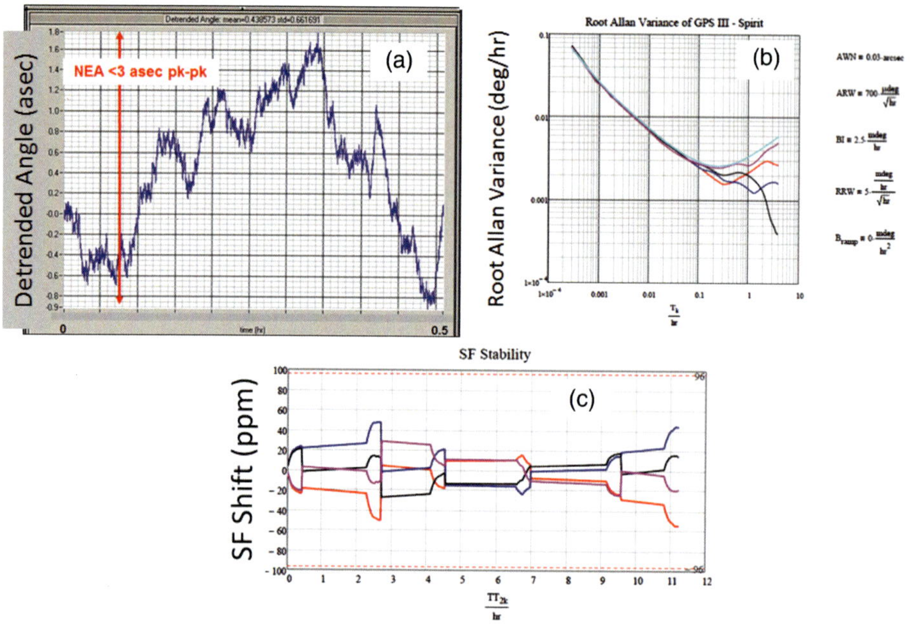

Figure 4.3 (a) Typical short-term angle drift performance. (b) Root-Allan variance of the typical short-term bias performance under a characteristic spacecraft thermal profile. (c) Typical short-term SF instability under the same thermal profile.

hazards. Figure 4.3(a) plots a typical detrended angle from these gyros, giving a noise-equivalent angle (NEA) of a few arc seconds—adequate short-term pointing performance for many commercial and other space applications. A root-Allan variance of typical bias performance, under spacecraft thermal conditions (typically ±10°C, over 10–20 h), is given in Fig. 4.3(b): all gyros have bias instability in the range of 0.002–0.004 deg/h (1σ). Finally, Fig. 4.3(c) presents the characteristic SPIRIT SF instability under the same thermal conditions; approximately 100 ppm pk-pk error is typical.

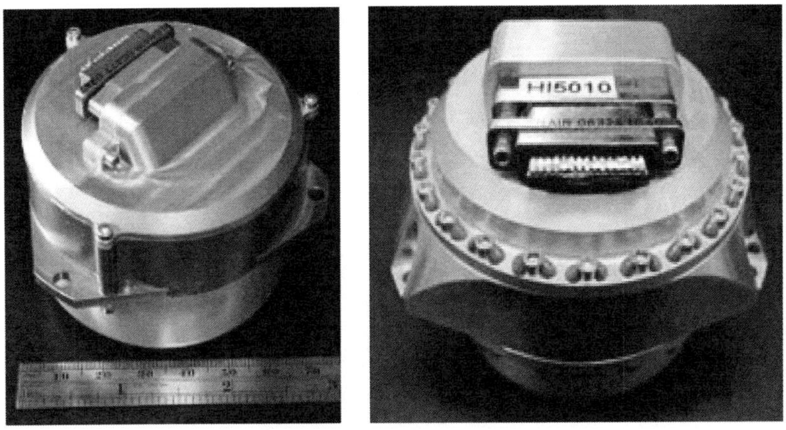

Figure 4.4 Family of depolarized, single-axis gyro sense heads in the performance-class SPIRIT IMU gyros.

IFOGs in the SPIRIT performance class are also available as single-axis sensors, as shown in Fig. 4.4. Individually shielded, sense heads provide very low magnetic sensitivity, guaranteed bias instability, and an ARW commensurate with the IMU performance discussed earlier. These gyros are in active production in quantities of ~10–20 per month.

As of 2017, Honeywell is also pleased to offer IFOGs in the SPIRIT high-performance (HP) class. These higher-performance gyros are based on polarization-maintaining (PM) fiber technology and are an interchangeable option in the SPIRIT IMU. They can also be available as single-axis sensors, as shown in Fig. 4.5. The PM fiber IFOG design is more robust with respect

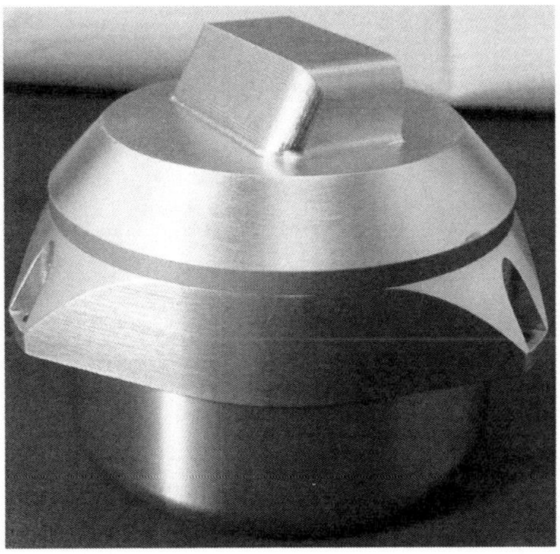

Figure 4.5 High-performance, polarization-maintaining, single-axis gyro sense head in the family of SPIRIT IMU gyros.

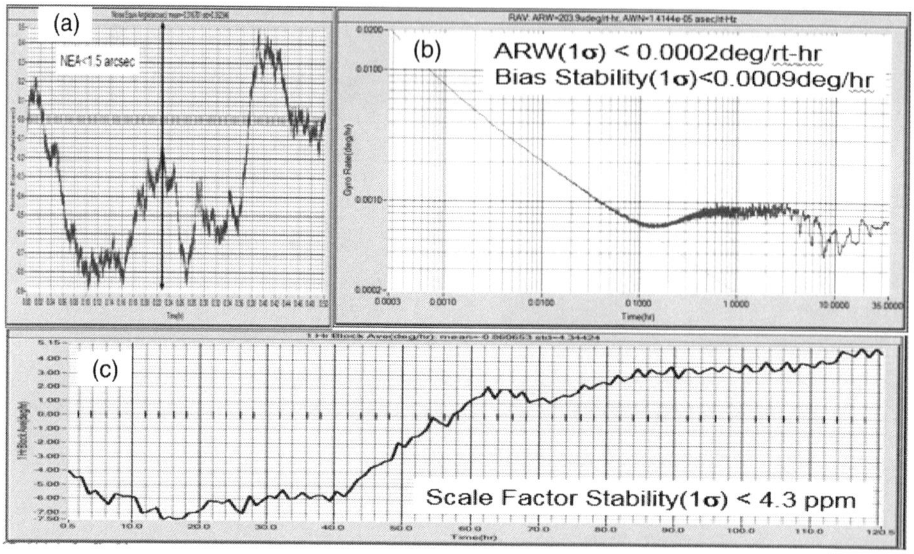

Figure 4.6 (a) Typical short-term angle drift performance. (b) Root-Allan variance of the typical short-term bias performance under a characteristic spacecraft thermal profile. (c) Typical short-term SF instability under the same thermal profile.

to optical errors, yielding better performance. Design has been simplified for reduced-touch labor and cycle time.

The HP SPIRIT IFOGs provide enhanced bias instability, ARW, and SF stability, as shown in Fig. 4.6. The bias stability can be as low as 0.001 deg/h (1σ), the typical ARW is 0.0002 deg/rt-h, and the short-term SF stability is 4–5 ppm. The HP SPIRIT gyros are in initial production now, with full qualification completed October 2018.

4.2.2 Strategic-grade IFOGs

The next performance range of interest for Honeywell IFOGs is strategic grade, i.e.,

$$0.0001 \text{ deg/h} \leq \textit{bias instability} \leq 0.001 \text{ deg/h},$$
$$0.00005 \text{ deg/sqrt(h)} \leq ARW \leq 0.0005 \text{ deg/sqrt(h)}.$$

This range of applications is addressed by two IFOG products that began development in the late 1990s and entered low-volume production in the early 2000s. One of these gyros is the High-Performance Space FOG (HPSFOG), serving the most-demanding space-pointing needs. A HPSFOG typically delivers short-term pointing instability ≤10 nrad, bias instability in the range of 0.0002–0.0006 deg/h over a typical diurnal thermal profile, and a total SF error in the single-digit PPM range. Serving a low-volume but highly demanding market segment, Honeywell strategic IFOGs are produced in volumes of roughly 5–10 per year.

4.2.3 Reference-grade IFOGs

Honeywell has developed a prototype IFOG in a class we call "reference grade," namely,

$$\begin{aligned} \mathit{bias\,instability} &\leq 0.0001\ \mathrm{deg/h}, \\ \mathit{ARW} &\leq 0.00005\ \mathrm{deg/sqrt(h)}. \end{aligned}$$

With the sensing loop shown conceptually in Fig. 4.7, this gyro is intended for a new range of possible applications in earth and scientific sensing, rather than as a compact unit for vehicle guidance and navigation. Figure 4.8 illustrates the reference-grade bias and ARW of this sensor: in a month-long segment of data, uncompensated bias instability is better than 0.00003 deg/h, and the ARW is about 0.000016 deg/sqrt(h) (16 μdeg/sqrt(h)). Figure 4.9 gives typical angle noise data for this gyro: the NEA is about 300 nrad over 30 minutes, and the angle white noise is less than 3×10^{-6} arcsec/sqrt(Hz). Some non-white angle noise is visible in the 60–120-Hz range, clearly an artifact of

Figure 4.7 Sensing loop assembly of a reference-grade gyro.

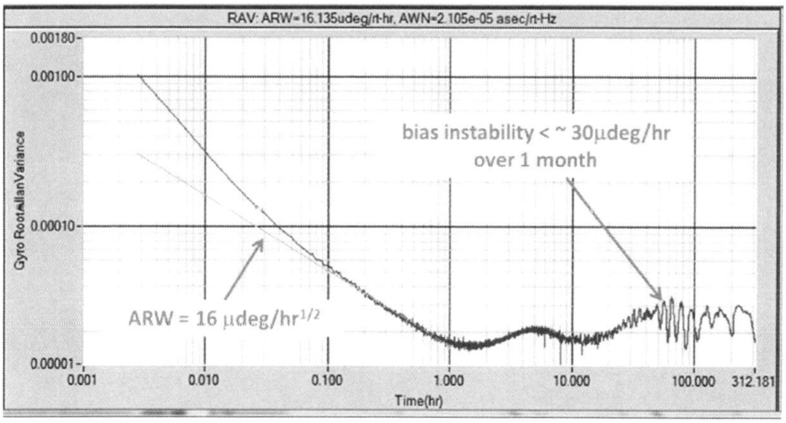

Figure 4.8 Root-Allan variance of a month-long run of a reference-grade gyro.

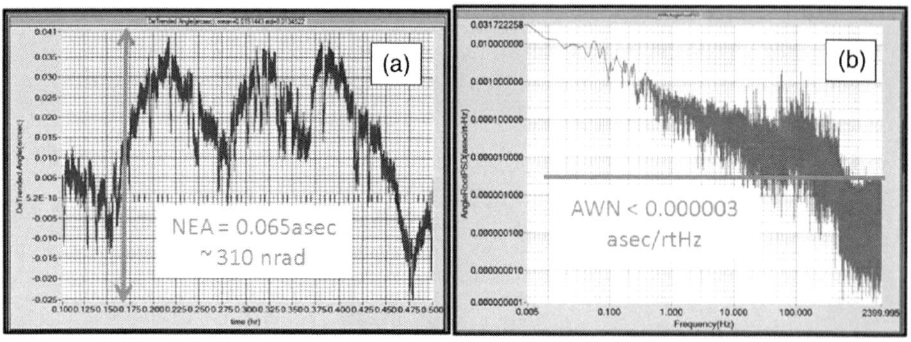

Figure 4.9 Typical reference-grade gyro data: (a) detrended angle over 30 minutes showing the NEA; and (b) angle root power spectral density (PSD) showing the angle white noise.

test-equipment contamination, which can be improved in straightforward ways. The aggregate performance of this gyro compares well to the best published performance (to our knowledge) of any practical laboratory instrument. For example, the ARW here is comparable to that of state-of-the-art atomic gyros,[12] but the IFOG is far more mobile and reproducible as a piece of engineering equipment. Similarly, the IFOG bias instability and angle white noise compare favorably to the best-ever published data for a hemispherical resonator gyro.[13] Hence, we believe the noise and bias performance of this reference-grade IFOG may be useful in a range of metrology, seismic, and structural sensing applications, as well as the calibration of inertial test equipment.

4.3 RFOG Development

The basic principle of the RFOG, based on the passive cavity technique,[3] is shown in Fig. 4.10. Light from a highly coherent laser at frequency f_0 is split

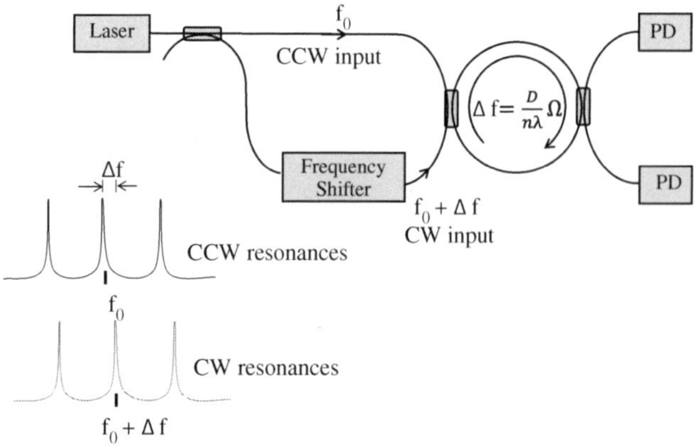

Figure 4.10 Basic principle of RFOG operation.

into two waves, one of which is frequency shifted by frequency Δf before each wave is introduced into a fiber ring resonator. To track the resonance frequencies of the resonator, optical frequencies f_0 and $f_0 + \Delta f$ are tuned to be centered on the ccw and cw resonances of the resonator using a standard phase-sensitive detection technique for sensing the line centers. Resonances are sensed in transmission through the optical ring resonator by two photodetectors (PDs). In the presence of rotation, the necessary frequency shift to track the resonance between the two waves is $\Delta f = D\Omega/n\lambda$, where D is the diameter of the coil, n is the index of refraction of the fiber, λ is the optical wavelength, and Ω symbolizes the rotation rate.

4.3.1 New RFOG architecture

The basic architecture of Fig. 4.10 has several challenging issues related to its implementation and performance. The use of a single laser plus a frequency shifter was originally chosen to minimize the relative frequency noise between cw and ccw inputs. In this way, any low-frequency laser jitter in the rotation band was common to both cw and ccw lightwave inputs and thus would not appear as rotation-induced resonant frequency changes. Laboratory designs have centered on the use of acousto-optic frequency shifters, which require high drive power and are expensive and large. Integrated optics, or serrodyne-type frequency shifters using lithium niobate, suffer from either poor sideband suppression, and from size and cost issues. This has been one factor leading Honeywell to the use of the multi-laser architecture,[14] as shown in Fig. 4.11. This RFOG configuration does not use frequency shifters to derive the different-frequency light waves needed to independently track cw and ccw resonance peaks, but rather separate semiconductor lasers. The architecture also facilitates the use of an improved method to reduce errors from optical backscatter by locking the lasers one free spectral range (FSR) apart.

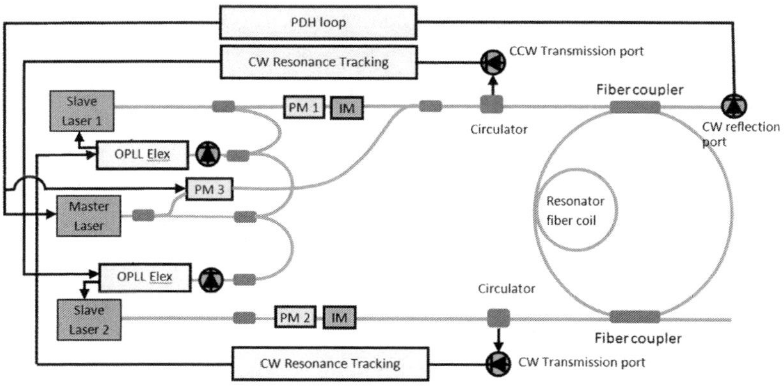

Figure 4.11 Improved RFOG configuration using phase-locked lasers.

In the architecture of Fig. 4.11, it is imperative to highly suppress relative frequency jitter in the sub-Hz to 100-Hz frequency region of the rotation sensing spectrum. Relative frequency jitter at the millihertz level is needed near DC so as to not add excess ARW that exceeds the requirement for navigation-grade sensing. This approach is now becoming realizable using ultra-high-gain–bandwidth phase lock loops to lock the lasers together. The figure shows three laser diodes in a multi-frequency laser source used in the RFOG approach. A master laser is used to perform a tight lock between the laser array and the resonator. This design uses feedback from the high-speed reflection port of the resonator, in combination with phase modulation applied by PM 3 is used in a servo to adjust the master laser frequency to a resonance of the resonator. This loop uses the well-known Pound–Drever–Hall technique to provide a high-gain–bandwidth loop. The slave lasers are then locked to the master laser by interfering them with a portion of the master laser to obtain a beat note and then maintaining the beat-note frequency at a tunable offset frequency. In each case, this is done by employing an optical phase lock loop (OPLL). The end result is that slave lasers 1 and 2 are now stabilized with respect to each other via electronic means instead of by optical frequency shifting.

Slave lasers 1 and 2 are introduced into the resonator in the cw and ccw directions, respectively. Each of them is sensed in transmission through the resonator and locked to the top of the resonance via a resonance-tracking loop. The resonance tracking loops rely on phase modulation imparted by phase modulators PM 1 and PM 2 to produce frequency modulation over each resonance lineshape. A standard phase-sensitive detection technique is used to detect the resonance line center. A different phase modulation frequency is applied to PM 1 and PM 2 to separate the intensity of backreflected waves from the main signal wave, as suggested by Iwatsuki et al.[15] The intensity of each slave laser is stabilized via an intensity modulator (IM) in a feedback loop (not shown) to reduce the effects of the optical Kerr effect.

Another significant challenge to realizing a navigation-grade RFOG that is addressed by the configuration of Fig. 4.11 is that of interference due to optical backscatter.[15–17] Backscatter from one wave into another causes beats at their frequency difference, which, at a low rotation rate, is an oscillatory error in the rotation rate band of interest. Without sufficient countermeasures, this error has been shown to produce dead-zone behavior (or the "lock-in" effect in closed-loop operation).[18] A RFOG arrangement that uses separate lasers allows this issue to be addressed; the slave lasers are offset by a large frequency difference to prevent the deleterious effects of backscatter from one direction into another. By locking to adjacent longitudinal resonances, as shown in Fig. 4.12, backscatter from one wave into the other causes its unwanted interference to be at a high frequency because the FSR is included

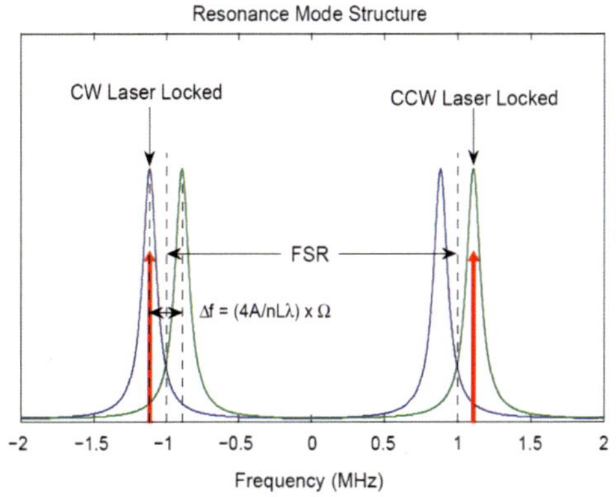

Figure 4.12 Ring resonator modes for clockwise (blue) and counter-clockwise (green) propagating directions. The cw laser frequency is locked onto the lower longitudinal resonance, whereas the ccw laser frequency is locked to the higher-frequency longitudinal resonance. The two resonances in each direction are separated by 1 FSR, and the cw and ccw resonances are shifted by Δf, due to the rotation rate.

in their frequency separation (e.g., 2 MHz for a resonator 100 m long). The phase-locked loops are controlled by the resonance-tracking feedback loops to tune the laser frequencies to resonance peaks separated by an FSR plus a variable frequency Δf (for rotation rate).

4.3.2 RFOG experimental results

An RFOG of the arrangement shown in Fig. 4.11 was built and tested for SF performance to verify its benefit to reducing the effects of backscattered light. The resonator was nominally 100 m in total length and consisted of two polarization-maintaining fiber couplers, a polarization-maintaining fiber coil, and two short strands of polarizing fiber (i.e., an **all-fiber configuration**). The polarizing fiber is used to eliminate light in the undesired polarization state of the fiber within the ring.[19] The FSR was roughly 2 MHz. The lasers were narrow-linewidth semiconductor lasers operating at a nominally 1.55-μm wavelength.

Performance data was taken for scale factor linearity, scale factor stability, and dead-band (i.e., linearity about a region near zero rotation rate); all of these performance characteristics are known to be degraded with backscattered light in the absence of effective countermeasures. It was also an objective to demonstrate the functionality and the range of the multi-frequency source. The gyro was mounted on a turntable as shown in Fig. 4.13(a). With the gyro's sensitive axis pointing along the applied rate, the gyro output versus rate was recorded across a rotation range of ±500°/s. As shown in Fig. 4.13(b), there is a very linear relationship between resonance frequency shift and rotation rate. To further quantify the SF linearity residuals, the gyro was pre-calibrated by an initial run forming a lookup table, and subsequent

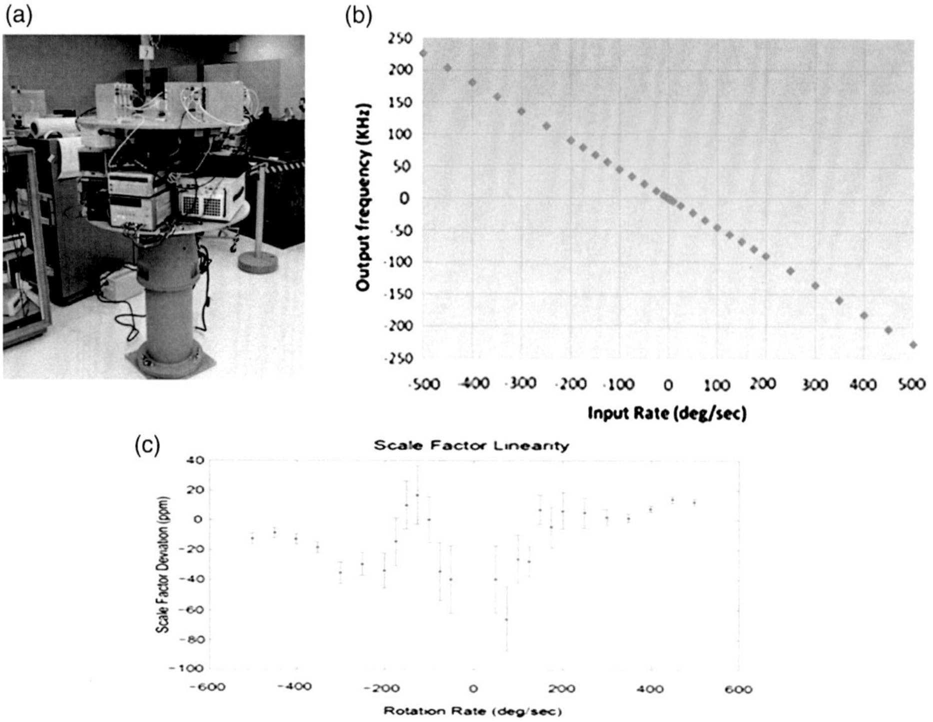

Figure 4.13 RFOG scale-factor linearity tests: (a) RFOG mounted on a rate table with electronic equipment and power supplies, (b) the gyro output frequency shift due to rotation as a function of rotation rate, and (c) the residual nonlinearity of the gyro scale factor (25 ppm).

runs were measured relative to the lookup table. The residuals in parts per million (ppm) versus rotation rate are shown in Fig. 4.13(c). The 1-sigma deviation of the residuals was computed to be less than 25 ppm.

An even more discriminating test for the effects of optical backscatter in the RFOG is that of linearity near zero rate, or a dead-band test. To do this the gyro's sensitive axis was placed perpendicular to the rate table axis. As the table was rotated at about 1 deg/s, the gyro's sensitive axis was swept through eastern, southern, western, and northern pointing directions measuring a component of earth rate that ranged from 0 deg/h to 12 deg/h. Data were taken over a 10-h period to assess repeatability and are plotted in Fig. 4.14. Note that in this test, the gyro axis was not perfectly perpendicular to the table axis, causing a 5 deg/h offset when the gyro axis was oriented east or west. The measured earth rate of 12.67 ± 0.11 deg/h was consistent with the true earth rate at our latitude. No dead-band was observed at zero rotation rate.

Scale-factor stability was also measured by mounting the gyro input axis parallel to that of the rate table and cycling the table input rate from 100 deg/s to –100 deg/s with a 15-s dwell at each rate. This process was repeated for 5 h.

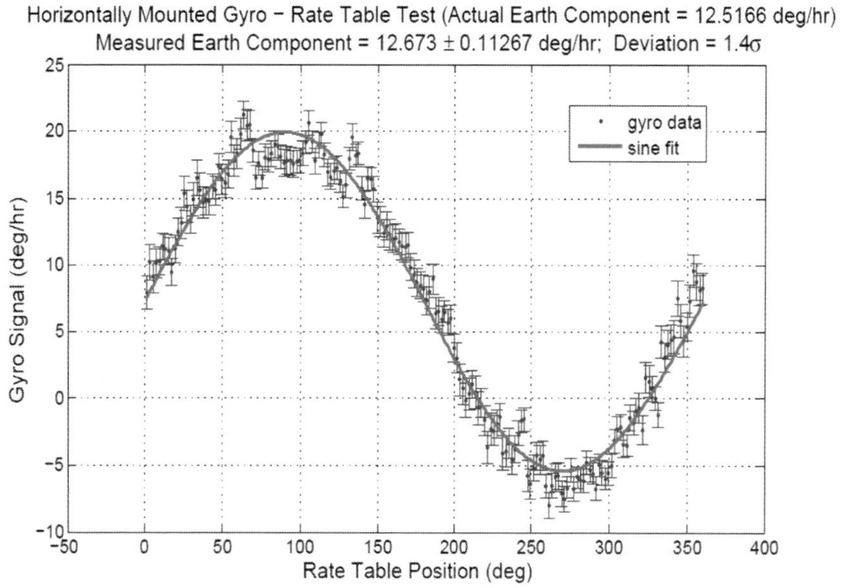

Figure 4.14 RFOG dead-band test. The gyro's sensitive axis is rotated to pick off a varying component of the earth rate. Within the error bars, no dead-zone is visible.

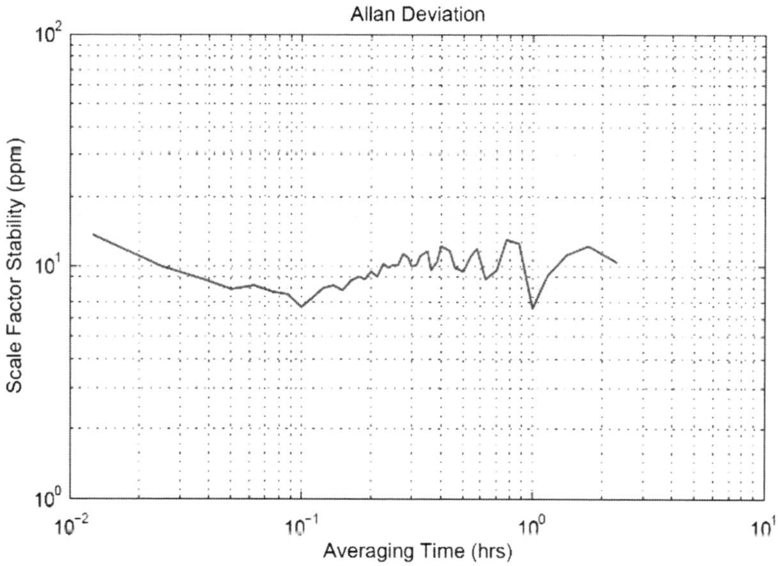

Figure 4.15 Allan deviation of SF variation.

The SF was computed by comparing the gyro output for positive and negative rates. The Allan deviation of the SF drift shown in Fig. 4.15 demonstrates an SF stability of about 12 ppm, i.e., stability that closely approaches that of navigation-grade performance.

While the promise of the RFOG approach is based on achievement of high performance in a small size, most of the effort to date has focused on realizing required performance using a relatively short fiber sensing coil. However, the other optics must be miniaturized and cost-reduced to take advantage of the benefits of the resonance-sensing coil architecture. This paper also reports some of our recent progress in RFOG development toward the realization of a next-generation compact device for commercial navigation applications by incorporating silicon optical bench (SiOB) technology to miniaturize resonator and input optics.[20] This arrangement is shown in Fig. 4.16. In this configuration, light propagates in cw and ccw directions around the multi-turn fiber optic coil consisting of 100 m of polarization-maintaining fiber. The ends of the fiber coil are placed in a v-groove etched in a silicon substrate forming the SiOB. By laying fibers in opposite ends of a v-groove with ball lenses in between, light is aligned and focused from one end of the loop fiber to the other. In the region between the ball lenses, light is collimated so that tiny beamsplitters can couple the light into and out of the resonator. Tiny polarizers are placed in between the resonator fiber ends to attenuate errors due to the propagation of light in the unwanted state of polarization within the resonator. In between the optics on the SiOB, light travels through free space. The SiOB plays a critical role in closing the resonator loop by connecting the two ends of the coil, as well as providing an input light path into and out of the resonator. The input paths also incorporate circulators on the chip to attenuate feedback to the lasers.

An example of the resonator-loop-closure region on a SiOB test device is shown in Fig. 4.17. In the photograph, the v-groove for the resonator runs from the upper left to the middle right of the photo. The input/output beamsplitters between the two ball lenses are used to collimate the light

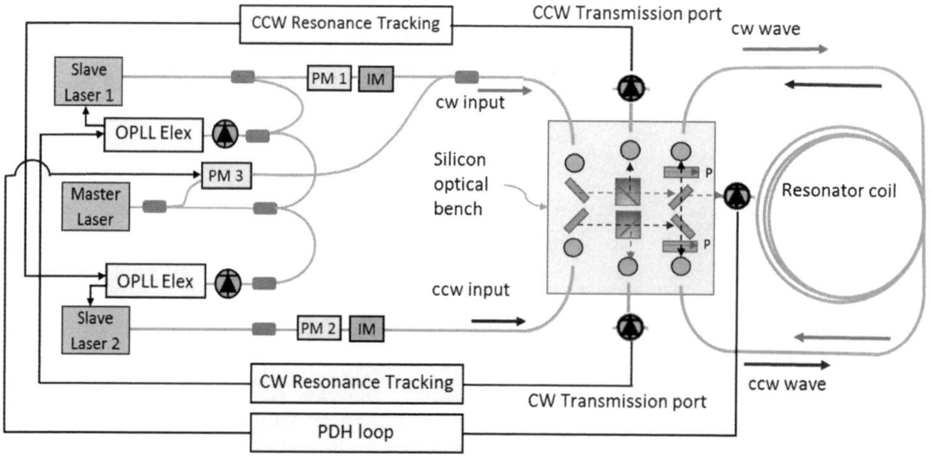

Figure 4.16 Resonator fiber optic gyro using miniaturized optics on a silicon optical bench.

Figure 4.17 Silicon optical bench photograph.

through the beamsplitters and focus the light into fibers. In the case of this test device, light was coupled into the resonator path by two fibers in parallel input v-grooves, followed by ball lenses.

To form the gyro shown in Fig. 4.16, the input laser light system consists of slave laser (SL) diodes and a master laser (ML) diode, very similar to that described in the arrangement of Fig. 4.11. Here, light from the ML is modulated in a phase modulator (PM3) before being directed to the SiOB, which directs it to propagate in the cw direction of the resonator. A portion of the ML light is detected in reflection from the resonator on the rightmost detector and demodulated at the frequency imparted to PM3. Again, each SL is locked to the master laser via an OPLL. Light from SL1 and SL2 are passed through a phase modulator (PM1 and PM2, respectively) prior to being directed to the SiOB and the resonator. Light from SL1 propagates in the cw direction of the resonator, whereas that of SL2 propagates in the ccw direction.

To measure the rotation rate, some of the cw wave from the resonator is coupled out via a resonator input/output beamsplitter, passed through a circulator located on the SiOB (middle features on SiOB) and coupled to a transmission port detector. The output of this detector is then demodulated at the phase modulation frequency applied to PM1 to form a discriminant for the CW resonance-tracking servo, which adjusts the frequency of SL1, via the OPLL, to a resonance condition in the cw direction. Likewise, the frequency of SL2 is adjusted to the resonance in the ccw direction.

Several key questions were addressed for the first time in this work.[20,21] Of specific importance in this advanced RFOG design is the attainment of

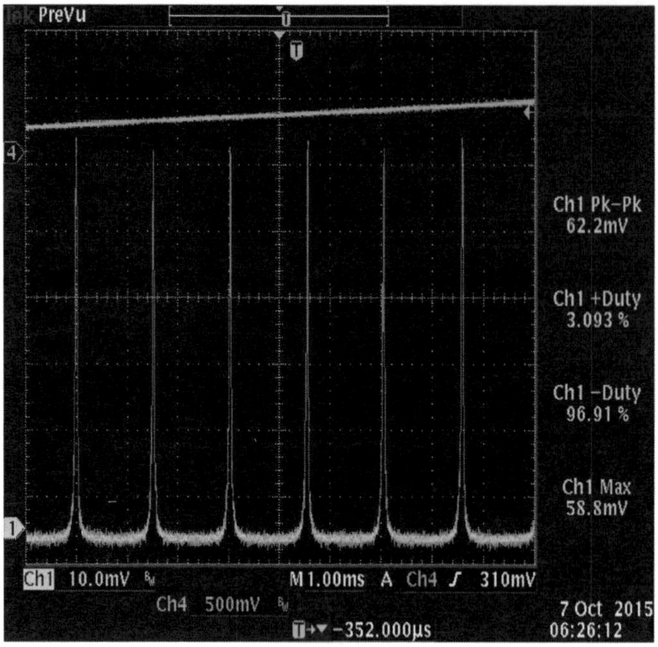

Figure 4.18 Scan of resonator lineshapes, showing a finesse of 35 with SiOB optics for fiber-to-fiber loop closure.

sufficiently high resonator finesse with an SiOB in the resonator loop, to achieve desirable ARW performance. A scan of the resonances in one direction of the resonator is shown in Fig. 4.18. As is shown, a finesse of 35 in a 2-inch-diameter coil of 100-m length has been realized. This result enabled a navigation grade ARW performance of about 0.0029 deg/rt-h. This performance is determined from Fig. 4.19, which plots the PSD of the gyro output. The white noise is given by the region on the right side (at 10 Hz), where the noise is flat and no longer influenced by $1/f$ noise.

Of critical importance is the assessment of the effect on bias performance of adding the SiOB with polarizer in the resonator.[22] Recent (Fig. 4.20) results show a bias stability of roughly 0.01 deg/h. These very encouraging early results are amongst the best reported for RFOGs, even though they are the first reported introduction of an SiOB into the loop. Note that the root-Allan variance of the data in Fig. 4.20 reflects a higher ARW than in Fig. 4.19, where antialiasing filters were employed and the gyro optical power was increased to suppress electronic noise.

4.3.3 RFOG component development and future implementation

Several emerging technologies that are envisioned to play a role in enhancing future RFOG products are being developed. These include low-loss hollow-core fiber (HCF),[23,24] integrated photonics technology for

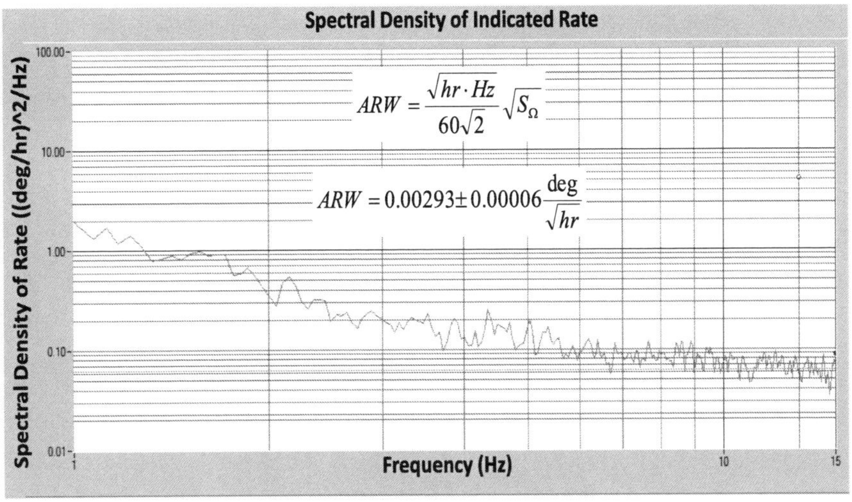

Figure 4.19 Power spectral density of RFOG noise showing a white-noise floor of 0.00293 deg/rt-h at 10 Hz for a 2-inch-diameter coil. The results are consistent with civil-navigation-grade performance.

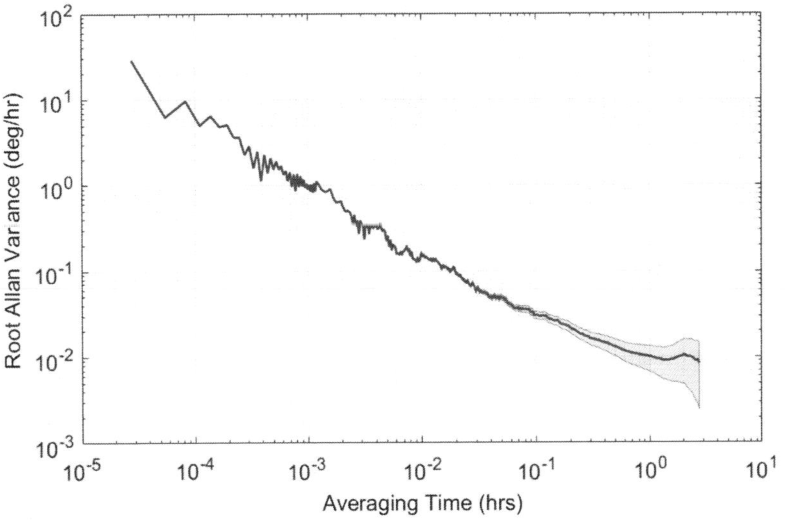

Figure 4.20 Root-Allan variance of RFOG data showing a bias stability floor of 0.01 deg/h for a 2-inch-diameter coil.

multi-frequency light sources (MFLSs), and further enhancement of the SiOB technology.

The RFOG platform is capable of utilizing sensing coils of polarization-maintaining solid-core fiber, polarizing solid-core fiber, or bandgap HCF with a length and diameter tailored to the exact application. HCF is a very attractive choice[25] in that it vastly reduces the size of the Kerr effect due to

Figure 4.21 Cross-section of hollow-core fiber.

optical power imbalances in the coil, and it dramatically increases the power threshold for stimulated Brillouin scattering (SBS). In short, it allows for the use of higher optical power to increase signal-to-noise sensitivity. The cross-section of a HCF is shown in Fig. 4.21.

The fiber in Fig. 4.21 has been developed to provide a high glass-free region and low loss for the propagation of light near the 1550-nm wavelength. It has special provisions to maintain polarization and attenuate higher-order spatial modes: both characteristics are needed to achieve the demanding bias-stability requirements of commercial navigation applications. This fiber shows encouraging performance characteristics at 1550 nm with 5.1 dB/km loss, and $h = 3.2 \times 10^{-5}$ m^{-1}. While more development is needed, formulation and demonstration of a fiber design of this type represents an important step toward readiness in an improved RFOG design.

A key element to the use of HCF for the resonator coil is the SiOB due to the difficulty of making suitably low-loss fiber couplers in HCF. Further, when the light wave is traveling between SiOB-mounted components, it is propagating in free space, allowing ease of coupling into and out of HCF with virtually no index mismatch. Not only is this a compact, low-cost platform for component integration, but it is also an efficient means for coupling light into an HCF. The SiOB is produced using many of the same process technologies used for MEMS technology. Specific attention is paid in the SiOB design to eliminate the need for active alignment. The technology shown in Fig. 4.17 is attractive in that it is capable of hosting fiber, lenses, mirrors, beamsplitters,

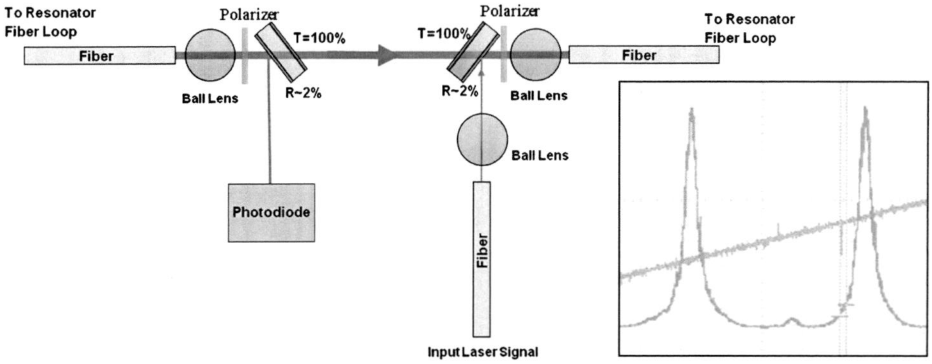

Figure 4.22 Schematic of an SiOB-based hollow-core fiber resonator and a scan of the resonances.

detectors, and circulators all on a single chip. A scan of one FSR of a resonator using HCF and an SiOB is shown in Fig. 4.22. It shows a finesse of 15 for this demonstration unit.[26]

To reduce the size and cost of the RFOG further, it is important to miniaturize optical components associated with the MFLS formed by the MLs, SLs, and power-combining optics. A compact package for the RFOG MFLS is shown in Fig. 4.23(a), which includes the entire MFLS and the many optical elements located on a silicon photonics (SiP) chip. A photo of an actual RFOG SiP chip is shown in Fig. 4.23(b). It provides dense packaging of over thirty components, including multiple photodetectors, a light splitter and combiners, and optical phase and intensity modulators. This is the second prototype, and further miniaturization is envisioned for the final product. Eventually, all MFLS electronics will be incorporated into a similarly sized package, or perhaps even smaller. (Graphics are provided by TeraXion, Honeywell's subcontractor.)

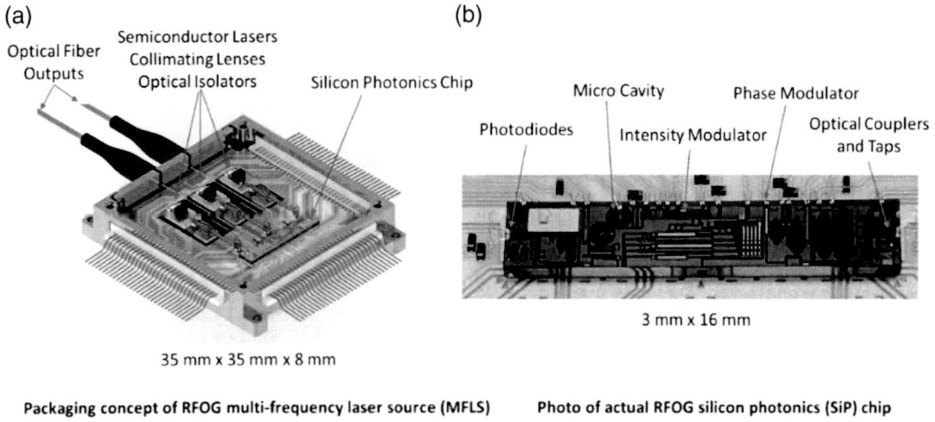

Figure 4.23 Low-cost, compact packaging technology under development for the RFOG.

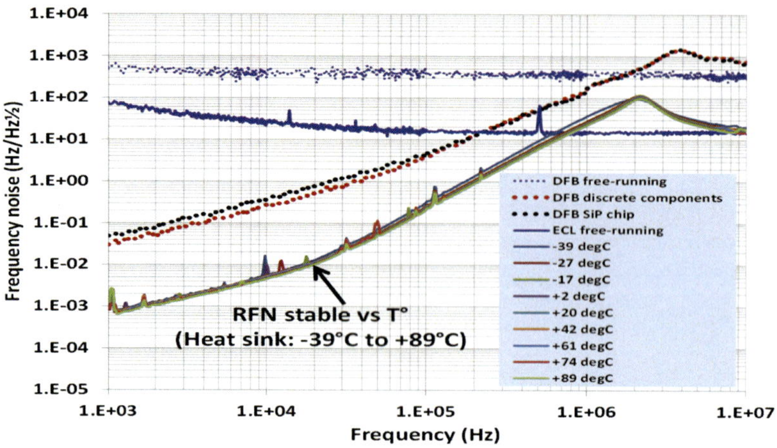

Figure 4.24 Relative frequency noise of an MFLS, and its stability over temperature.

Initial test results[27] of the relative frequency noise of the two SLs stabilized by an early-generation SiP module are shown in Fig. 4.24. The results are shown over a –39°C to +89°C temperature range. The top two plots show free-running frequency noise between the DFB and ECL lasers. The very bottom plots (all coinciding) show the stabilized-laser relative frequency noise for various temperatures. The frequency noise at a low frequency is suppressed into the mHz/sqrt(Hz) regime and is invariant with temperature. This shows exciting progress toward an MFLS assembly for our RFOG navigation applications.

Finally, we also report our first tests of temperature stability of a SiOB test device. This is particularly significant because any misalignments and loss variation in the fiber-to-fiber coupling within the resonator loop can adversely affect the finesse. As a first trial, we subjected a SiOB-based resonator device with a finesse of around 20 to a variation in temperature, as shown in Fig. 4.25. The SiOB temperature was varied over a non-condensing temperature range of 20°C to 85°C (the unit was not packaged or sealed). As depicted in Fig. 4.25, the finesse demonstrated remarkable stability, showing negligible change over temperature.

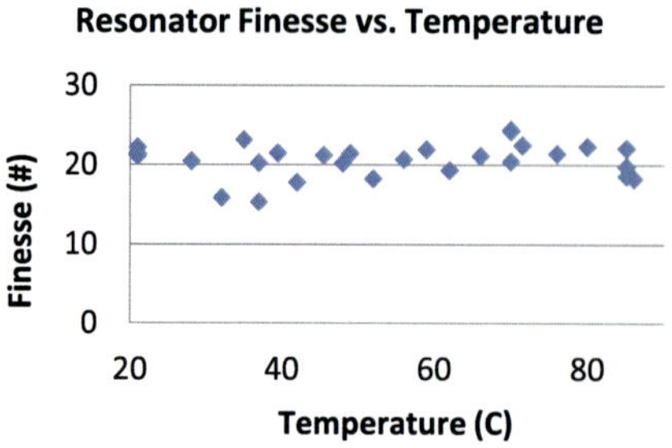

Figure 4.25 Resonator finesse vs. temperature using SiOB technology.

4.4 Summary

Honeywell's commitment to the fiber optic gyro remains strong, with active production of numerous IFOG products for navigation- and strategic-grade applications, as well as new developments in the "reference-grade" performance range. In parallel, RFOG development is underway for next-generation applications with both rigorous performance requirements and reduced size allowance.

An improved RFOG approach has been presented here using two phase-locked lasers to probe adjacent resonant frequencies in a ring resonator. This approach enables countermeasures for optical backscatter effects and is expected to eliminate the need for expensive and power-consuming frequency shifters. The SF stability of 12 ppm over 5 h, the 25-ppm linearity up to ±500 deg/s rotation rate, and the dead-band-free operation represent the state of the art in RFOG SF performance, to our knowledge. Taken together, these results demonstrate the functionality of the new approach and its attractiveness toward backscatter-error mitigation. A design based on a silicon optical bench is presented as a means of moving toward more miniaturized product designs. Presented ARW and bias stability performance of 0.0029 deg/rt-h, and 0.01 deg/h, respectively, represent the best in RFOGs to date. Future designs involving HCF, SiOB technology, and a miniaturized phase-locked laser system are under development as an attractive product approach of the future. These developments represent a significant step toward navigation-grade RFOGs for civil aircraft applications.

References

[1] H. C. Lefevre, *[The Fiber-Optic Gyroscope], Artech House, Boston and London*, 30–32, (1993).

[2] H. C. Lefevre, P. Graindorge, H. Arditty, S. Vatoux, and M. Papuchon, "Double closed-loop hybrid fiber gyroscope using digital phase ramp." *Proc. SPIE MS* **8**, 444–447 (1985).

[3] S. Ezekiel and S. Balsamo, "Passive ring resonator laser gyroscope," *Appl. Phys. Lett.* **30**(9), 478–480 (1977).

[4] D. Shupe, "Fiber resonator gyroscope: sensitivity and thermal nonreciprocity," *Appl. Opt.* **20**(2), 286–289 (1981).

[5] S. Ezekiel and H. Arditty, "Fiber-optic rotation sensors" in *[Fiber-Optic Rotation Sensors and Related Technologies]*, Springer-Verlag, Berlin & Heidelberg, 2–26, (1982).

[6] G. Sanders, B. Szafraniec, R. -Y. Liu, M. Bielas, and L. Strandjord, "Fiber-optic gyro development for a broad range of applications," *Proc. of SPIE* vol. 2510, 1995.

[7] G. Sanders, B. Szafraniec, L. Strandjord, R. Bergh, A. Kaliszek, R. Dankwort, C. Lange, and D. Kimmel, "Progress in high performance fiber optic gyroscopes," , *Proc. OFS* 12, (1997).
[8] S. Sanders, L. Strandjord, and D. Mead, "Fiber optic gyro technology trends– a Honeywell perspective", *Proc. OFS* 15, (2001).
[9] R. DeMaio, S. Sanders, N Giere, and J. Sewell, "Fiber optic gyro - inertial measurement Unit (IMU) designed for the industry's demand for a high-Quality, high-reliability, and long life commercial IMU", *Proc. GNC2011, 8th* International ESA Conference on Guidance, Navigation, and Control Systems (2011).
[10] S. Sanders, A. Taranta, S. Mosor, M. Alden, L. Hendry, R. DeMaio, N. Giere, and J. Sewell, "Fiber optic gyros in a high-quality, high-reliability inertial reference unit for commercial satellites", *Proc. OFS* 22, (2012).
[11] B. Szafraniec, J. Feth, R. Bergh, and J. Blake, *Proc SPIE* **2510**, 37–48 (1995).
[12] D. Durfee, Y. Shaham, and A. Kasevich, "Long-term stability of an area-reversible atom-interferometer Sagnac gyroscope," *Phys. Rev. Lett.* **97**, 240801 (2006).
[13] D. Rozelle, Proc. *ION Joint Navigation Conference*, 324–357 (2015).
[14] J. Wu, M. Smiciklas, L. Strandjord, T. Qiu, W. Ho, and G. Sanders, "Resonator fiber optic gyro with high backscatter-error suppression using two independent phase locked lasers," *Proc. of OFS-24*, 2015.
[15] K. Iwatsuki, K. Hotate, and M. Higashiguchi, "Effect of Rayleigh backscattering in an optical passive resonator gyro," *Appl. Opt.*, **23**(21), 3916–3924 (1984).
[16] G. Sanders, M. Prentiss, and S. Ezekiel, "Passive ring resonator method for sensitive inertial rotation measurements in geophysics and relativity," *Opt. Lett.* **6**(11), 569–571 (1982).
[17] T. Kaiser, D. Cardarelli, and J. Walsh, "Experimental developments in the RFOG," *Proc. SPIE 1367*, Fiber Optic and Laser Sensors **VIII**, 121–126, (1990).
[18] F. Zarinetchi and S. Ezekiel, "Observation of lock-in behavior in passive resonator gyroscope", *Opt. Lett.* **11**, 401–403 (1986).
[19] K. Iwatsuki, K. Hotate, and M. Higashiguchi, "Eigenstate of polarization in a fiber ring resonator and its effect in an optical passive ring-resonator gyro," *Appl. Opt.* **25**(15), 2606–2612 (1986).
[20] E. Benser, G. Sanders, M. Smiciklas, J. Wu, and L. Strandjord, "Development and evaluation of a navigation grade resonator fiber optic gyroscope," Proc. of **ISS**, 2015.
[21] G. Sanders, L. Strandjord, J. Wu, W. Williams, M. Smiciklas, M. Salit, C. Narayanan, E. Benser, and T. Qiu, "Development of Compact

Resonator Fiber Optic Gyroscopes," 4th IEEE International Symposium on Inertial Sensors and Systems, March 2017.

[22] G. Sanders, L. Strandjord, J. Wu, W. Williams, M. Smiciklas, M. Salit, C. Narayanan, E. Benser, and T. Qiu, "Improvements of Compact Resonator Fiber Optic Gyroscopes", *DGON* **ISS**, September 2017.

[23] B. Mangan, L. Farr, A. Langford, P. Roberts, D. Williams, F. Couny, M. Lawman, M. Mason, S Coupland, R. Flea, H. Sabert, T. Birks, J. Knight, and R. Phillip, "Low loss (1.7 dB/km) hollow core photonic bandgap fiber," *Proc of OFC*, p. PD24, 2004.

[24] J. Fini, J. Nicholson, R. Windeler, E. Monberg, L. Meng, B. Mangan, A. DeSantolo, and F. DiMarcello, "Low-loss hollow-core fibers with improved single-modedness," *Opt. Express*, vol. **21**, 5, pp 6233–6242, 2013.

[25] G. Sanders, L. Strandjord, and T. Qiu "Hollow core fiber optic ring resonator for rotation sensing," *Proc of OFS-18*, 2006.

[26] L. Strandjord, T. Qiu, J. Wu, T. Ohnstein, and G. Sanders, "Resonator fiber-optic gyro progress including observation of navigation-grade angle random walk," *Proc. OFS* 22, (2012).

[27] S. Ayotte, D. Faucher, A. Babin, F. Costin, C. Latrasse, M. Poulin, E. G-Deshenes, M. Pelletier, and M. Laliberte, "Silicon photonics-based laser system for high performance fiber sensing," *Proc of OFS-24*, 2015.

Glen A. Sanders is a Sr. Fellow in Honeywell's Aerospace Advanced Technology Group, and the Chief Systems Engineer of the Strategic Sensors Group of Honeywell Defense and Space located in Phoenix, AZ. Glen received his Ph.D. in Physics from MIT in 1983 and has worked at Honeywell a total of 32 years in precision optical measurements, fiber optics, and sensing. At Honeywell, he led efforts in high performance fiber optic gyros for applications in space platform stabilization, submarine navigation, and other guidance areas. Glen was also a founder of NxtPhase, Inc., commercializing fiber optic current sensors for the electric utilities. The team received the R&D 100 Award for one of the top new products of 2002. Recent development activity at Honeywell has focused on next generation optical-resonator sensors, including resonator fiber optic gyros. Glen has chaired and co-chaired many conferences, including serving as Technical Chair of the 18th Intl. Conference on Optical Fiber Sensors (OFS). He has served as a member of the OFS Technical Program Committee, OFS International Steering Committee, and the OFS International Honorary Committee for 25 years combined. He has over 90 inventions and 50 publications.

Steven J. Sanders is Director of Engineering at Honeywell Quantum Solutions. He attended U.C. Berkeley for a B.S. in Engineering Physics, and then went on to complete a Ph.D. in Applied Physics at Caltech in 1997. After working in Japan for a year as a Henry Luce Foundation Fellow at Mitsubishi's Advanced Technology R&D Center, Steve joined the Honeywell Technology Center in 1998. He spent the first 18 years of his Honeywell career as a Research Scientist, Manager, and Engineering Fellow in the Strategic Sensors group; there he led programs in fiber-optic gyroscopes, guidance and navigation systems, and fiber telecommunication technology. Along the way, Steve earned a 6σ Master Black Belt and 16 patents. More recently he has joined Honeywell's efforts to develop advanced quantum computing technology and full quantum information systems.

Lee K. Strandjord, is a Fellow in the Strategic Sensors Group of Honeywell's Defense and Space located in Phoenix, AZ. He received a B.S. degree in Physics from the University of Minnesota in 1987 and has worked at Honeywell for 32 years. From 1989 to 1990, he was a resident engineer at the Charles Stark Draper Laboratory for technology transfer of the U.S. Army's Solid State Inertial Measurement Unit (SSIMU) RFOG program. From 2001 to 2007, he was the Principal Investigator and lead Systems Engineer of a strategic grade fiber optic gyro being developed for the U.S. Navy's Trident submarine navigator. From 2007 to present, his focus has been on resonator fiber optic gyros, semiconductor laser stabilization, sensor signal processing, and laser based displacement sensors. Mr. Strandjord has published over 20 research papers and holds 80 patents.

Tiequn Qiu obtained his B.S. from the Physics Department of Fudan University, Shanghai, China. He got his Ph.D. degree in Lasers and Nonlinear Optics at Regensburg University, Germany, in 1997. From 1998 to 2003, he worked as a research associate or laser/optical engineer in CUNY, Quantronix Corp. and CIENA Corp., where joined the development of nonlinear optical material, high power lasers and ultra-high density fiber optical communication systems. From 2003 to 2005 he worked as a research scientist in the Optical Science Center at University of Arizona, successfully developed high power lasers using short rare earth doped fibers. In 2005, Dr. Qiu joined the Strategic Sensor Group at Honeywell, where he worked as a principal engineer in design, test and numerical simulation of fiber optical gyros. He currently holds more than 40 patents in fiber optical sensors and laser sources.

Jianfeng Wu received his B.S. and M.S. in Physics from Fudan University in 1998 and 2001, and Ph.D. degree in optics from the University of Arizona College of Optical sciences in 2005. From 2004 to 2010, he was an Optical Engineer with NP photonics. Since 2011, he has been a Staff Scientist with Honeywell International in Phoenix, AZ. His research interests include fiber lasers, fiber optical gyroscope, fiber optical sensing, and Silicon photonics.

Marc Smiciklas is a Principal R&D Scientist in the Advanced Technology division of Honeywell Aerospace. Marc's background is in Atomic, Molecular and Optical (AMO) Physics, with a focus on precision measurements. He received his Ph.D. in Physics from the University of North Texas in 2010. Following his Ph.D., Marc spent 4 years as a Post Doc under Michael Romalis at Princeton University. There he worked with atomic magnetometers and performed experiments testing fundamental symmetries. This included a test of Lorentz Symmetry performed at the Amundsen-Scott South Pole Station in Antarctica. He joined the Advanced Technology group at Honeywell in 2014 where he develops RFOG technology.

Derek Mead is currently a senior staff engineer for inertial navigation guidance and controls systems working with Honeywell for over 27 years. Mead received a bachelor's of applied physics and his master's in electrical engineering with an emphasis in fiber optic technology. Mead is currently responsible for managing new inertial navigation systems designs using interferometric fiber optic gyros. Over the years, Mead has introduced over dozens of inertial navigation systems for avionic, terrestrial, underwater and space applications. Mead provided excellence in system engineering through product concept definition, design specifications, development, end of life analyses and design qualification. In addition to his leadership in system engineering, Mead has previously worked in pressure sensors used for altimeter and airspeed indicators for both commercial and business commuter aircraft. Mead developed hardware and software designs for aircraft Air Data controllers, when aircraft systems were converted from analog to digital design. Mead has spent over ten years working in the Honeywell research lab in the initial development of the interferometric fiber optic gyro. In this period of time, Mead received seven patents, six outstanding engineer awards and one technical achievement award. Mead has also worked in the accelerometer sensors divisions for the Avionics, Space and Defense markets. Mead was on a large team that designed, developed and verified the world's most accurate interferometric fiber optic gyro mentioned in this paper. Mead was the lead technical director for the design, development and qualification for the Honeywell high performance Spirit fiber optic gyro mentioned in this paper, which is an industry game changer. Mead's future pursuits is to develop a state of the art digital fiber optic gyro to reduce navigation controls systems size, weight and power.

Sorin Mosor is a Senior Optical Engineer in the Strategic Sensors Group at Honeywell Aerospace Engineering & Technology. Sorin received his Ph.D. (2005) in Optical Sciences from University of Arizona, with a dissertation on semiconductor quantum optics; he has co-authored seven papers in this field, published in major scientific journals. Sorin has been at Honeywell for over 12 years, working on fiber optic gyroscopes. He has led the optical design effort on multiple FOG programs, being a major contributor to all efforts to improve gyro performance. During this time he was co-author on 4 patents and received the Honeywell Technical Achievement Award and Excellence in Innovation award (2015).

Alejo Arrizon is the Principal Electrical Engineer in the Launch Vehicle Division at Northrop Grumman. He received his B.S and M.S. degree in Electrical Engineering from Arizona State University in 2011 and 2015, respectively. From 2011 to 2017, he worked at Honeywell in the Strategic Sensors Group as a test and electrical design engineer where he was highly involved in developing test systems for various Fiber Optic Gyroscope (FOG) programs.

Waymon Ho is a Test Engineer with a decade of experience in new test equipment development, script automations, and software programming in both Labview FPGA and Labview software. He has a Bachelors in Electrical Engineering and a Masters of Business Administration degree at Arizona State University. Waymon is also a proud father of two kids. Waymon is currently working as a Staff Engineer in Collins Aerospace.

Waymon worked as a Senior Engineer for 15+ years at Honeywell. He was given an opportunity building data acquisition system that prototype various different feedback loops, writes firmware, and software to demonstrate Resonator Fiber Optic Gyro (RFOG) performance. Waymon was part of a team that turned the RFOG technology from research into advanced development phase.

Mary Salit is a Principal Optical Physicist with Honeywell Advanced Connected Sustainability Technologies. She received her Ph.D. in Physics from Northwestern University in 2009, and joined Honeywell the same year. She was previously a principal investigator for an optical waveguide gyroscope program, designed integrated photonics devices to achieve sub-diffraction-limited resolution in telescopes, and was a Silicon Optical Bench task manager for Honeywell's RFOG program.

Neil A. Krueger is a Research Scientist in the Advanced Sensors and Microsystems Research Technology Center of Honeywell Aerospace Advanced Technology in Plymouth, MN. He received a B.S. in Engineering Physics from Rose-Hulman Institute Technology with a concentration in Micro-/Nanotechnology (2011), where he was also a four-year member of the varsity baseball team and named a Capital One/CoSIDA Academic All-America in 2011. Neil then obtained his Ph.D. in Materials Science & Engineering from the

University of Illinois at Urbana-Champaign under the guidance of Professor Paul V. Braun, earning a National Defense Science & Engineering Graduate Research Fellowship. Since joining Honeywell in July 2016, Neil has supported resonator fiber optic gyro development by helping to oversee the continued development of the silicon optical bench. In addition, he has led and been a part of various optical material and integrated photonics developments including large-area optical metasurfaces, optical frequency combs, optical phased arrays, silicon photonics manufacturing, and high-efficiency fiber coupling for optical interconnects. Overall, he has authored more than 10 peer-reviewed journal articles and is a named inventor on more than 20 invention disclosures and patent applications.

Clarence Laskoskie received his M.S. in Electro-Optics from University of Houston – Clear Lake. Recently retired, he worked with Honeywell as a research scientist/engineer from 1985 to 2018 developing, designing and building optical sensors and associated test stations and procedures. Mr. Laskoskie authored/co-authored 13 patents on optical sensors and devices.

Chellappan Narayanan has over 20 years of R&D experience in the field of photonics, with progressively increasing levels of responsibility. He received his M.S. and Ph.D. in Electrical Engineering from the University of Texas at Austin, in 1990 and 1994, respectively. From 1994–1996, he was at AT&T Bell Laboratories, Murray Hill, NJ, where he was engaged in passive optical component research. He moved to the Optoelectronics Center of Bell Labs, Lucent Technologies in 1999, and became a Bell Labs Distinguished Member of the Technical Staff in 2000. He is presently a senior engineer at Honeywell involved in the design of FOG-based inertial navigation systems. He has published over 30 journal papers, hold 20 patents.

Wes Williams is a Staff Scientist with Honeywell Aerospace Advanced Technology and the Portfolio Manager for Honeywell Resonator Fiber Optic Gyroscope Development. He has over 20 years of experience developing fiber optic sensors. He has a Master (1998) and Bachelor (1996) of Science in Mechanical Engineering from the Massachusetts Institute of Technology.

Chapter 5
Fiber Optic Gyros from Research to Production*

George A. Pavlath
Northrop Grumman Corp.

5.1 Abstract

Fiber optic gyros are a great success story for a new inertial measurement technology that successfully transitioned from the laboratory in 1975 to production in 1992. This chapter will review their research, advanced development, product development, and production transfer. The focus of the chapter will be this cycle from Stanford University to Northrop Grumman.

5.2 Research

Stanford University began the development of fiber optic gyros in 1975 when this author started there as a first year graduate student in Applied Physics. The first fiber gyro was built on a rotating table originally used for a Michaelson and Morley type ether drift measurement in the 1920s. This gyro was operated at 633-nm wavelength with a HeNe laser and had 600 meters of fiber enclosing a 1-m^2 square area. It clearly demonstrated rotation over approximately ±13 fringes in 1996.

The next Stanford fiber gyro[1–5] was built using a semiconductor laser source operating at 820 nm. It demonstrated numerous all fiber optic components: couplers, polarizer, polarization controller, and phase modulator. It was eventually able to measure down to 0.01 deg/h rotation rates with angle random walks around 0.001 deg/rt-h. It demonstrated mitigation techniques for amplitude modulation, polarization nonreciprocity, Shupe

---

*Chapter adapted from G. A. Pavlath, "Fiber optic gyros from research to production," *Proc. SPIE* **9852**, 985205 (2016) [doi: 10.1117/12.2228562].

effect, Faraday effect, and Kerr effect. All of these components, performance, and mitigations were demonstrated by 1982.

Due to the limited funding and number of students working on the fiber gyro project, the following areas were not able to be addressed: performance over wide temperature ranges (–55 to +75 °C), shock and vibration environments, long-term performance stability, scale factor stability, and dynamic range.

5.3 Development

In 1982, Litton Guidance and Control Systems took over the Stanford Fiber Optics Research Program from Atlantic Richfield (ARCO). At the same time, this author graduated from Stanford with his Ph.D. and was hired by Litton to develop the fiber optic gyro.

The first task was to reproduce the Stanford all-fiber-optic gyro but in a more compact and rugged format. This was the X-1 gyro, which was 4 inches in diameter and 4 inches in height (Fig. 5.1). This gyro achieved the same performance as the larger Stanford all-fiber gyro. With additional development, this gyro could have been used on local level-platform-based systems.

Unfortunately in the early 1980s, the customer base was moving away from platform-based systems to strapdown systems. These strapdown systems require scale factor (SF) accuracy and stability on the order of a few parts per

Figure 5.1 First fiber optic gyro built at Litton (now Northrop Grumman) in 1983.

million (ppm) along with large dynamic ranges of 10^8 to 10^9. To achieve these performance levels requires closed-loop operation of the fiber optic gyro. The all-fiber gyro from Stanford University was an open-loop gyro, and an all-fiber feedback component did not exist.

Closed-loop operation of fiber gyros was achieved by Cahill and Udd[6] at McDonnel Douglas in the late 1970s using Bragg cells. This gyro could achieve the scale factors needed, but the Bragg cells were bulk optics that were difficult to interface with and power hungry. Researchers and engineers had built fiber gyros in the 1980s using integrated optic phase shifters driven by linear voltage ramps. Again, this permitted closed-loop operation, but SF accuracies were only 0.1 to 1.0%, a long way away from the few ppm needed for strapdown operation.

In 1985, Arditty and Lefevre[7] of Thompson CSF published their paper on the serrodyne dual-loop-closure method for closed-loop operation of the fiber gyro. The linear voltage ramp was replaced with a digital staircase. The step size commanded by the first loop closed the loop and permitted a wide dynamic range. The second loop, the SF loop, permitted controlling the reset of the voltage staircase ramp to be exactly 2π with accuracy of 1 ppm, which permitted accurate and stable scale factors. Their method with numerous modifications is the basis for all closed-loop fiber optic gyros in production.

The first closed-loop gyros at Litton were the EDM-2 and EDM-3 gyros shown in Figs. 5.2 and 5.3 respectively. They were 4 inches in diameter and 1 inch in height. These gyros were prototypes for a fiber-optic-gyro-based inertial navigation system.

Figure 5.2 EDM-2 (Engineering Developmental Gyro) utilizing a separately packaged light source and detector.

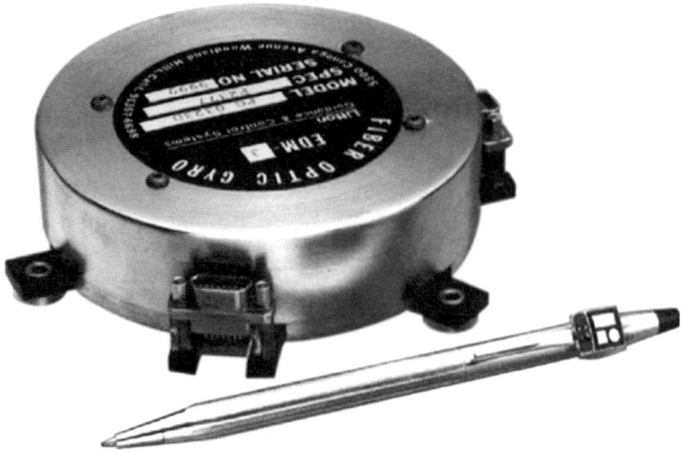

Figure 5.3 EDM-3 gyro containing a light source and detector.

5.4 Productionization

In 1989, product development began on a 1 deg/h-class tactical IMU, the LN200 shown in Fig. 5.4 along with some key physical parameters. The schedule called for production start in January 1992. The IMU had to operate from –54 to +71°C, operate in 15-G rms random vibration, and operate through shocks of 1500 G. In addition, production-cost targets had to be achieved.

A cost model was developed that flowed the product cost down into assembly labor, test labor, and component costs. The component costs were used by the engineers along with the environmental requirements to develop the necessary components, vendors, and fabrication and assembly processes.

Many vendors were developed to produce environmentally robust components that achieved their cost targets along with all of their physical and optical performance requirements. Vendors were developed for the pigtailed optical source, pigtailed photodetector, and fiber optic couplers.

The cost of the polarization-maintaining (PM) optical fiber was a major cost driver in 1989. This was addressed by making long-term, large-quantity

- LN200
    - 3.5" ϕ x 3.35" H
    - 1.65 lbs
    - < 15 W
- Gyros:
    - 1.0 º/hr bias
    - 100 ppm SF
    - 0.07 º/rt-hr ARW
- Silicon Accelerometers:
- 300 ug bias
- 300 ppm SF
- 35 ug/rt-Hz noise

Figure 5.4 LN200 IMU and key parameters.

purchasing agreements with the fiber vendor. The price of the PM fiber today is well below the cost target set back in 1989.

Another major cost item was the multifunction integrated optics chip (MIOC, see Fig. 5.5). This was addressed by building the appropriate-class clean rooms at Litton for production of these MIOCs. These in-house production facilities have built over 10,000 MIOCs a year and have improved the 1989 cost target for the MIOC.

Two major assembly processes that did not have solutions in 1989 by the fiber optics industry were the winding of the fiber optic sensing coil and the pigtailing of the MIOC. These processes had to be automated in order to meet the cost targets. Litton automated these processes by developing coil-winding robots and pigtailing robots, as seen in Fig. 5.6. These robots also improved the 1989 cost targets.

The last major assembly process was fusion splicing of optical fibers to form the fiber gyro circuit. The fiber optic industry had by 1990 developed automatic fusion-splicing systems along with automated jacket-stripping and rejacketing systems. These systems were purchased from commercial vendors and used to great success.

The LN200 started production in January 1992 and remains in continuous production. Over 30,000 LN200 systems utilizing three fiber gyros and three silicon MEMS accelerometers have been delivered to nearly 200 customers.

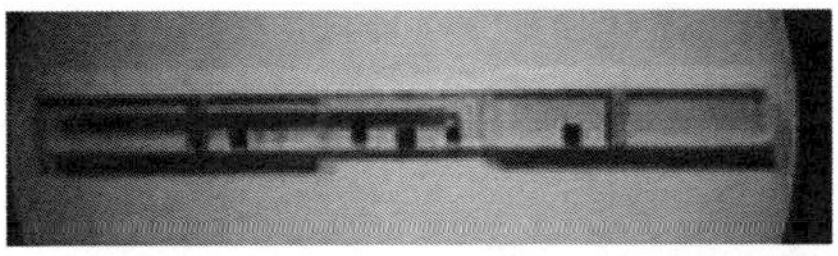

Figure 5.5 Multifunction integrated optics chip (MIOC) for Northrop Grumman fiber gyro.

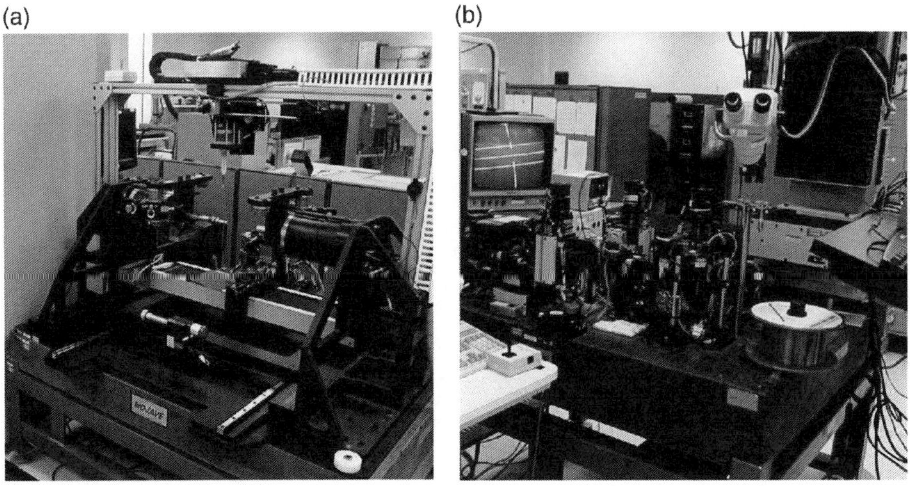

Figure 5.6 (a) Coil-winding robot, and (b) MIOC pigtailing robot.

They are used in a wide variety of applications, from undersea to outer space, and have an excellent reliability record in the field.

In 2000, Northrop Grumman purchased Litton Industries. In 2002, the LN251 and the LN270 (Fig. 5.7) began low-rate production and in a few years transitioned to high-rate production. The LN251 is qualified as the AN/ZSN-1 by the USAF as a 0.8 nautical mile per hour (nmph) navigator, and the LN270 is qualified by the US Army as the AN/VSN-12 for land-navigation and pointing systems. Over 3000 of these systems have been delivered to Northrop Grumman's customers.

The components that were used in the LN200 were reused for LN251/LN270. The coil-winding and pigtailing robots were modified to handle the larger, longer fiber coil and the higher-accuracy pigtailing requirements for the MIOC.

Around 2005, Northrop Grumman repackaged the LN251 in a different form factor and with different interface cards and software to be a replacement for the USAF Standard Navigation System. This is designated the LN260 and is shown in Fig. 5.8. Over 2000 have been delivered.

(a) (b)

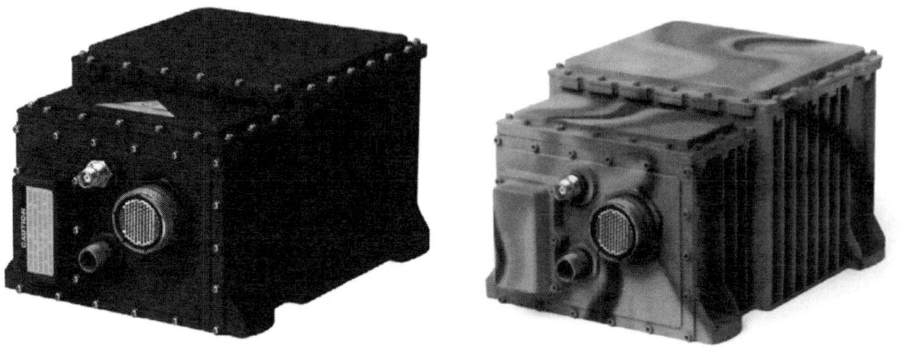

Figure 5.7 (a) LN251 airborne inertial navigator, and (b) LN270 inertial land navigator.

Figure 5.8 LN260 standard inertial navigator.

5.5 Summary

The fiber optic gyro is the great success story about the transition of a fiber optic product from research into high-volume production. All of the developmental and production challenges such as size, weight, cost, environmental ruggedness, and reliability have been achieved. Northrop Grumman has delivered over 100,000 fiber optic gyros in numerous inertial products to hundreds of customers. Northrop Grumman's fiber-optic-gyro-based products remain in forward production at high rate and will continue for at least two decades.

References

[1] C. C. Cutler, S. A. Newton, and H. J. Shaw, "Limitation of rotation sensing by scattering", *Optics Letters* Vol. **5**(11), 488–490 (1980).

[2] K. Bohm, P. Marten, K. Petermann, and E. Weidel, "Low-drift Fibre Gyro Using a Superluminscent Diode", *Electron. Lett.*, Vol **17**(10), pp 352–353, 1981.

[3] H. C. Lefevre, R. A. Bergh, and H. J. Shaw, "All-fiber gyroscope with inertial-navigation short-term sensitivity", *Opt. Lett.*, vol. **7**(9), pp 454–456, 1982.

[4] R. A. Bergh, H. C. Lefevre, and H. J. Shaw, "All-single-mode fiber-optic gyroscope with long term stability", *Opt. Lett.*, vol. **6**(10), pp 502–504, 1981.

[5] K. Bohm, P. Russer, E. Weidel, and R. Ulrich, "Low-noise fiber-optic rotation sensing", *Opt. Lett.*, vol. **6**(2), pp 64–66, 1981.

[6] E. Udd and R. F. Cahill, "Phase-nulling fiber-optic laser gyro", *Opt. Lett.*, vol. **4**(3), pp 93–95, 1979.

[7] H. C. Lefevre, J. P. Bettini, S. Vatoux, and M. Papuchon, "Progress in Optical Fiber Gyroscopes using Integrated Optics", AGARD CPP-383, 9A/1-13, 1985.

Chapter 6
Technological Advancements at Al Cielo Inertial Solutions

Meir Rosilio, Lisa Koenigsberg, Noam Pasternak, and Arnon Arbel
Al Cielo Inertial Solutions, Ltd

6.1 Introduction

For over ten years, Al Cielo Inertial Solutions (ACIS) has been supplying Israeli and international customers with high-performance, reliable inertial solutions based on closed-loop FOGs. This technology has been proven over the past few decades as a growing number of IMU systems based on FOG technology have been successfully deployed. FOG systems are found in tactical and in strategic IMUs, providing a range of solutions for navigation and control. IMUs, as well as systems-gyroscopes-only inertial reference units (IRUs), are also used for high-precision attitude referencing. This chapter is most, but not exclusively, relevant to this application.

Customer specifications are usually apportioned into several figures of merit that together allow for specific qualification and acceptance tests under known inertial input conditions. For example, separate tests are often used to test the small signal errors, i.e., bias, and the large signal errors, i.e., scale factor. This approach has many advantages, particularly when the system is designed for a variety of customers and applications. However, as it will be shown, for some applications the standard control loop, used in the closed-loop FOG, may lead to errors. These errors are unique in that they are not usually accounted for in the overall navigation budget, and in turn standard acceptance tests are not designed to quantify these errors. Moreover, even if these errors were observed, they cannot be corrected or compensated for as is normally practiced for other errors. Therefore, a new parameter called "navigation accuracy" was defined to describe FOG performance vis-à-vis this error.

The goal of the work herein described was two-fold. Firstly, an optimized control loop was custom designed to meet the particular profile and conditions of interest. Secondly, a multilevel acceptance test was specified to prove compliance. The first goal, the design, was achieved using standard concepts and simulations of feedback loops incorporating control blocks. The second goal, the verification, required hardware-in-the-loop (HITL) simulations and special test equipment.

Section 6.2 reviews the standard closed loop for a FOG, with emphasis on the particular analysis approach used (it will be analyzed here as a deadbeat controller). The closed-loop error for particular profiles is shown as computed by using either Monte Carlo or HITL simulation. Section 6.3 describes an optimized closed-loop controller and presents substantially improved results. Various methods of verification are used to increase confidence in the optimized controller. The chapter concludes with some experimental results in Section 6.4.

6.2 Standard Control Loop

6.2.1 Control model

The all-digital closed-loop approach for a single-axis FOG[1] is well known, as well as the enabling technology of a multi-integrated optical circuit (MIOC) most commonly made from lithium niobate. A MIOC provides an electro-optic phase modulator featuring high bandwidth and linearity. Several previous works[2–6] presented and analyzed the gains and dynamics of the closed-loop system. Most previous designs constitute a single-loop path that contains two integrals in series. Where an additional integral loop is discussed,[3] analysis and results proving the benefits are not given.

Our particular model of the standard logic circuit is shown in Fig. 6.1. Additional loops are often present, for example, a loop that tracks changes in the gain of the feedback voltage. In addition, some randomization scheme is often incorporated to eliminate dead-band and improve resistance to vibration. This work is only concerned with the rotation-rate closed loop. Since this scheme is commonly known, only a brief description will now be given to serve as a basis for further adaptations.

It has been shown[3] that for conditions of input-step and steady-state error of zero the deadbeat control filter is

$$H(z) = G2 \cdot \frac{1}{1 - z^{-1}} \cdot \frac{1}{1 - z^{-k}}.$$

Several simplifications are inherent to this model. Although there are analog electronic elements that create additional time delays that are smaller than the inverse of the eigenfrequency, the dominant delay in the system is due to digital processing. Digital electronics delays, commonly 2–4 times the inverse of the eigenfrequency, are generated at the A2D and the digital processing unit. The requirement to synchronize the modulator and the demodulator in order to achieve a lock-in-amplifier function is also a factor. The smaller delays of the

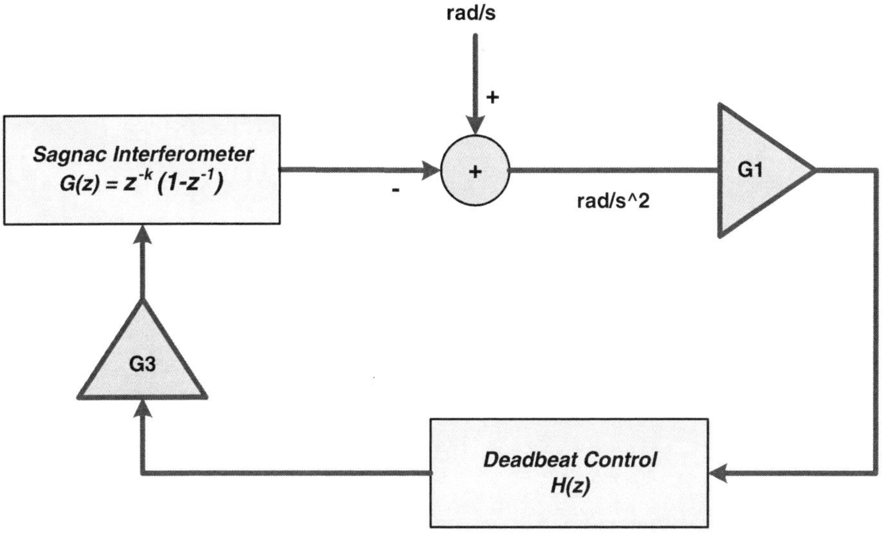

Figure 6.1 Block diagram of a standard closed-loop FOG control loop.

analog electronics can, therefore, be safely ignored. Also, some analog electronic blocks function as high-pass or low-pass filters. Here, as well, only the dominant blocks that affect the rate signal and noise are considered. As in previous treatments,[3] the modulation and demodulation stages can be removed from the model since they effectively cancel each other out. Also, lumped gains are used when practical. For example, we can combine the gains due to the optical transfer function of the Sagnac effect with a two-signal interferometer (analog and digital gains, as well as gains due to domain transfer), the signal domains being optical, analog, and digital. The nonlinearities of the gains can also be approximated and incorporated into the model in a Monte Carlo simulation.

6.2.2 Sub-specifications and verifications

Gyro specifications are commonly derived using navigation-error equations.[8,9] This way, given an overall navigation-error requirement, the sub-specifications for the gyro or system are allocated. For example, the error parameters for a single-axis gyro may be apportioned into noise, bias, scale factor, and misalignment. These parameters are then commonly measured using standard procedures.[9] During a calibration process, these error parameters are modeled and compensated for at the digital section of the flow that follows the closed-loop circuit depicted in Fig. 6.1. This standard method is sufficient for a large variety of applications and is most practical for products targeting a generic customer base.

The previous method, however, fails to consider errors generated by the control loop itself. These errors are usually not modeled in the navigation-error equations. They are not apportioned as a sub-specification to be verified by either a qualification test plan or an acceptance test plan. Consequently, typical verification tests are not designed to observe these errors. For example, SF error measurements usually compare a FOG measurement to a known input. However, this measurement is usually performed during the input of a constant rotation rate. The measurement is the integral of the rotation rate over some applied rotation-profile, and the measurements is taken while the FOG is actually measuring earth-rate inputs. Also, control-loop errors due to acceleration and deceleration often cancel each other out, and thus the FOG ability to track changes in the rotation rate or track the relative angle during angular accelerations and deceleration is often not observed. The end result is that excellent SF results may be measured, when in fact the customer's overall requirements are not met.

6.2.3 Navigation accuracy sub-specification

To complement the ordinary-FOG sub-specifications, an additional parameter (referred to as navigation accuracy) was identified to answer to the aforementioned problem. Several rotation-rate profiles were found to be relevant to a particular application. The navigation accuracy is the error of the integral of the rotation rate or the relative angle at a critical inquiry time during the profile. In general, at this inquiry time, the rotation rate is not constant. As an example, for a particular application, the customer requirement for navigation-accuracy error was on the order of several μrad.

6.2.4 Monte Carlo simulation

A Monte Carlo simulation, implemented with Simulink, was used to analyze the standard-control-loop filter for stability and general behavior. For given coil dimensions and open-loop gains, the gain margin was 30.9 dB, and the

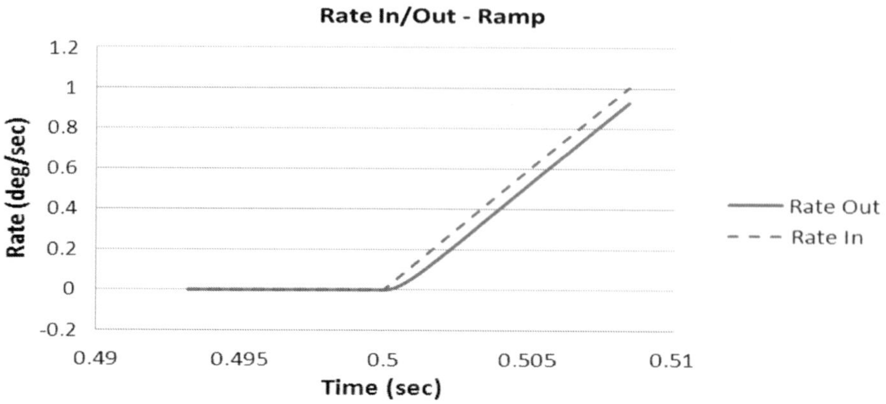

Figure 6.2 Standard-control-loop ramp response.

phase margin was 87.6 deg. The response to a rate ramp input is shown in Fig. 6.2. The rate error and angle error are given in Fig. 6.3. For a standard control loop, the angle error is unbounded and will reach prohibitive values for ramps that span several minutes.

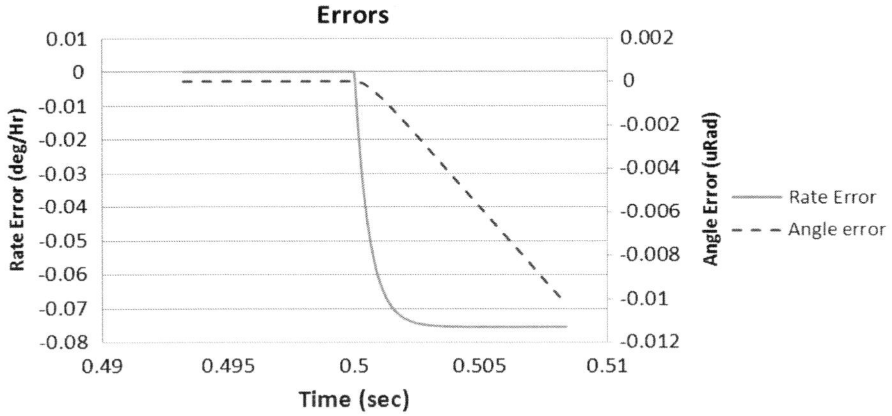

Figure 6.3 Standard-control-loop ramp errors.

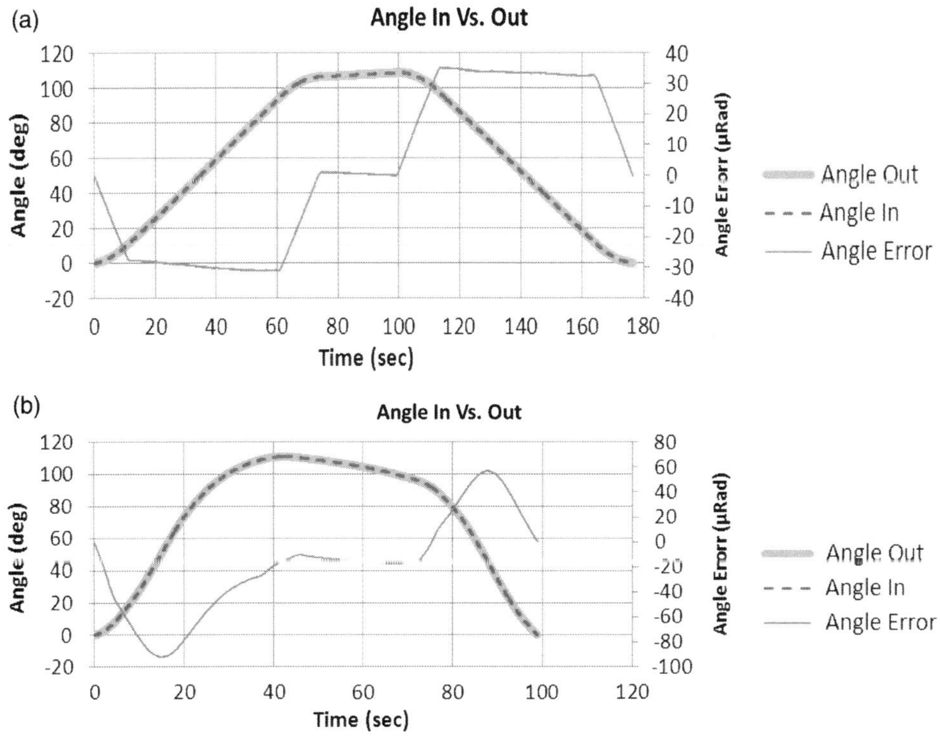

Figure 6.4 Relative angle error during two user-defined profiles in HITL simulation; (a) and (b) depict two separate profiles.

6.2.5 HITL simulation

A HITL simulation was built into the digital section of the FOG. The HITL simulation is designed to allow the FOG to close the loop while injecting artificially generated phase disturbances according to a user-defined profile. This profile could be a sine wave used to verify the signal bandwidth or a step function or ramp to verify convergence and tracking. In addition, any particular relevant profile as defined by a customer could be tested. The advantages of a HITL simulation are many. HITL is much more robust than a typical Monte Carlo simulation, as most of the simplifications and approximations are avoided. Also, the simulation contains all of the random noise sources, including quantization noise arising from all domains: optical, analog, and digital. In addition, most of the gain nonlinearities in the system are incorporated into the simulation. Figure 6.4 shows the measurement error of the relative angle during specific user-defined profiles using the HITL simulation.

The underperformance of the standard control loop is made apparent. The peak-to-peak error of the absolute angle can exceed 100 μrad, whereas only several μrad are allowed. These loop errors were shown to exist under several different user-defined profiles, which confirmed the conclusion implied by analysis as well as by Monte Carlo simulation.

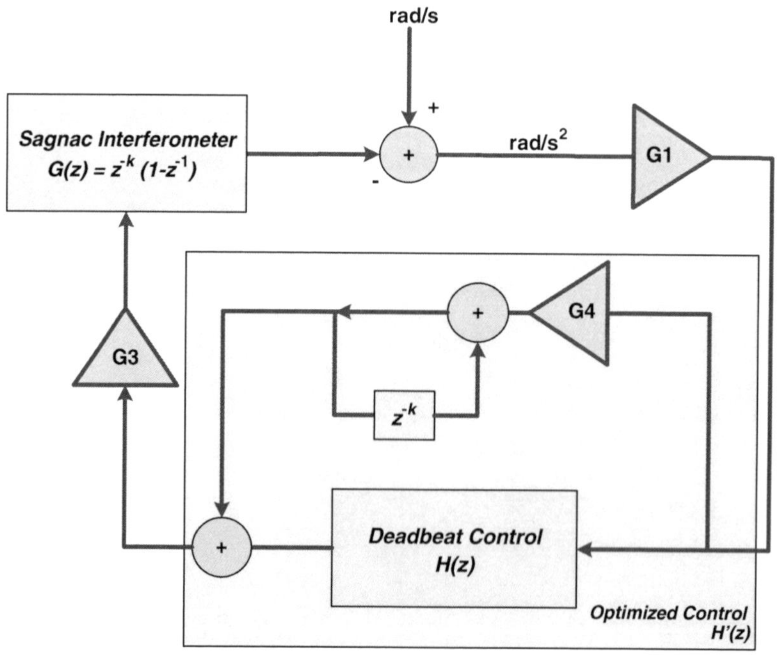

Figure 6.5 Block diagram of an optimized closed-loop FOG control loop.

6.3 Optimized Control Loop

6.3.1 Control block

A control block was designed to improve the accuracy of the relative angle measurement during acceleration profiles. It was found that an additional integral path in parallel to the series loop will meet the stability requirements, which can be described as

$$H'(z) = G2 \cdot \frac{1}{1 - z^{-1}} \cdot \frac{1}{1 - z^{-k}} \cdot \left(1 + G4 \cdot \frac{1}{1 - z^{-k}}\right).$$

The margin analysis was repeated for this new control block; the block diagram is shown in Fig. 6.5. Several classic tuning methods were used to determine the relative gains of the two paths until the loop was optimized for all requirements.

6.3.2 Monte Carlo simulation

Monte Carlo simulation was used to analyze the optimized control-loop filter for stability and general behavior. For the same coil dimensions and open-loop gains from Section 6.2.4.1, the gain margin was 26.6 dB, and the phase margin was 60.3 deg. These stability margins are lower than those of the standard control loop in Section 6.2.4.1, although still high enough for the majority of applications. However, using this filter in a more-generic-application product requires further study to ensure that there is no practical penalty. The response to a rate-ramp input is shown in Fig. 6.6, with the errors in rate and angle given in Fig. 6.7.

For the optimized control loop, the rate error is transient, and the angle error is small and bounded. This control loop exhibits the characteristics that are needed for the specifications of navigation accuracy.

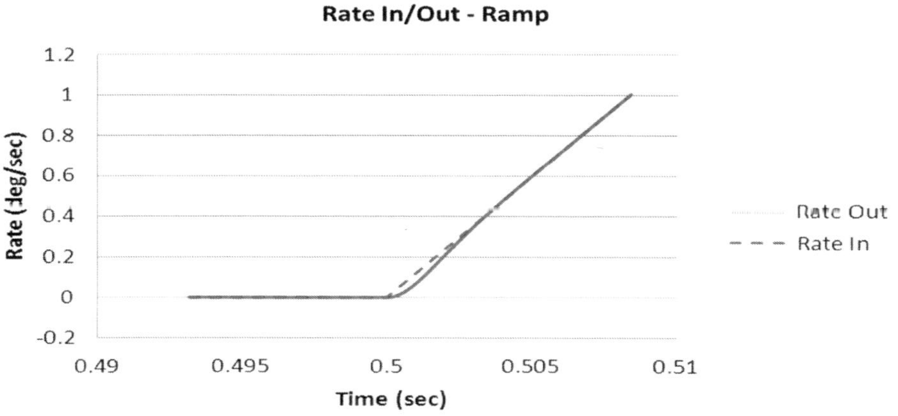

Figure 6.6 Optimized-control-loop ramp response.

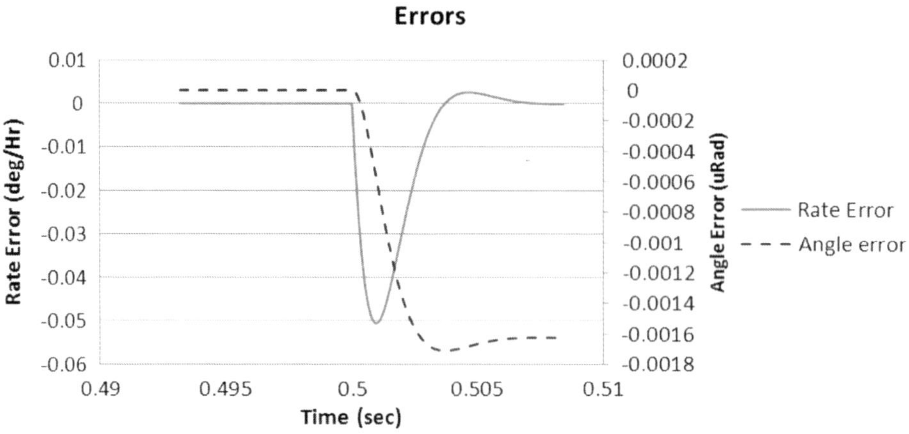

Figure 6.7 Optimized-control-loop ramp errors.

6.3.3 HITL results

Applying the HITL simulation to the new control loop with the same user defined profile as in Fig. 6.4, yields the results in Fig. 6.8. The peak-to-peak error of absolute angle is less than 1 μrad during the entire profile. For any critical inquiry time, the control loop meets, and even surpasses, the customer

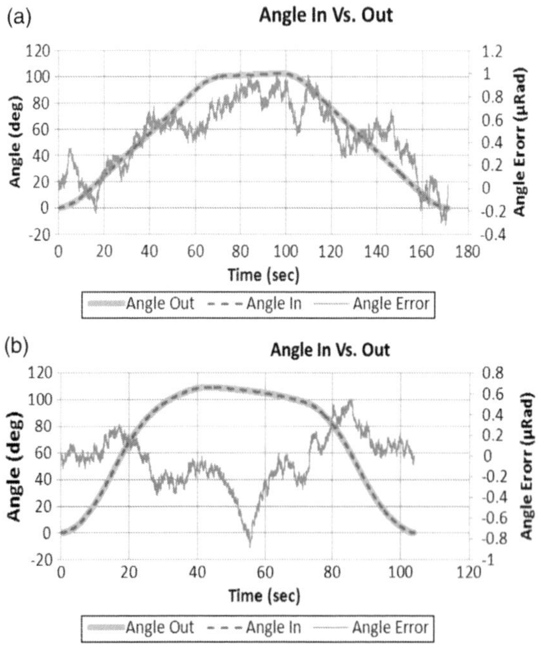

Figure 6.8 Relative angle error during two user-defined profiles in HITL simulation with an optimized control loop. (a) and (b) depict two separate profiles.

requirement of several μrad. A substantial improvement was seen for several user defined profiles, demonstrating the robustness of the solution. Improvements of up to three orders of magnitude in navigation accuracy were observed.

6.4 Inertial Measurements

The ultimate verification of the design is a comparison of true angles to the measurement of the FOG angles during rotation. This measurement has several similarities to the standard SF measurement. In both tests, the FOG sensitive axis is aligned parallel to a rate-table rotation axis, and the rate table and the FOG are rotated together. Similarly, the angle measurements of the FOG, after subtracting the angle measured due to the earth rotation, are compared to the rate-table angle. However, in the new protocol, the comparison between the FOG and rate-table angle is performed during rotation, not only at the initiation and the termination of the test. This is achieved by passing a pulse, created following completion of a one degree of rotation of the rate-table. The pulse is transmitted to the FOG electronics, which in turn calculates its angle measurement synchronized with the rate-table pulses. A user-defined profile containing angular accelerations and decelerations can be applied to the rate table.

Results for such experiments are shown in the following figures. Two profiles similar to those used in the simulations were applied, and the two control schemes were compared. First, it is shown that there was no deterioration of rate measurements due to the change in control scheme (see Figs. 6.9 and 6.10). Figure 6.11 shows a comparison of the angle errors for one profile, and Fig. 6.12 shows a similar comparison for another profile.

As can be seen from the results, significant improvements in the angle error have been achieved. However, these results are not as good as those predicted by the simulation. Initial analysis points to the complexity of the experiment as well as to some inaccuracies of the rate table. Further work is being done to investigate this discrepancy, as well as on-going comparison of prediction to operational mission testing.

6.5 Conclusion

Customized controller analysis leading to an optimized closed-loop filter for the FOG has been shown to feature substantially improved performance. The benefits of fulfilling the specification of navigation accuracy were confirmed under realistic profile conditions. Three orders of magnitude improvement was demonstrated with an optimized controller block over the standard deadbeat controller. Preliminary results also point to improvements under dynamic conditions such as high vibration profiles. Further work is needed to ensure that there is no practical penalty for the use of a generic-application product.

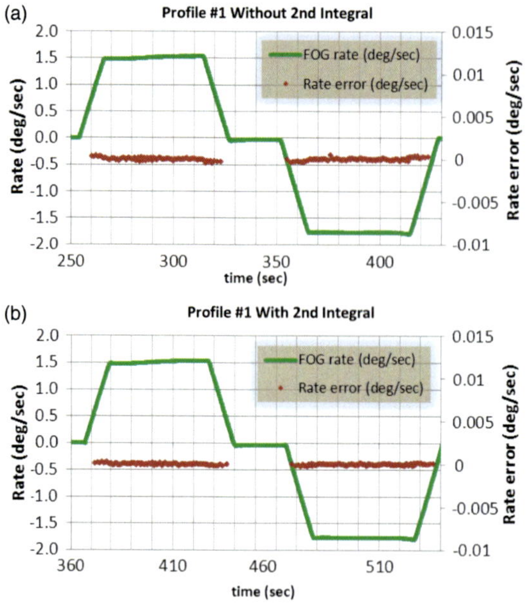

Figure 6.9 Rate error during a specific user-defined profile on a rate table comparing the two control schemes.

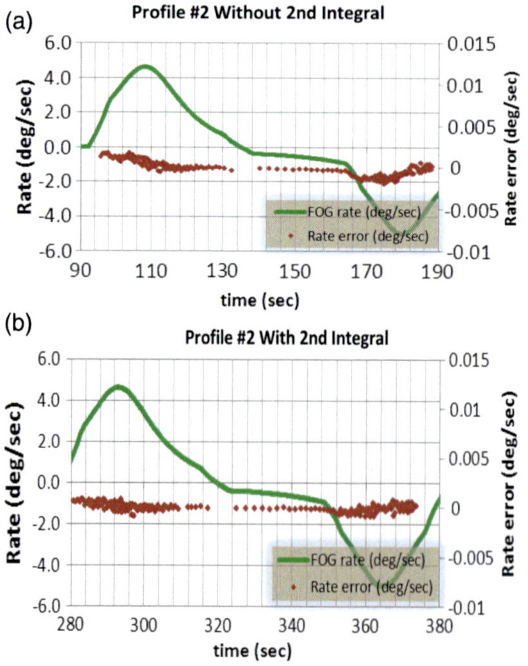

Figure 6.10 Rate error during a different user-defined profile on a rate table comparing the two control schemes.

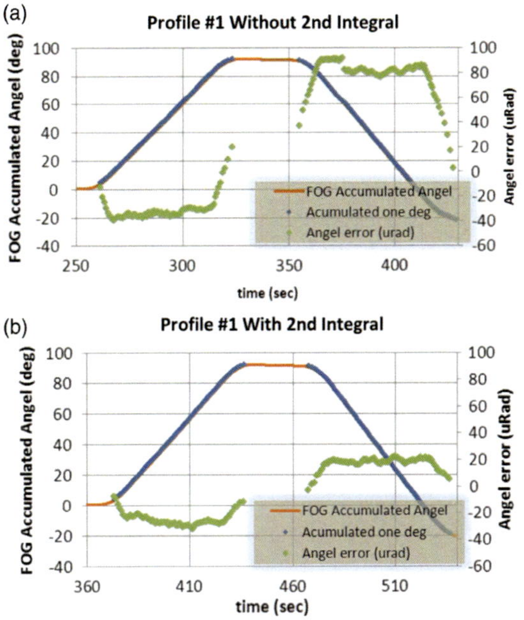

Figure 6.11 Relative angle error during one user-defined profile on a rate table comparing the two control schemes.

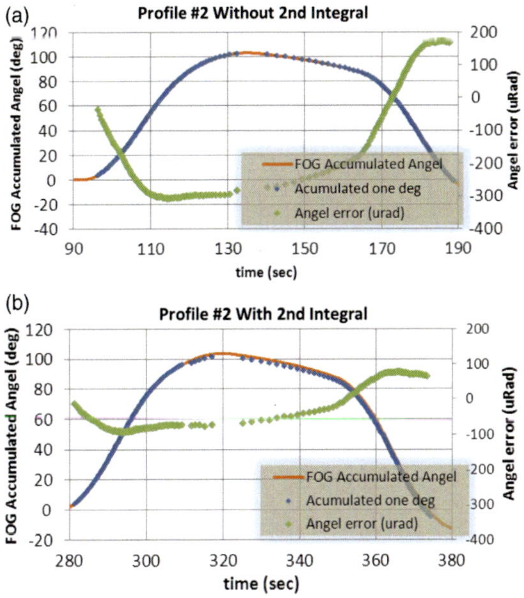

Figure 6.12 Relative angle error during a second user-defined profile on a rate table comparing the two control schemes.

Acknowledgement

The authors would like to acknowledge the fruitful collaboration with our customers who identified the problem and the solution and were full partners in realizing the analysis, implementation, and verification.

References

[1] H. C. Lefevre, *[The Fiber-Optic Gyroscope Second Edition]*, Artech House, Boston & London, p. 134 (2014).
[2] M. S. Bielas et al., "Stochastic and dynamic modeling of fiber gyros," *Proc. SPIE* **2292**, 240–254 (1994).
[3] G. Spahlinger et al., "Error Compensation via Signal Correction in High Precision Closed-Loop Fiber Optic Gyros," *Proc. SPIE* **2837**, 218–227 (1996).
[4] G. A. Pavlath, "Closed-loop fiber optic gyros," *Proc. SPIE* **2837**, 46–60 (1996).
[5] A. Noureldin et al., "Computer Modeling of Microelectronic Closed Loop Fiber Optic Gyroscope," *Proceedings of the IEEE Canadian Conference on Electrical and Computer Engineering*, 633–638 (1999).
[6] H. Junliang et al., "Modeling and Simulation of Digital Closed-loop Fiber Optic Gyroscope", *IEEE Proceedings of the 6th World Congress on Intelligent Control and Automation*, pp. 1659–1663 (2006).
[7] P. G. Savage, *[Strapdown Analytics Second Edition Part 2]*, Strapdown Associates, Inc., Minnesota, Chapter 13 (2007).
[8] D. Titterton and J. Weston, *[Strapdown Inertial Navigation Technology, 2nd Edition]*, IEE, Chapter 12 (2004).
[9] *IEEE Standard Specification Format Guide and Test Procedure for Single-Axis Interferometric Fiber Optic Gyros*, 952-1997(R2003).

Chapter 7
Current Status of Fiber Optic Gyro Efforts for Space Applications in Japan

Shinji Mitani, Tadahito Mizutani, and Shin-ichiro Sakai
Japan Aerospace Exploration Agency (JAXA)

7.1 Current Status of FOGs for Space Applications

Intermediate-grade interferometer fiber optic gyroscopes (IFOGs) have already been used in various space applications. The IFOG made by Japan Aviation Electronics Industry Ltd. (JAE) was used in the NASDA TR-1A rocket (1991–1998). This experiment, starting in 1991, marked the world's first space application of the IFOG.[1] Figure 7.1(a) shows the appearance of JAE's IFOG. The inertial-grade (0.05 deg/h) IFOG from Mitsubishi Precision Company Ltd. (MPC) (Fig. 7.1(b)) was used in the ISAS S-520 #1, M-3SII #6/7, and M-V #1 rockets (1997–present).[1] In 1997, the M-V #1 rocket successfully launched the radio astronomical satellite "HARUKA" (MUSES - B) into orbit (Fig. 7.1(c)). For HARUKA, the JAE inertial-grade 0.05-deg/h grade IFOG was installed for attitude control, and an initial observation exceeding a lifetime of three years was carried out. In small-satellite application, the JAE intermediate-grade IFOG named JG-34 was used in the JAXA (NASDA) microsatellite μ-Labsat (2002–2006), and the same product with a modified coil length named JG-34F (Fig. 7.1(d)) was used in the JAXA (ISAS) microsatellite REIMEI (2005–present).[2] The weight of JG-34F was 150 g/axis, and index angle random walk (ARW) was approximately 0.05 [deg/$\sqrt{}$h]. On-orbit data evaluation shows that accurate attitude knowledge coincides with the expectations based on gyro performance. The same FOGs on the REIMEI satellite were used in JAXA's spacecraft IKAROS, which was launched in 2010 and became the first spacecraft to successfully demonstrate solar-sail technology in interplanetary space by successfully deploying a 20-m sail.[3]

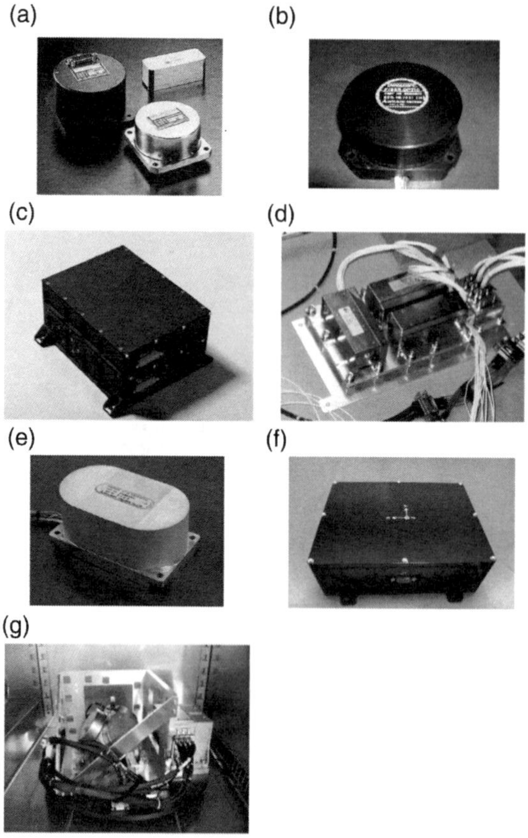

Figure 7.1 Appearance of IFOG models for space applications in Japan: (a) JAE IFOG in the NASDA TR-1A rocket (1991–); (b) MPC IFOG in the JAXA (ISAS) M-V rocket (1997–); (c) MUSES-B (HARUKA) IRU (1997–); (d) JAE JG-34F for REIMEI satellite (2005–); (e) MPC MFG-200 MOD-II on SERVIS-I (2003); (f) TSC TA7584 for microsatellites (2014); and (g) engineering model IRU for ASTRO-G satellite (2008).

In another effort, the MPC IFOG (MFG-200 MOD-II, shown in Fig. 7.1(e)) was demonstrated on the Space Environment Reliability Verification Integrated System (SERVIS-1) satellite developed by the Institute for Unmanned Space Experiment Free Flyer (USEF) and launched in 2003.[4] The angular random walk (ARW) was approximately 0.005 [deg/$\sqrt{}$h]. The three-axis FOG unit named TA7584 (Fig. 7.1(f)) was built by Tamagawa Seiki Co., Ltd. (TSC) for microsatellites being developed with the Next-Generation Space Systems Technology Research Association (NESTRA). The TA7584 series was widely used in satellites weighing less than 100 kg, such as the HODOYOSHI satellite series, Shindai-Sat (GINREI),[5] UNIFORM-1, and PROCYON in 2014.

High-precision FOGs have also been developed in recent years. A random walk 0.0005 [deg$\sqrt{}$h] level of the engineering model (EM) inertial reference unit (IRU) for ASTRO-G was developed.[6] Figure 7.1(g) shows the appearance of the EM IRU. Although the ASTRO-G program has been canceled, we continue to develop the high-accuracy FOG-IRU in order to satisfy these requirements and cover the specifications of future earth-observation and science missions.

7.2 Activities for Improving Coil Performance

This section describes the work that has gone into the development of high-precision IFOGs. Particularly with respect to the high accuracy of the coil, key technologies have been found over several years. Thermal strain significantly affects the stability of IFOG performance. Therefore reducing temperature-induced drift is one of the challenges in meeting the high-precision specification for a bias stability of 0.002 deg/h (1σ). We describe some practical examination results, such as the coil-winding method and a thermal structure study.

7.2.1 Symmetrical winding

When the temperature changes in the coil of an IFOG, bias fluctuation (called the Shupe effect) affects the output rate by a non-reciprocal effect of light. Regarding the distribution of the rate of temperature change across a sensing fiber $\delta T/\delta t$ (l) (function of l) of the coil outside T_D and inside, assuming a linear distribution of 0 in the case of solenoid winding, the bias drift rate θ_E is as follows:[7]

$$\dot{\theta}_E = \frac{n_c N_T \pi}{6}\left(\frac{\Delta n_c}{\Delta T} + n_c\alpha\right) \cdot T_D = S_0 \cdot T_D, \tag{7.1}$$

where n_c is the fiber-core refractive index, $\delta n_c/\delta T$ is the temperature change rate of the core refractive index, α is the coefficient of linear thermal expansion (CTE) of the fiber, $\delta T/\delta t$ is the time rate of change of temperature, c is the speed of light in vacuum, and N_T is the total number of coil turns Assuming $n_c = 1.45$, $\delta n_c/\delta T = 10^{-5}$ /°C, $\alpha = 0.5$ ppm/°C, $T_D = 10$°C/h, and $N_T = 7{,}000$, Eq. (7.1) leads to 3.3 (deg/h)/(°C/h).

Therefore, some devised elaborate winding patterns are typically used, with fibers arranged on each side at the same distance from the center position to cancel the resulting nonreciprocal phase error. This is referred to as symmetrical winding. Table 7.1 lists the bias-rate analytical solution of each symmetrical winding technique as compared with solenoid winding.[8] Shupe sensitivity is calculated as $\dot{\theta}_E/T_D$, where m is the turn number of layers. The simple analytical solution is a case of ignoring the fiber diameter. $\dot{\theta}_E$ is inversely proportional to m in the case of dipole winding, inversely proportional to m^2 in the case of quadrupole winding, and zero in the case of octapole winding. Therefore, we try to develop an octapole winding

Table 7.1 Primitive analysis of the effect of reducing symmetrical winding[8] (ignoring fiber diameter and with $m = 48$).

Winding Pattern	Gyro Bias Drift $\dot{\theta}_E$	Shupe Sensitivity
Solenoid	$\left(1+\frac{1}{m}\right)\cdot S_0\cdot T_D$	3.3(deg/h)/ (°C/h)
Dipole	$[\frac{3}{2(m-1)}]\cdot S_0\cdot T_D$	0.11(deg/h)/ (°C/h) (3.2%)
Quadrupole	$[\frac{3}{m(m-1)}]\cdot S_0\cdot T_D$	0.0044(deg/h)/ (°C/h) (0.13%)
Octapole	0	0(deg/h)/ (°C/h) (0%)

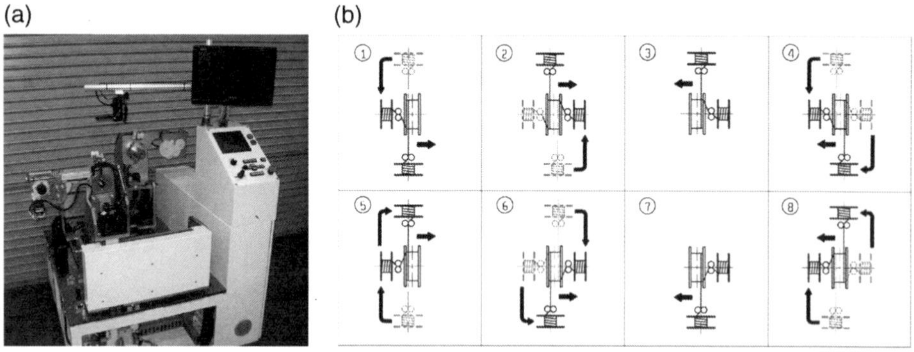

Figure 7.2 (a) Developed octapole winding machine made by Kawamasa Industry Ltd., and (b) octapole turns of the winding-step procedure. The center is the main bobbin, and two delivery bobbins move so as to surround the main bobbin. The eight-step movement is repeated, making a symmetrical winding.

machine that can wind fiber precisely and symmetrically. Figure 7.2 shows the developed octapole winding machine made by Kawamasa Industry Ltd. and the octapole turns of the winding-step procedure. We fabricated some trial coils with this winding machine in order to verify the effects of coil parameters, such as on-coil turn-number error or winding flaws, bobbin materials, and fiber

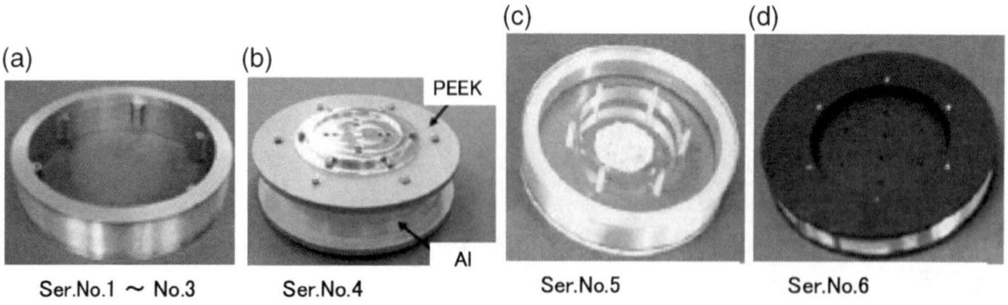

Figure 7.3 Fabricated optical fiber sensing coils: (a) aluminum bobbin, (b) PEEK and aluminum-compounded bobbin, (c) quartz bobbin, and (d) CFRP bobbin. Total fiber length is about 3.4 km using an octapole winding pattern. Table 7.2 lists more information about the coil parameters.

winding tension. Figure 7.3 shows the appearance of the fabricated optical fiber coils. The detailed results are explained in a later subsection.

7.2.2 Thermal conductivity and strain attenuation

The use of a low-thermal-expansion bobbin such as glass or CFRP is effective to suppress Shupe sensitivity.[9] In order to confirm its effectiveness, a prototype coil of a glass bobbin was evaluated in a vacuum environment. Figure 7.3(c) shows the appearance of the quartz bobbin coil. Figure 7.4 shows the result of comparing the evaluation value with the thermal-analysis simulation values. The thermal analysis result showed a sensitivity of 1×10^{-4} (deg/h)/ (°C/h), while the test result achieved 2×10^{-4} (deg/h)/ (°C/h). Given this demonstrated result of a prototype coil of a glass bobbin, the experimental values were confirmed to exhibit relatively good agreement with the analytical values.

The development of thermal strain in a lightweight carbon fiber-reinforced plastic (CFRP) FOG under a thermal vacuum condition simulating the space environment was also investigated.[10] Several materials for IFOG bobbins were then quantitatively compared using experimentally validated finite element analysis (FEA) from the aspects of thermal strain and weight to illustrate the clear advantage of CFRP.

To understand the mechanism of strain distribution in detail, it was subjected to thermal strain numerical analysis by FEA (Nastran® static analysis). Figure 7.5(a) shows the results of strain numerical analysis. We found there is a high stress value in the vicinity of the bobbin flange.[11] This result is presumably due to discontinuity of the linear thermal expansion coefficient of different materials (aluminum and glass). The bobbin inner wall providing stress relief for the winding of the dummy fiber (discarded winding) was considered valid. Figure 7.5(b) also plots strain analysis values over the entire fiber length based on the structure of the octapole winding. In this

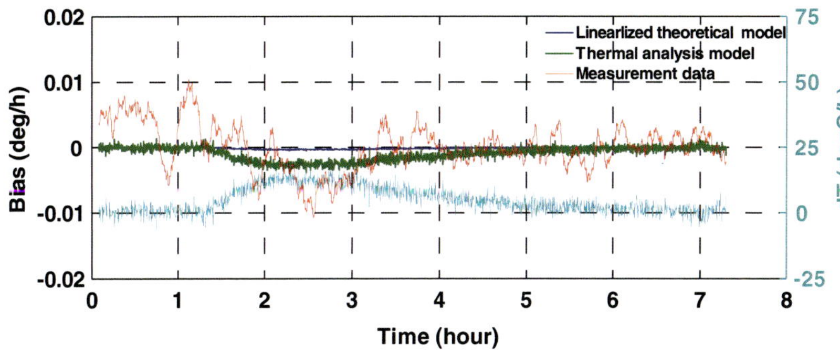

Figure 7.4 Temperature rate – gyro rate bias-time history in a quartz bobbin (Fig. 7.3(c)). The blue line represents simple analysis assuming that the *dT*/*dt* distribution is linearized. The green line represents detail analysis based on the constructed thermal analysis model.

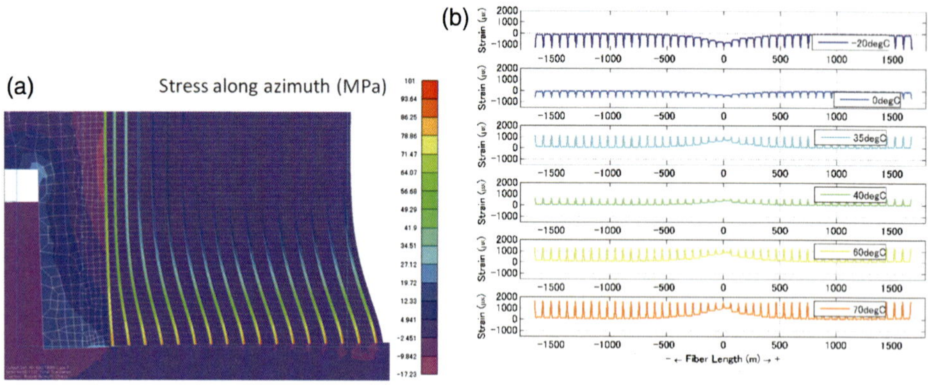

Figure 7.5 Stress numerical analysis results: (a) cross-section of FOG bobbin (enlargement of around fiber coil), and (b) strain change along the entire fiber length during temperature change between –20°C and 70°C.

simulation result, a sharp strain waveform appears at the winding turn-up position near the bobbin flange portion, and the peak and valley of the sharp point exhibited reverse behavior along with the temperature change.

In order to verify the analysis, we utilized pulse-pre-pump Brillouin optical time-domain analysis (PPP-BOTDA, NBX-6040, Neubrex Co., Ltd.). NBX-6040 is a commercially available, high-spatial-resolution measurement system that can obtain the Brillouin gain spectrum along an optical fiber with a spatial resolution of a few centimeters. Figure 7.6 shows the results of strain measurement by PPP-BOTDA. Based on the structure of the octapole winding, the diagram is displayed across the strain analysis value along the total fiber length. The distributed strain of the aluminum bobbin and 40-gf tension winding (Fig. 7.6(a)) shows some interesting features as FEA

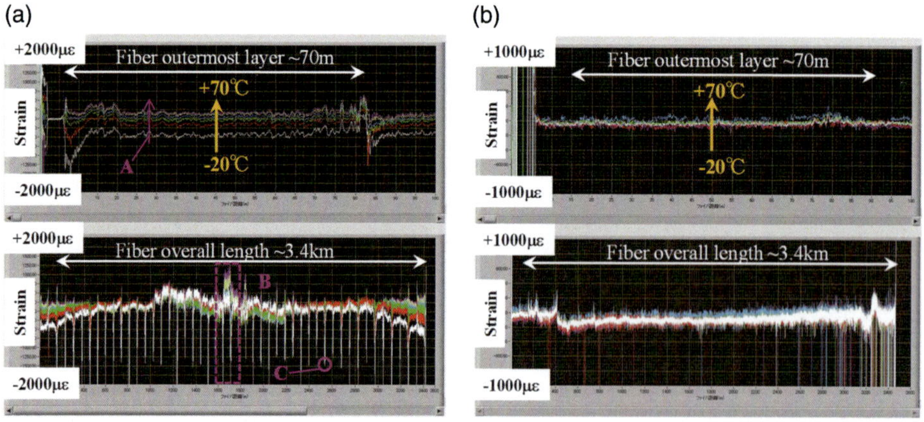

Figure 7.6 Strain of winding coil measurement results with PPP-BOTDA: (a) strain distribution of Ser. No. 4 coil, and (b) strain distribution of Ser. No. 5 coil.

simulation can be observed, such as the increasing level of strain as the temperature rises (Point A), high strain along the inner bobbin despite the winding of dummy fiber (Point B), and steep distortions that appear in the winding turn-back position near the vicinity of the flange portion (Point C). Conversely, the distribution of the CFRP bobbin and 15-gf tension winding (Fig. 7.6(b)) clearly indicates that these negative features are effectively depressed. This result shows the importance of the coefficient of expansion in bobbin material and fiber winding tension.

7.2.3 Zero-sensitivity winding design

A novel coil parameter design method based on Shupe error analysis was proposed.[11] Once the coil parameters such as fiber diameter, including coating thickness, fiber length, and the number of layers, are fixed, only one degree of design freedom remains with respect to the remaining two parameters (coil diameter D and number of turns N_T). If all of the coil parameters are determined, the Shupe sensitivity can also be uniquely obtained by the Shupe error expression in Ref. 7. Figure 7.7(a) shows thermally induced drift for every eight layers for the coil parameters of Ser. No. 4 and No. 5. Once all phase errors in each eight layers are summed up, then thermally induced drift error in this coil is obtained. Now consider that the other parameters except the (D, N_T) parameters are not changed. The (D, N_T) parameters realize thermally induced drift sensitivity (integrated phase error value over each layer) (Fig. 7.7 (b)). The coil of Ser. No. 6 was designed with these parameters. Conventionally, the (D, N_T) parameters were decided based on the volume

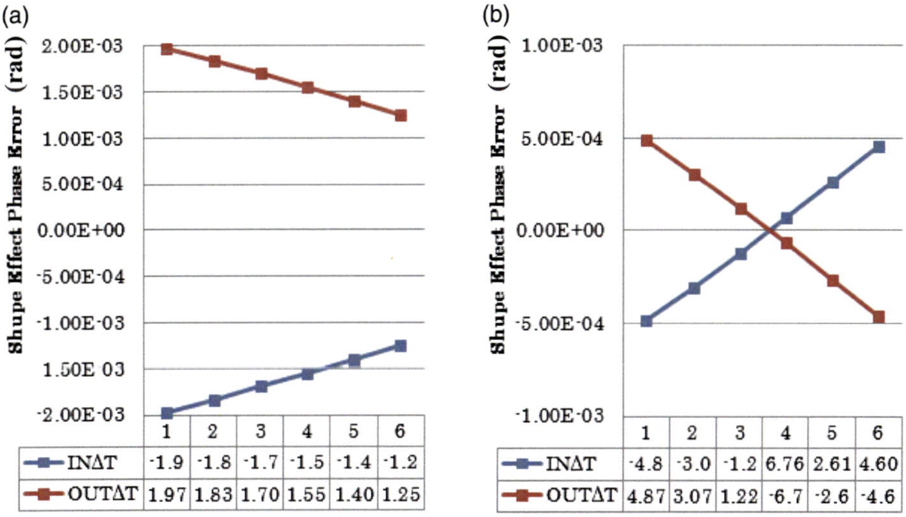

Figure 7.7 Thermally induced drift error for every eight layers: (a) Ser. No. 4 and No. 5 coil parameters, and (b) Ser. No. 6 coil parameters (including proposed zero-sensitivity parameter).

constraints of products and other factors. However, the above procedure becomes a new design guideline for high-precision FOGs.

7.2.4 Summary of activity results

This subsection summarizes the verification coil mentioned in this study. Figures 7.8(a) and (b) show the winding coil turn error of Ser. No. 3–6 as the appearance of the coil is already shown in Fig. 7.3. Figures 7.8(c) and (d) show the Shupe sensitivity of each coil. Table 7.2 summarizes the coil design parameters of Ser. No. 4–6. We find that Shupe sensitivity critically depends on winding flaws, making it absolutely necessary to produce a perfectly symmetric coil with not one turn of winding error. In order to allow for the correct value of Shupe sensitivity and enable thermal analysis by a thermal mathematical model, it is necessary to take into account the stress-induced section in addition to the heat-induced section. In consideration of these points and given the result of a prototype coil of a glass bobbin, the experimental values were confirmed to exhibit relatively good agreement with the analytical values. Under the care of a precision coil design and winding process, some low-sensitivity coils suitable for high-precision specifications were achieved.

7.3 Conclusion

This chapter introduced examples of Japanese products that are applicable to space use. We also described some efforts toward high-grade navigation use in Japan. Under the care of a precision coil design and winding process, some low-sensitivity coils suitable for high-precision specifications were achieved.

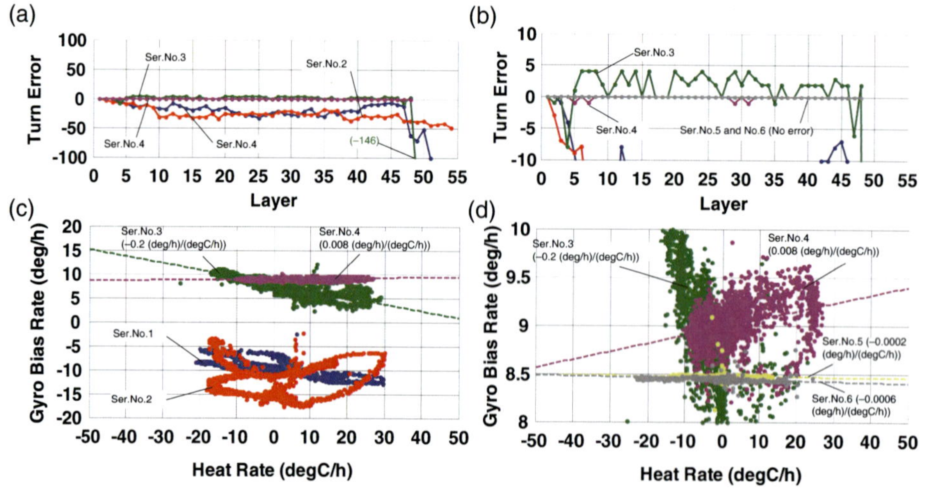

Figure 7.8 (a) Coil turn error, (b) coil turn error (magnified), (c) Shupe sensitivity of each coil, and (d) Shupe sensitivity of each coil (magnified).

Table 7.2 Summary of parameters for each fabricated coil.

	Coil Parameter	Ser. No. 4	Ser. No. 5	Ser. No. 6
1	Sagnac Coefficient	5.8 s (length: 3.371 km, coil diameter: 130.05 mm)	←	7.4 s (length: 3.371 km, coil diameter: 166.15 mm)
2	Flange Material	PEEK	Same as bobbin material	←
3	Bobbin Material (CTE and thermal conductivity)	Aluminum ($\alpha \sim 23$ ppm/K, $\lambda \sim 200$ W/m/K)	Electric melting quartz ($\alpha \sim 0.5$ ppm/K, λ 1.5 W/m/K)	CFRP($\alpha < 1$ ppm $\lambda < 10$ W/m/K)
4	Dimension Tolerance	R < 0.08 mm Width tolerance ±0.1 mm	←	←
5	Sensing Fiber	FIBERCORE SM-1500 (6.4/80), acrylic-coating diameter of 170 μm	←	←
6	Winding Accuracy	48 octapole winding layers, symmetry per 4 layers	48 octapole winding layers, zero winding flaws	32 octapole winding layers, zero winding flaws
7	Inner Dummy Winding	Done (8 layers)	←	←
8	Adhesive	Acrylic resin	UV cure epoxy resin AT9390 (NTT-AT Co.)	←
9	Tension	40 ± 5 gf	←	15 ± 0.5 gf

We will continue our study toward achieving practical application of high-precision, space-ready FOGs.

Acknowledgement

The authors wish to express their great appreciation to Prof. K. Hotate of the University of Tokyo for his beneficial advice and support.

References

[1] A. Ohno, A. Kurokawa, T. Kumagai, S. Nakamura, and K. Hotate, "Applications and Technical Progress of Fiber Optic Gyros in Japan," Optical Fiber Sensors, 2006.

[2] S. Sakai, Y. Fukushima, A. Ohno, and H. Saito, "In-Orbit Performance Evaluation of Temperature Controlled Small Fiber Optical Gyro on Microsatellite REIMEI," Optical Fiber Sensors, 2006.

[3] S. Sakai and H. Saito, "Fiber Optical Gyros in the Spacecraft Control Applications," FOG Workshop, Optical Fiber Sensors, 2012.

[4] M. Kojima, S. Hirata, K. Hama, and S. Nakamura, "Development and Demonstration of Fiber-Optic Gyro-Inertial Reference Unit," *Proc. of Conference on Space Science and Technology*, 2006 (in Japanese).

[5] Y. Shimizu, H. Kumagai, H. Hashimoto, H. Sugiyama, and A. Nakajima, "Operation Status of Attitude Control System for R&D on Visible Light Communication Experimental Satellite GINREI," *Proc. of Conference on Space Science and Technology*, 2014 (in Japanese).
[6] S. Mitani, Y. Ishijima, H. Suzuki, S. Sakai, T. Toda, T. Hashimoto, A. Ohno, Y. Miyahara, and H. Okikura, "Development of High Performance Fiber Optical Gyro Inertial Reference Unit (FOG-IRU) for Satellite Applications," 7th International ESA Conference on Guidance, Navigation & Control Systems, 2008.
[7] D. Shupe, "Thermally induced nonreciprocity in the fiber-optic interferometer," *Applied Optics*, Vol. **19** (5), pp. 654–655, 1980.
[8] N. Frigo, "Compensation of Linear Sources of non-reciprocity in Sagnac Interferometers," *Proc. of SPIE* Vol. **412**, pp. 268–271, 1983.
[9] T. Mizutani, "Precise Sensing Utilizing Optical Fiber for Spacecraft," 23rd International Conference on Optical Fiber Sensors, 2014.
[10] S. Minakuchi, et al., "Thermal Strain in Lightweight Composite Fiber-Optic Gyroscope for Space Application," *J. Lightwave Technol.*, Vol. **33** (12), pp. 2658–2662, 2015.
[11] S. Mitani, T. Mizutani, K. Shinozaki, and Y. Sasaki, "Reduction of Thermally Induced Drift on Interferometric Fiber Optic Gyroscope," *Proc. of Conference on Space Science and Technology*, 2014 (in Japanese).

Shinji Mitani received a M.S. from the Department of Aeronautics and Astronautics in Kyoto University in 2004. He has been with JAXA Tsukuba Space Center for 14 years. While enrolled in JAXA, he entered Kyoto University and received a Ph.D. in electrical engineering in 2013. His current position is a senior researcher, research unit I, Research and Development Directorate. He has been engaged in the R & D of Advanced Sensors and Actuators for Spacecraft Guidance Control (Fiber Optic Gyro, etc.), R & D of Advanced Attitude and Orbit Control System for Future Formation Flight missions and high-precision pointing missions etc..

Dr. Mitani is a member of the Japanese Society for Aeronautical and Space Sciences (JSASS) and the Society of Instrument and Control Engineers (SICE).

Chapter 8
Fiber Optic Gyro Development at Fibernetics

Ralph A. Bergh
Fibernetics LLC

8.1 Introduction and Past Development

Fibernetics started in 1983 with the small-scale manufacturing of fiber optic directional couplers, polarizers, and polarization controllers based on concepts first demonstrated at Stanford University.[1–3] Sales of these components and small systems such as fiber resonators were largely to research laboratories. In parallel with this manufacturing effort, we conducted research on FOGs. We licensed the intellectual property that developed from this research to Honeywell International, Inc., which began a long-term relationship between the two companies in FOG development. The Honeywell–Fibernetics collaboration slowed down in 2004, but Fibernetics continued its development effort through government contracts and internal funding. Now Fibernetics is poised to manufacture IMUs to serve the high-performance-inertial-system market.

This chapter begins with a short overview of our current development, including a phase-modulation scheme and a source-sharing scheme. The so-called occasional-calibration phase-modulation scheme is a combination of a digital phase ramp and dual-ramp phase modulation.[4–6] The digital phase ramp performs the function of closed-loop operation, and the dual ramp performs two functions: closed-loop operation, and calibration of the phase modulation with respect to the interferogram for accurate SF determination. This design thus provides precise digital representation of the phase difference caused by rotation. The chapter then presents our novel source-sharing system, which makes efficient use of light from a single light source to power all three Sagnac interferometers. It is efficient because light that would normally be thrown away by the first coupler is used to power other Sagnac interferometers. It involves a single light source and three detectors, one for each FOG. The directional couplers are connected in series rather than in parallel, as in other source-sharing schemes. The coupling ratios of the directional couplers can be adjusted to equalize the optical power incident on the various photo-detectors at the output of the FOGs. A calculation of the coupler-splitting ratios that equalize the optical powers at the various photo-detectors are presented for three or more FOGs. Although equalizing these optical powers is not critically important in the IMU, it does help balance the SNRs and the control-loop gains of the three FOGs. Finally, a novel FOG architecture is presented that uses a multicore fiber in an effort to further reduce the size of the FOG.

8.2 Current Development

Recently, Fibernetics began testing its prototype IMU designed for navigation-grade application. Early results demonstrate promising FOG-noise and bias-stability characteristics. The fiber coils are wound in-house with our custom winder designed to provide precisely wound coils that improve the thermal stability of the FOGs. Post-winding processing reduces stress within the coils. The power dissipation of the IMU is reduced significantly by using a single light source instead of three, and by using low-power high-performance analog electronic components and low-power high-density digital electronic components. Compared to other source-sharing systems, this design delivers about 50 percent more of the available light to each of the photo-detectors. This efficient source-sharing architecture means that less source power is required to achieve the same SNR. When less power is used to drive the source, less heat is generated, which puts less demand on the light-source thermoelectric cooler at high temperatures, further reducing the total power dissipation.

8.3 Basic FOG Design

A discussion of phase-modulation waveforms can benefit from reviewing the basic architecture of a high-performance FOG (Fig. 8.1). While it was easy to describe the Sagnac effect as two counter-propagating waves that arrive back at the common input/output port of a Sagnac interferometer with a phase difference that is proportional to the rotation rate, actually developing a high-performance FOG design proved to be much more challenging. The Sagnac effect is very small. Optical fiber offers a way to amplify the effect by providing a long optical path length that can be wound into an acceptable-sized coil using many turns of fiber. This increases the signal but also increases possible errors in phase difference if the FOG is improperly built. It is important even with a relatively static environment that the optical path be

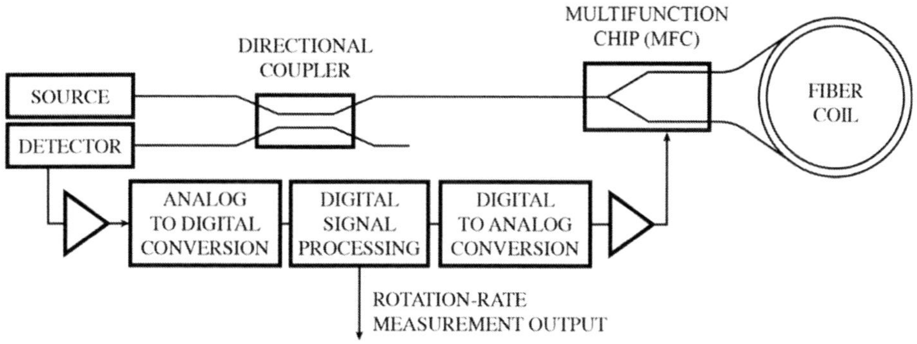

Figure 8.1 High-performance closed-loop FOG with a multifunction chip.

the same in each direction around the interferometer. While optical fiber provides the optical path for the necessary sensitivity enhancement, in order to make nano-radian phase-difference measurements after light has propagated through giga-radians of phase delay in glass it is crucial that the light travel exactly the same optical path in opposite directions from input to output through the interferometer. To accomplish this, the FOG must be designed to take full advantage of the fundamental principle of reciprocity. It took four years after the first publication of a demonstrated FOG in 1976 to develop the optical architecture upon which all high-performance FOGs are based.[7–9] A single-mode fiber, together with a polarizer at the common input/output port, is used to create the well-documented reciprocal design that has been called the "minimum configuration" FOG.[8–11] If a reciprocal design were not used, then impossibly small variations in the environment could cause unacceptably large phase-difference variations. With a reciprocal design, there is a chance of making a FOG with adequate bias stability. Perfect reciprocity would result in zero average phase shift between the interfering waves.

Phase modulation, typically used to phase-bias the interferometer, creates a necessary deviation from perfect reciprocity. This biasing phase modulation performs four functions: (1) it ensures that the signal returning from the interferometer is sensitive to small, nonreciprocal phase shifts and therefore small rotation rates; (2) it creates a way of determining the direction of rotation; (3) it modulates the FOG output signal at some relatively high frequency that makes it possible to discern this signal from numerous DC offsets that are likely present in the electronics; and (4) it reduces the noise in the detected signal by heterodyning the rotation-induced signal to a high frequency and by providing a means (through the use of a lock-in amplifier) to reduce the detection bandwidth and void excess low-frequency noise.[12]

In this design, a directional coupler splits light from the source, directing a portion of the light to the Sagnac interferometer (multifunction chip (MFC) and fiber coil) and typically discarding the rest of the light. The MFC performs three functions. First, it polarizes the light by guiding only one state of polarization if using proton exchange waveguides in lithium niobate. Second, it splits the light from one waveguide into two waveguides. Those two waveguides are connected to the two ends of the coiled fiber. Light coupled into the ends of the fiber propagate through the fiber in opposite directions, pass through each other, and return to the MFC. The splitter then recombines the light and returns the recombined light to the directional coupler. The optical power of the recombined light is sensitive to the phase difference between the constituent light waves according to the interferogram discussed below. The third function of the MFC is to provide a means to effect the high-speed phase modulation necessary for accurate closed-loop modulation schemes. The latter is the main reason for using a MFC.

An electro-optic phase modulator allows us to consider stepped waveforms that are popular because of performance improvements. Rise times and fall times can be just a few nanoseconds. An example of a stepped waveform is the "digital phase ramp" published by Lefevre et al.[4] The most convenient way to make use of an electro-optic phase modulator is with a waveguide. And if we are already using a waveguide to perform the task of phase modulation, then we might as well create additional functions of the splitter/combiner and polarizer. Putting all of the components on the MFC, including the first coupler, would have obvious benefits, but it would also create problems of inadequate filtering that would compromise the ability to make full use of the principle of reciprocity. The input/output length of fiber ahead of the Sagnac interferometer (between the directional coupler and the MFC in Fig. 8.1) is the best single-mode spatial filter with outstanding rejection of higher-order spatial modes. This length of single-mode fiber, together with the polarizing capability of the MFC, makes an outstanding truly single-mode filter.

A stepped waveform lends itself to making excellent feedback systems that not only null the phase shift due to rotation but also make 2π adjustments to keep the electronics driving the phase modulator from reaching saturation. A stepped feedback waveform works together naturally with a square wave for bias modulation. Other waveforms are possible, but a square-wave modulation works well for biasing the Sagnac interferometer, and it also works well with digital electronics. Figure 8.2 shows this waveform, and Fig. 8.3 shows how it works with the interferogram. This chapter assumes a bias modulation at the "proper frequency," a term used to denote $1/2\tau$ Hz, as shown in Fig. 8.2.

Phase modulation in the Sagnac interferometer converts the raised cosine DC signal (see Fig. 8.3(a)) into an AC signal the amplitude of which is proportional to the sine of the rotation-induced phase shift.[8–11] Examples of zero phase difference and $\pi/8$ phase difference are shown in Figs. 8.3 and 8.4, respectively. The AC signals mentioned are in Fig. 8.3(c) with zero amplitude and Fig. 8.4(c) with nonzero amplitude.

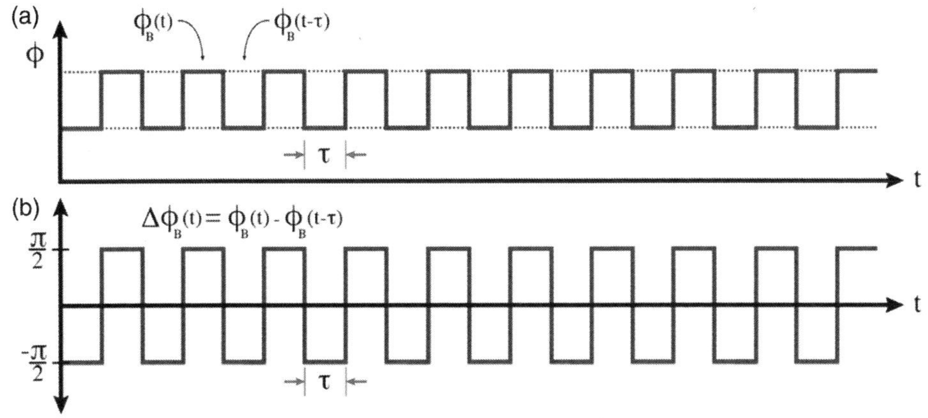

Figure 8.2 Square-wave phase modulation $\phi_B(t)$, minus its delayed version $\phi_B(t - \tau)$, shown using a dotted line in (a) to create the phase-difference modulation shown in (b).

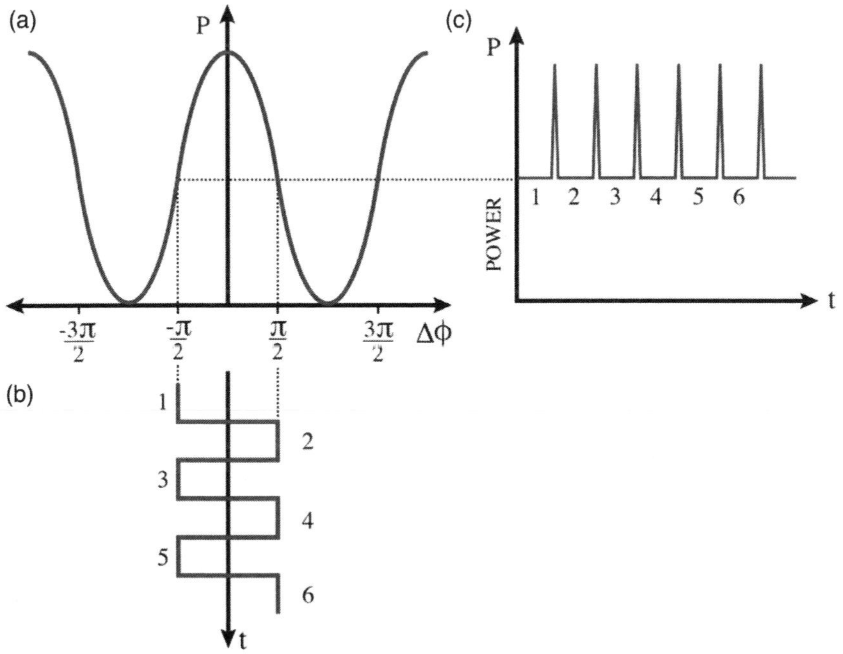

Figure 8.3 The (a) interferogram maps the (b) square-wave phase-difference modulation, shown with zero rotation rate, into a (c) flat power vs. time graph separated by glitches.

Figure 8.3(a) shows detected optical power versus phase difference between interfering waves of identical amplitude and polarization. This is the classic interferogram of a two-wave interferometer. The square-wave bias phase-difference modulation (Fig. 8.3(b)) with no added nonreciprocal phase difference due to a rotation, maps into a zero-amplitude detected optical power signal in Fig. 8.3(c). The spikes in the optical power versus time shown

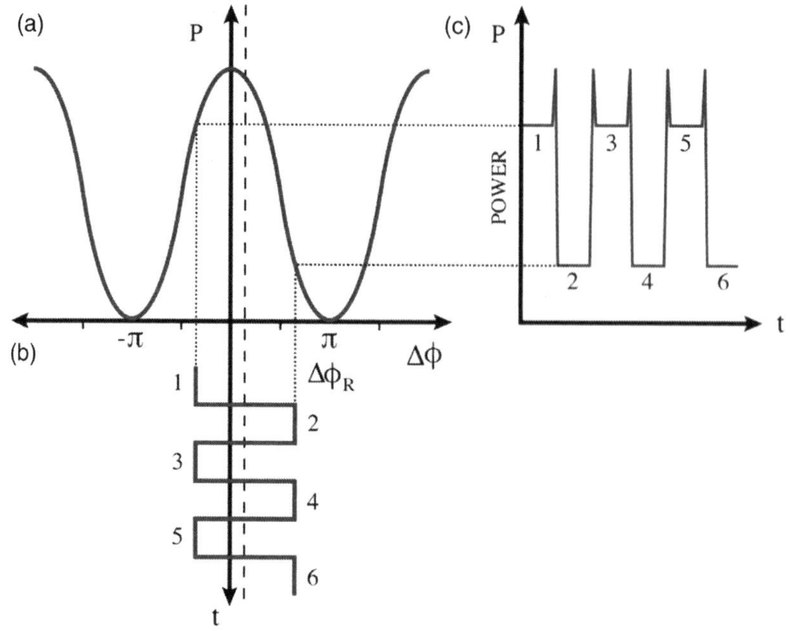

Figure 8.4 A rotation-induced phase difference of $\pi/8$ is added to (b) square-wave phase-difference modulation and mapped by an interferogram (a) to (c) the square-wave detected optical power.

in Fig. 8.3(c) are the result of transitioning from one phase difference to another for the short time during which the phase difference passes through zero. These spikes, or glitches as they are called, are gated out and do not contribute to the signal.

When a rotation rate is applied, the square-wave bias phase-difference modulation maps into the nonzero detected optical power shown in Fig. 8.4(c). The odd segments of the square-wave phase modulated output signal on the left in Fig. 8.4(b) map to the odd segments at the top of the power versus time output curve in Fig. 8.4(c). Likewise, the even phase difference segments on the right in Fig. 8.4(b) map to the lower segments in Fig. 8.4(c). This would be opposite if the rotation rate were in the other direction, in which case the nonreciprocal phase difference would be negative.

The detected optical power signal is demodulated at the bias-modulation frequency to create a so-called "error" signal. In a closed-loop FOG, this error signal is nulled by adding a stepped phase ramp to the phase modulation (see Fig. 8.5). This stepped ramp has also been called a "digital phase ramp."[4] Such a ramp is limited in amplitude, so a 2π step is used to keep the phase modulation from getting out of range. This 2π step does not affect the signal so long as this value is sufficiently accurate. The accuracy of the amplitude of the phase-modulation waveform is guaranteed by adequate calibration, which is the subject of the next section.

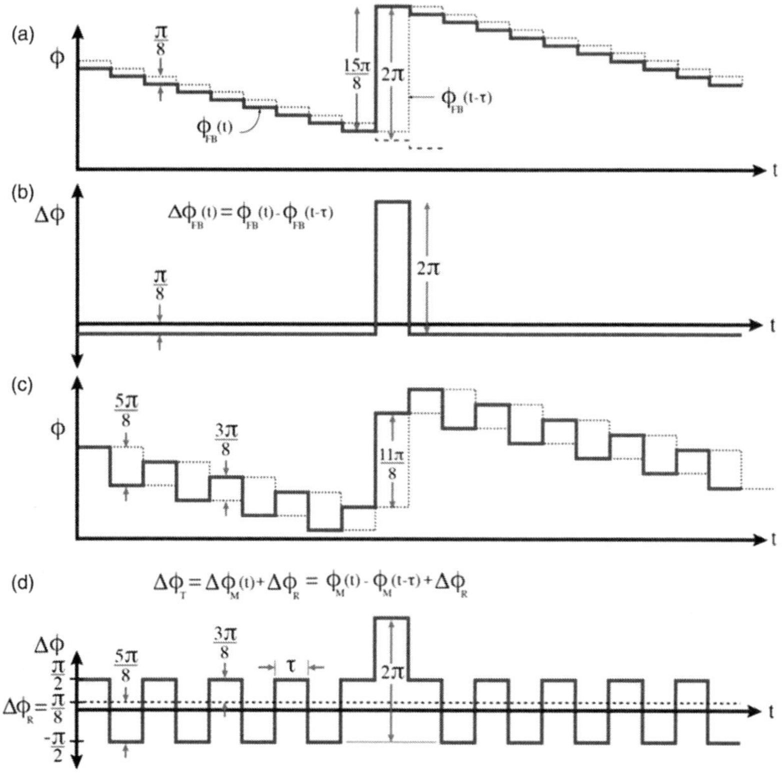

Figure 8.5 (a) Stepped phase ramp with steps of $\phi_M = -\pi/8$ every τ, and (b) an adjustment step of plus $\phi_M = 15\pi/8$ to make a transparent $\Delta\phi_M = 2\pi$ step in phase difference shown. (c) Total phase modulation is the sum of the stepped phase ramp and square wave bias modulations. (d) The total phase difference is the sum of the phase-difference modulation $\Delta\phi_M$ and phase difference due to rotation $\Delta\phi_R$.

The stepped waveform shown in Fig. 8.5(a) in a solid line is intended to create an offset-bias phase difference that is equal to and opposite the phase difference due to rotation.[4] The dotted line is the same as the solid line but delayed in time by τ the transit time of light through the fiber in the fiber coil. The solid line represents the phase imparted to the counter-propagating waves as they exit the Sagnac interferometer, and the dotted line represents the phase that was imparted to those same waves at an earlier time when they entered the Sagnac interferometer. The phase difference between the interfering waves upon exiting the interferometer resulting from this modulation is shown in Fig. 8.5(b). The phase difference in Fig. 8.5(b) is simply the solid line minus the dotted line in Fig. 8.5(a). The phase difference in Fig. 8.5(b) would be a constant $-\pi/8$ if the staircase phase modulation could go on forever, but the phase modulator has a limited range, and a 2π step is used to stay within this range. This 2π step in phase manifests itself as a deviation from the constant $-\pi/8$ phase difference by a 2π pulse for a time of τ. This 2π pulse in phase difference is transparent once it is mapped by the interferogram to optical power. The waveforms in Fig. 8.5(c) are the same as those in Fig. 8.5(a) but with the addition of a square-wave bias modulation. The same way the phase modulation in Fig. 8.5(a) led to the phase-difference modulation in Fig. 8.5(b), the phase modulation in Fig. 8.5(c) leads to phase difference in Fig. 8.5(d), with the exception that in Fig. 8.5(d) a constant $\pi/8$ phase difference due to rotation was added to the modulation-induced phase difference to create the total phase difference that is displayed.

8.4 Dual-Ramp Phase Modulation

Phase modulation, as mentioned previously, biases the interferometer and supplies feedback to null the rotation-induced phase shift. Additionally, phase modulation is used to calibrate the digital electronics with respect to the phase difference reported by the interferogram. This calibration ensures that the rotation measurement is not dependent upon the transfer function between the digital electronics and the phase modulation of the light in the interferometer. To compensate for the rotation rate and to perform the calibration of this compensation, two control loops are needed: one to add a phase shift that is equal to and opposite the Sagnac phase shift, and one to eliminate the signal dependence on the transfer functions of the digital-to-analog conversion, the gain of the electronic amplifiers driving the MFC, and the MFC phase modulators. Each of these transfer functions varies somewhat with the environment, but because the effect of these changes is measured, the calibration loop makes the signal independent of these changes.

All successful closed-loop schemes applied to the FOG have used dual control loops. There are similarities and differences between these various control schemes. The differences usually manifest in the calibration control loop. The evolution of the calibration control loop developed at Fibernetics is discussed next.

8.4.1 Low-frequency approach

The dual-ramp scheme was originally envisioned as a low-frequency approach to effect a dual-closed-loop FOG.[5] It was first demonstrated using PZT phase modulators.[13] A triangle-wave modulation amplitude of many tens of π can be generated with a PZT phase modulator. The control loop is locked about $+\pi$ during the ramping upslope portion of the waveform and about $-\pi$ during the downslope portion of the waveform, both of which last for many τ, i.e., many loop-transit times. Neither control loop is locked during the transition from upslope to downslope at the peak of the modulation waveform or from downslope to upslope at the bottom of the modulation waveform (see Fig. 8.6).

The peak and valley of this waveform required transition times in excess of τ, the light transit time through the fiber coil. During this transition time,

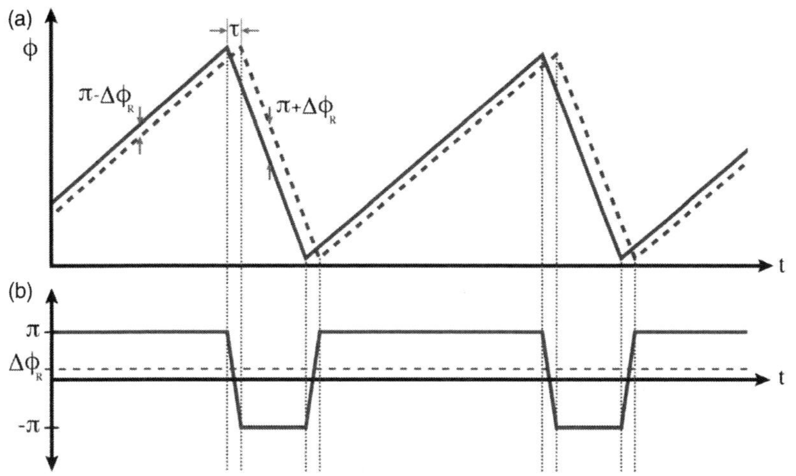

Figure 8.6 Original analog dual-ramp modulation and associated phase-difference shifting the total phase difference from $\Delta\phi_R$ to π and $-\pi$.

the signal stayed gated out and the phase modulation remained uncontrolled. Although this PZT design achieved stable control, it did not achieve stellar performance. While work on the PZT modulator version of dual-ramp control of the FOG was being performed in the 1980s and 1990s, work also progressed on a stepped version of this waveform. Some of this work was performed in collaboration with Honeywell after they had purchased a nonexclusive license to Fibernetics patents. The dual-staircase waveform requires a high-speed phase modulator but only a few nanoseconds of gating per τ with no additional gating at the peaks and valleys of the waveform. The success of the dual-staircase approach made the low-frequency approach less compelling, and so the work was abandoned.

8.4.2 High-frequency approach

Two descriptions of the dual-ramp phase modulation scheme result in the same waveform. As it was originally conceived, the idea was to lock the Sagnac interferometer alternately to $+\pi$ and $-\pi$, which leads to the component waveforms shown in Figs. 8.7(a) and 8.7(b). Figure 8.7(c) shows the addition of these two waveforms. The waveform in Fig. 8.7(a) shifts the phase difference to $+\pi$ on the up steps and to $-\pi$ on the down steps. An even number of up steps and an even number of down steps are used so that an integer number of bias-modulation periods occur on each one.

The second description might be considered slightly more mainstream. The concept here is more aligned with locking the phase difference to zero with 2π steps thrown in for calibrating the waveform and thereby the electronics. Figure 8.8(a) shows the 2π steps, and Fig. 8.8(b) shows the bias-modulation waveform. The difference between Figs. 8.7(b) and 8.8(b) is the

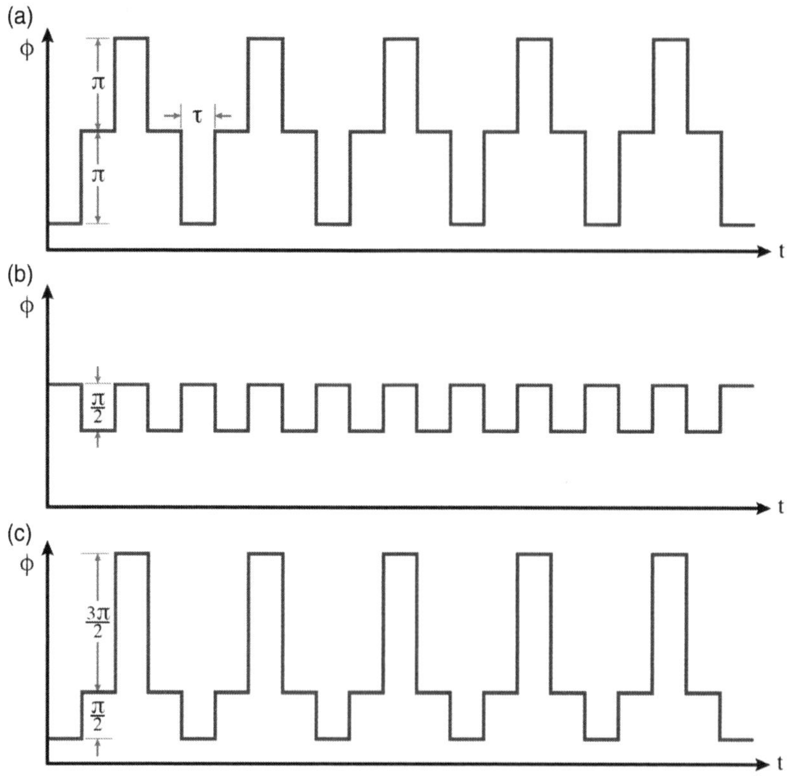

Figure 8.7 (a) Stepped triangle-wave and (b) square-wave bias modulation are summed to produce (c) dual-ramp phase modulation.

phase of the bias-modulation waveform. Note the 180-deg phase shift between them. Note also that Figs. 8.7(c) and 8.8(c) are identical.

The dual-ramp modulation shown in Fig. 8.9(a) leads to the phase-difference modulation shown in Fig. 8.9(b) and the detected power shown in Fig. 8.9(c). Figure 8.9(b) shows that the high-speed dual-ramp modulation has four phase-difference values, which has been called "four-states" modulation.[14]

Note that the glitch pattern contains a component at the bias-modulation frequency. If the glitches are not well gated and some of the glitch contaminates the signal, this can cause a bias offset. This problem and its remedy will be discussed later in this section. Figure 8.10 shows how the interferogram maps phase-difference modulation versus time into optical power versus time.

The detected optical signal resulting from this type of modulation waveform is basically flat with short-duration glitches as the interfering optical waves transition from one phase difference to another. If the system were operated open loop and if the rotation induced a phase shift of $\Delta\phi_R = \pi/8$, then a square-wave optical signal would occur at the bias-modulation frequency, as shown in Fig. 8.11.

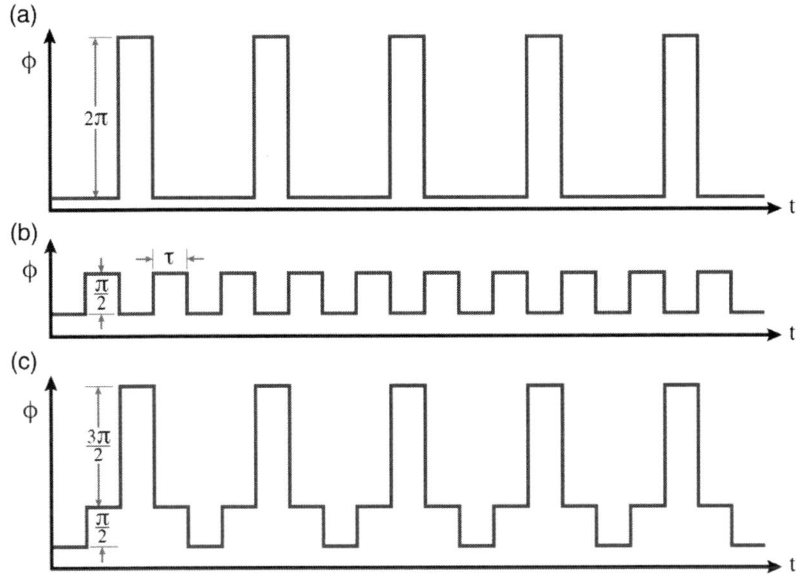

Figure 8.8 The (a) 2π steps and (b) square-wave bias modulation combine to produce (c) dual-ramp phase modulation.

To close the rotation-rate control loop, a $-\pi/8$ ramp is added to the dual-ramp waveform, as shown in Fig. 8.12. It is not easy to see the ramp and reset in the figure once it is added to the dual-ramp waveform. The reset occurs in Fig. 8.12 when taking a small up step instead of a large down step. Then the dual-ramp waveform continues with another large up step. In the dual-ramp waveform, each step involves either a change in direction or a change in step size. A large up step precedes either a large down step (change in direction) or a small up step (change in step size). A decision whether to change direction is made after every pair of steps, or in other words, every bias-modulation period. The duration of each step is τ.

The phase-difference modulation shown in Fig. 8.12(b) assumes four positions. Consider this as locking to two points on the interferogram, e.g., π and $-\pi$. The bias modulation is then added to $+\pi$ to create the phase differences of $\pi/2$ and $3\pi/2$, and to $-\pi$ to create the phase differences of $-\pi/2$ and $-3\pi/2$. This was much more evident when large-amplitude phase modulation was employed as in Fig. 8.6. The reason is that the modulation would lock to $+\pi$ for a relatively long period of time (i.e., many bias-modulation cycles) and then switch to $-\pi$ for another long period of time. In other words, the four-position modulation allows locking to zero phase difference and to the amplitude of 2π. This type of modulation offers independent measurements of the phase difference due to rate and modulation amplitude compared with 2π according to the interferogram. This is not the case for ramp-and-reset modulation, or digital-phase-ramp modulation, when

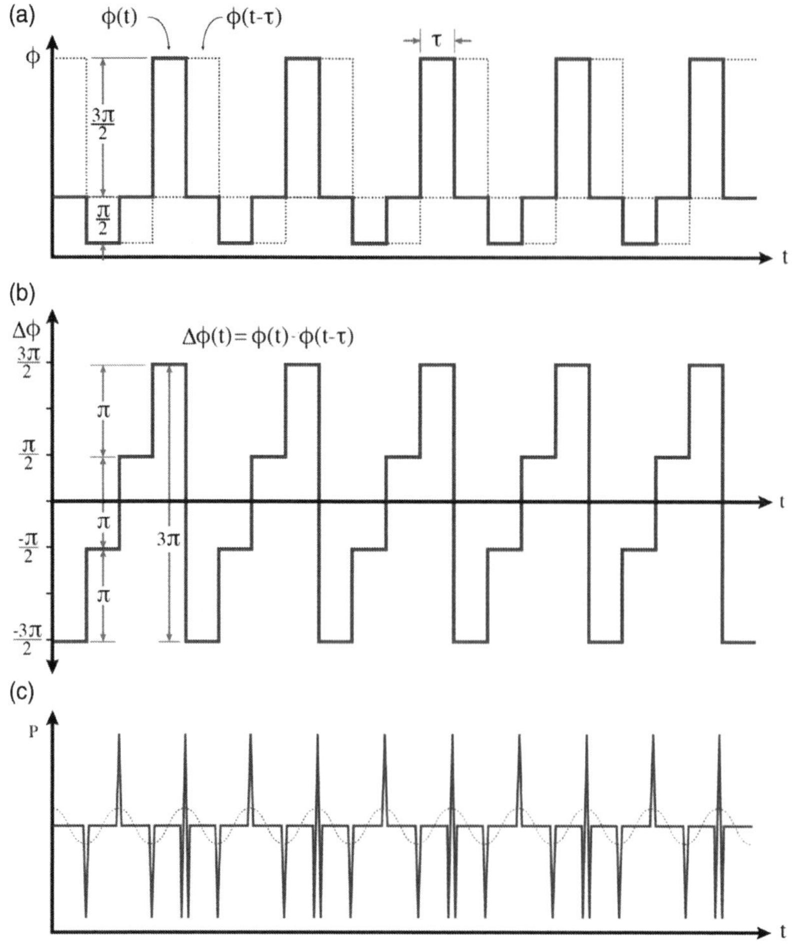

Figure 8.9 (a) Dual-ramp phase modulation leads to (b) phase-difference modulation and (c) optical power.

the reset portion of the modulation is used to calibrate the phase-modulation amplitude. Modulation-amplitude error in both of these cases is mixed with the rate loop error. This is why dual-ramp modulation and four-point digital phase-ramp modulation are so powerful.[5,6,11]

If the waveform has an incorrect amplitude, as seen in Fig. 8.13, the result would be an error signal similar to Fig. 8.13(c). The detected power contains a square wave synchronized to the dual-ramp modulation. If the modulation amplitude is too large, the large steps increase the detected optical power and the small steps reduce the detected optical power. The result is a signal that is orthogonal to the rate-induced phase difference signal. In other words, an amplitude error signal does not generate a bias error signal. The two loops remain independent. Looking more closely at Fig. 8.13, the bias error signal created by the detected power signal of

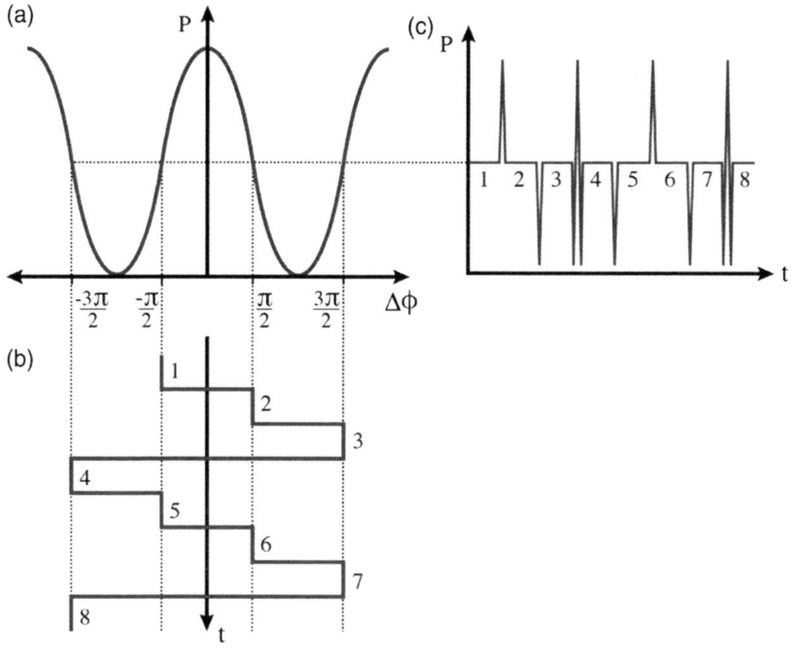

Figure 8.10 (a) An interferogram maps the (b) phase-difference modulation to (c) optical power vs. time.

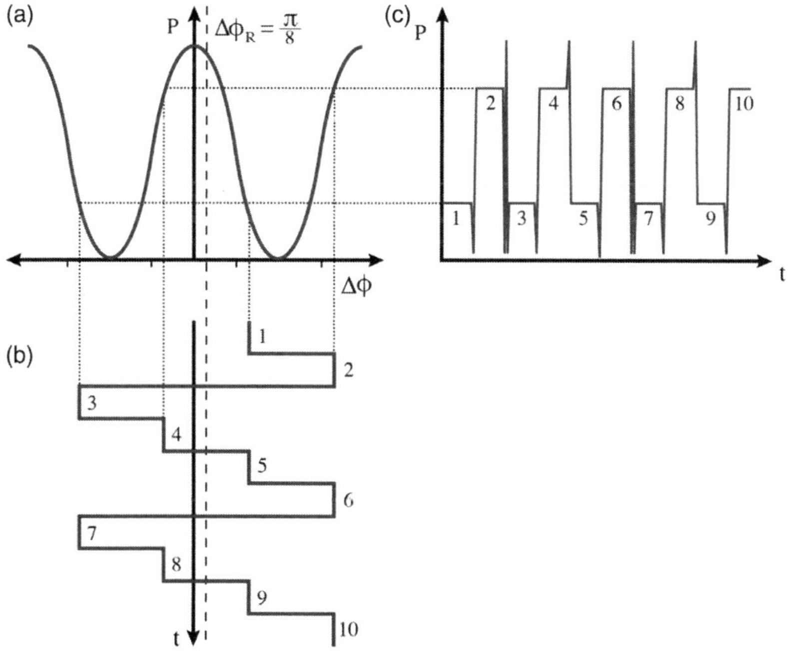

Figure 8.11 (a) Rotation-induced phase difference of $\Delta\phi_R = \pi/8$ is added to (b) the phase-difference modulation, and the result is (c) a square wave in the optical power at the bias-modulation frequency. This demonstrates the optical signal created by a rotation rate.

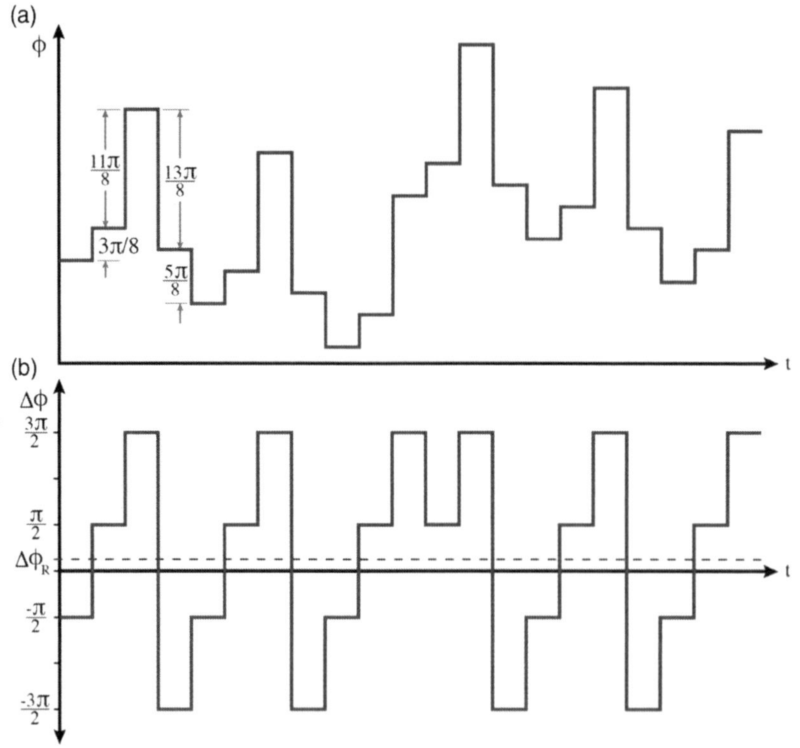

Figure 8.12 (a) Adding a $\phi_M = -\pi/8$ per step-down ramp creates (b) a minus $\Delta\phi_M = -\pi/8$ offset in the phase-difference modulation that compensates for the $\Delta\phi_R = \pi/8$ due to the rotation rate.

step 1 minus step 2 is compensated for by the equal and opposite bias error created by step 3 minus step 4.

The dual-ramp modulation has several preferable features in comparison to other modulation schemes. Most importantly, it locks to four points on the interferogram, allowing for two distinct signals: one for the phase difference due to rotation, and one for amplitude control of the modulation waveform. The latter effectively calibrates the transfer function from digital signal to phase difference. This transfer function is a slowly varying function, and it is not necessary to perform the calibration at high speed, except perhaps at startup. Another feature is that it makes one transition per τ; thus, there is only one glitch per τ to gate out of the signal.

Unfortunately, there is an issue with the glitch pattern in the dual-ramp modulation. This is evident when looking at Fig. 8.9(c). At low rotation rates, the glitch pattern contains a component at the bias-modulation frequency. If the glitches are not completely gated out, this small component causes a small bias offset. One solution is to modify the waveform to reduce the update rate of the calibration. The bias error is small and can be made completely negligible by decreasing the update rate by two or more orders of magnitude at low rotation rates. Figure 8.14 shows an occasional calibration waveform

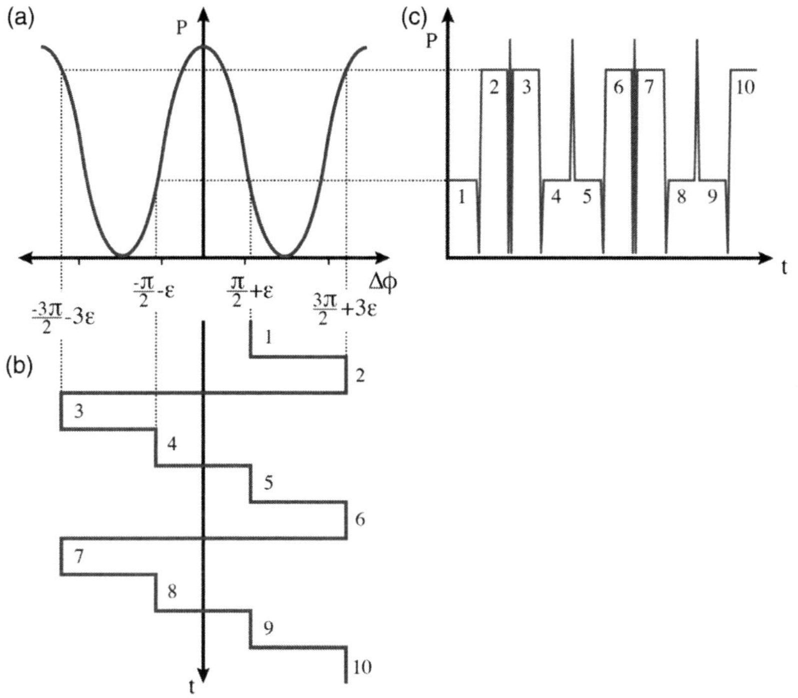

Figure 8.13 (a) Waveform with an error, (b) The phase-difference modulation waveform amplitude is too large, and (c) the detected power has a square-wave component at half the bias-modulation frequency.

with an update rate reduced by a factor of 2. Note that this does not affect the bandwidth of the rate control loop of the FOG.

Alternatively, the sign of the calibration step can be changed, as shown in Fig. 8.15. This sign switching removes any component at the bias-modulation frequency.

In summary, for this phase-modulation section, all successful closed-loop schemes applied to the FOG have used dual control loops. There are similarities and differences between these control schemes, and our phase-modulation approach has some advantages and disadvantages. A dual-ramp closed-loop modulation scheme has been presented that closes the rate loop and calibrates the digital electronics with respect to the interferogram. The Fibernetics approach differs from other schemes by having a step size of τ instead of $\tau/2$. Thus, there is only one glitch per τ, resulting in half the glitches that must be gated out of the signal. Problems that arise from these glitches can be fixed by using an occasional-calibration waveform, as described previously.

8.5 Three-Axis Source-Sharing Design

The use of a single source for three FOGs is desirable because of the reduction of electronics and optical components. Only one set of drive electronics is

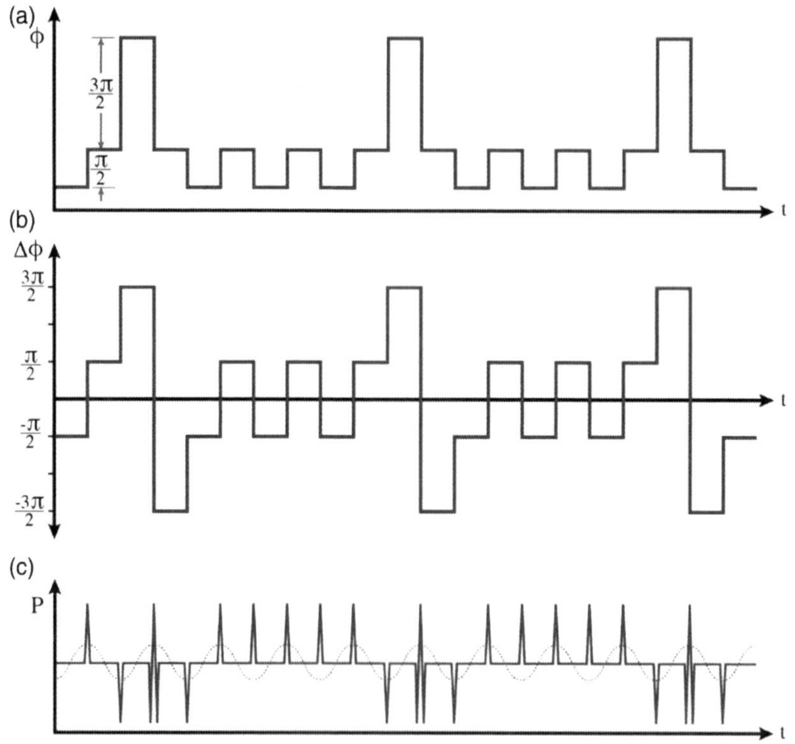

Figure 8.14 Occasional calibration waveform with an update-rate reduced by a factor of 2. (a) Phase modulation and (b) corresponding phase-difference modulation creates (c) detected optical power and associated glitches.

required, as well as only one set of optical components, which is especially compelling if a fiber source with all its complexity is employed. Thermal management is also simplified.

The conventional way to create a source-sharing design uses a directional coupler to split the source light from the source into three or more parts. Each of the outputs of that directional coupler is used as a substitute for the source shown in Fig. 8.1. Figure 8.16 shows three FOGs with their sources replaced by a single source and a 1×3 directional coupler.

Fibernetics is doing something different involving three directional couplers and two isolators instead of three directional couplers and a 1×3 splitter.[15]

In this design, light is distributed to the fiber directional couplers in series: first coupler #3, then coupler #2, and then coupler #1 (as depicted in Fig. 8.17). Each of those couplers takes a portion of light and sends it to its respective Sagnac interferometer. Light from the each Sagnac interferometer returns to the respective coupler, which then directs a portion of that light to the respective detector. While it is important to generally equalize the optical power on each detector, the coupler splitting ratios are not critically important

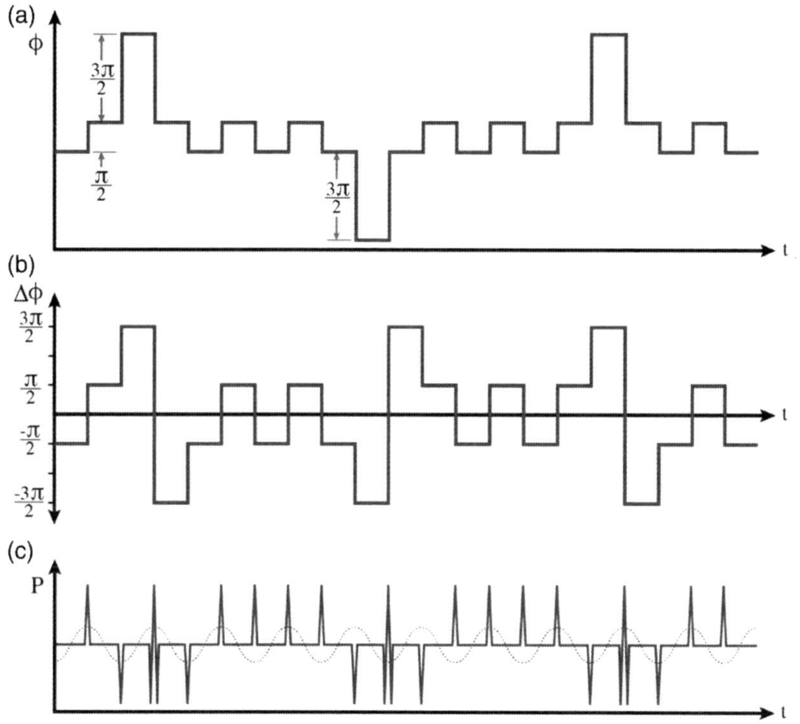

Figure 8.15 Occasional-calibration waveform in which (a) the sign of the calibration step is changed, along with the (b) corresponding phase-difference modulation and (c) detected power with glitches. Note that the glitch pattern associated with the calibration modulation changes sign with respect to the superimposed sine wave.

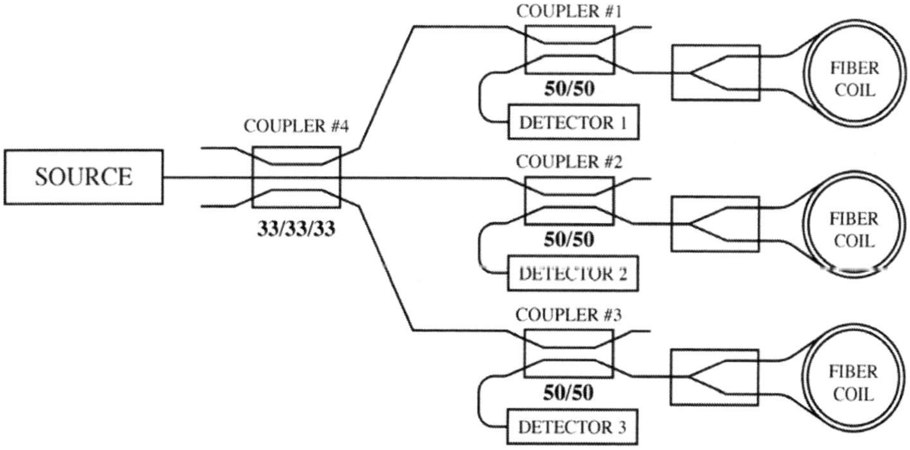

Figure 8.16 Conventional source-sharing approach with single light source split by a 1×3 directional coupler to power three independent FOGs.

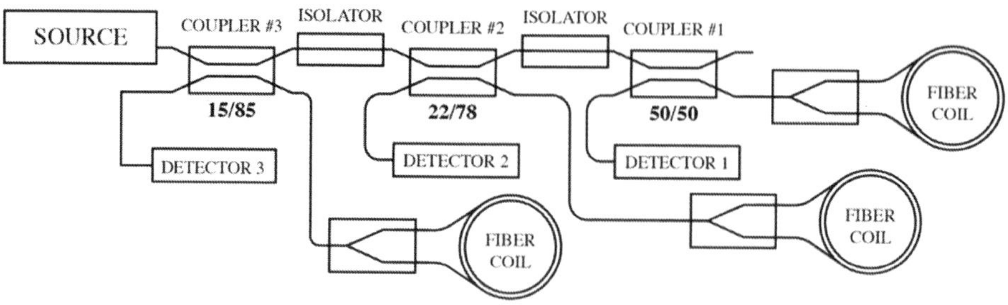

Figure 8.17 Source-sharing three-axis FOG with a single light source and high optical efficiency.

and can deviate somewhat from the values mentioned here. A 50/50 coupler #1 will maximize the light on the last detector in the series (detector 1), just like the single-FOG case. To select an equal power on detector #2, a coupler is used that taps a much smaller fraction of the power and directs it toward the second Sagnac interferometer. To demonstrate the concept, a calculation is made assuming lossless components. Later, loss is introduced to show how it affects coupler ratio selection. For even more efficient use of the source light, as shown in Fig. 8.17, a scheme was devised whereby the otherwise wasted light from coupler #3 is used to power the other FOGs downstream, and the otherwise wasted light from coupler #2 is sent to coupler #1. This source-sharing system increases optical efficiency over other source-sharing systems, thus reducing required power and cost.

Under the assumption of lossless components, the calculation to determine the optimal coupling ratio for the coupler #2 with the goal to maximize detected power and equalize the power received by detectors #1 and #2 is

$$P_2 = P_{2source} K_2 (1 - K_2), \tag{8.1}$$

$$P_1 = P_{2source} (1 - K_2) K_1 (1 - K_1), \tag{8.2}$$

where P_1 and P_2 are the detected optical powers at detectors #1 and #2, respectively, and K_1 and K_2 are the coupling ratios of couplers #1 and #2. If the powers are chosen to be equal, then

$$K_2 = K_1 (1 - K_1). \tag{8.3}$$

Now $K_1 = 0.5$ to maximize the power on detector 1. Then $K_2 = 0.5$ $(1–0.5) = 0.25$, independent of the coupling ratio of coupler #3.

Next, calculate the value of coupler #3 for equal power directed to detector #3:

$$P_3 = P_{3source} K_3 (1 - K_3), \tag{8.4}$$

$$P_2 = P_{3source}(1 - K_3)K_2(1 - K_2), \tag{8.5}$$

$$K_3 = K_2(1 - K_2) = \frac{1}{4} \cdot \frac{3}{4} = \frac{3}{16}. \tag{8.6}$$

Likewise, if a fourth axis were added, its splitting ratio would be

$$K_4 = K_3(1 - K_3) = \frac{3}{16} \cdot \frac{13}{16} = \frac{39}{256} \cong 0.15. \tag{8.7}$$

Thus, for an arbitrary number of FOGs in series the relationship is simply

$$K_n = K_{n-1}(1 - K_{n-1}) \tag{8.8}$$

for lossless components. Note that adding a fourth FOG would only reduce the light power on all of the detectors by 15%. Note also that the choice of the splitting ratio of the first directional coupler K_1 is completely unconstrained to keep the optical powers equal. It will, however, affect the choice of splitting ratios of the directional couplers upstream for equal power.

In the case of three FOGs, the amount of light on the detectors is

$$\frac{P_1}{P_{3source}} = (1 - K_3)(1 - K_2)K_1(1 - K_1) = \frac{13}{16} \cdot \frac{3}{4} \cdot \frac{1}{2} \cdot \frac{1}{2} = \frac{39}{64} \cdot \frac{1}{4} \cong 0.61 \cdot \frac{1}{4}, \tag{8.9}$$

compared with 1/4 for a single FOG. Thus, we have 61% of the light of a single FOG on each of the three detectors. This can be compared to 33% of the light for a conventional source-sharing system. The result is 61/33 = 1.85, which shows that for a lossless system the light on the detector is nearly doubled.

Now consider the case where each isolator imparts 0.5 dB of loss; the power delivered to detector #1 would be

$$P_1 = P_{2source}(1 - K_2)I_1K_1(1 - K_1), \tag{8.10}$$

and together with Eq. (8.1) and equalizing P_1 and P_2, the coupling ratio formula would become

$$K_2 = I_1K_1(1 - K_1), \tag{8.11}$$

where the isolator loss I_1 is included. This would drop the ideal coupling coefficient of coupler #2 because more light would be required downstream to make up for the loss in the isolator. Likewise, the formula for the third coupler would be

$$K_3 = I_2 K_2(1 - K_2). \tag{8.12}$$

This can be generalized as above to

$$K_n = I_{n-1} K_{n-1}(1 - K_{n-1}), \tag{8.13}$$

where the addition of downstream loss reduces the required splitting ratio of the couplers upstream to retain equal powers on all detectors.

As an example, consider the case of a 0.5-dB loss for each isolator:

$$K_2 = I_1 K_1(1 - K_1) = 0.89 \cdot \frac{1}{2} \cdot \frac{1}{2} = .2225 \tag{8.14}$$

and

$$K_3 = I_2 K_2(1 - K_2) = 0.89 \cdot .2225 \cdot .7775 \cong .154 \tag{8.15}$$

The optical power at the detector is then

$$\begin{aligned} \frac{P_1}{P_{3source}} &= I_2 I_1(1 - K_3)(1 - K_2)K_1(1 - K_1) \\ &= 0.89 \cdot 0.89 \cdot 0.846 \cdot 0.7775 \cdot \frac{1}{2} \cdot \frac{1}{2} = 0.52 \cdot \frac{1}{4}. \end{aligned} \tag{8.16}$$

This shows that the source-sharing design has 0.52 times as much light at each detector compared with a single-axis design, which is greater than 50% more light than the 0.33 from the conventional approach.

The isolators in Fig. 8.17 are included in our early designs to make sure that the detected signals and noise of the FOGs on the right do not contaminate the detected signals of the FOGs to the left. We are using them in early designs to be conservative. It is not at all clear that they are necessary. To improve isolation without isolators, some tricks could be played. The modulation frequencies of the FOGs could be made sufficiently different from each other. The coupling ratio of detector 1 could be reduced from 0.5 with very limited impact on the signal level.

The comparison of the technique described above (shown in Fig. 8.17) to the more standard source-sharing scheme that uses a 1×3 coupler and three 2×2 directional couplers, one 2×2 coupler attached to each of the outputs of the 1×3 coupler (see Fig. 8.16) shows a significant power savings. One might question whether using circulators at the outputs of the 1×3 coupler in place of the 2×2 couplers would provide an even better efficiency.[12] It is true that circulators offer improved efficiency compared to directional couplers and might offer the greatest performance for FOGs. The reason that we began using directional couplers is that their performance over temperature is well

Table 8.1 Relative power delivered to each detector for a variety of source-sharing schemes and the same source power.

Design	Relative Power at Detectors
3 × 3 coupler with 2 × 2 couplers	0.08
Fibernetics design with isolators	0.13
Fibernetics design without isolators	0.15
3 × 3 coupler with circulators (specifications for 0–70°C)	0.17–0.20

established, whereas circulator specifications typically cover the limited temperature range of 0 to +70°C. It would be interesting to test the circulators over a wider temperature range to determine if this approach can provide greater efficiency at temperature extremes. The loss comparison between various designs is shown in Table 8.1.

8.6 Future Development

Further improvements to the gyro design are possible that have the potential of achieving navigation-grade performance in an IMU that is the same size as today's tactical IMUs (e.g., Honeywell's 33-cubic-inch HG 1700 IMU). The use of multicore fiber is one development path that will help accomplish this feat.

8.6.1 Multicore fiber

Multicore fiber has the potential to reduce the overall size of the largest component in the FOG, the fiber coil, without reducing the length of the optical path. This is possible by making better use of the volume of the fiber, using four (possibly more) cores instead of just one.[16] The concept is to create a fiber ring by splicing the two ends of a multiwaveguide, or multicore fiber, together and simultaneously align the first and second ends of the fiber such that waveguide #1 of the second end of the fiber is aligned with waveguide #2 on the first end of the fiber, waveguide #2 of the second end is aligned with waveguide #3 of the first end, and so on. Thus, light in the fiber will travel through waveguide #1, then waveguide #2, then waveguide #3, and so on until it passes through all of the waveguides in the fiber. Light will propagate through the entire ring N times, where N is the number of waveguides in the fiber. A low loss splice of this type is possible in a four-core fiber manufactured by Fibercore Ltd. See Fig. 8.18 for a cross-section of multicore fiber with four waveguides.

There are two architectures to employ multicore fiber in the fiber coil of a FOG. Although the one-directional-coupler design shown in Fig. 8.19 is elegant, it has the danger of transmitting unwanted light to the photodetector if the directional coupler does not couple 100% of the light from the standard fiber to the multicore fiber. Note that the directional coupler might

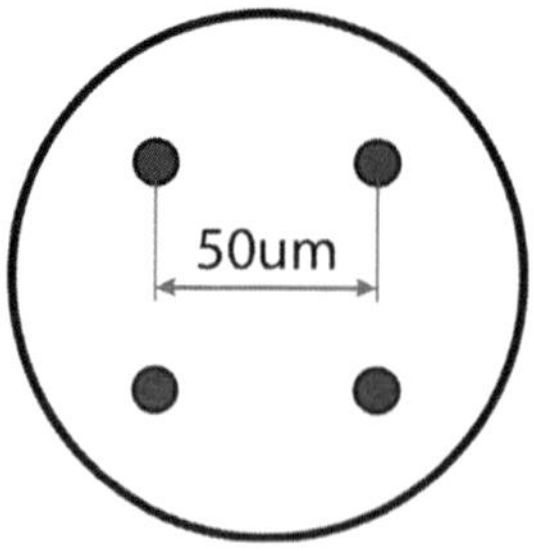

Figure 8.18 Cross-section of multicore fiber with four waveguides.

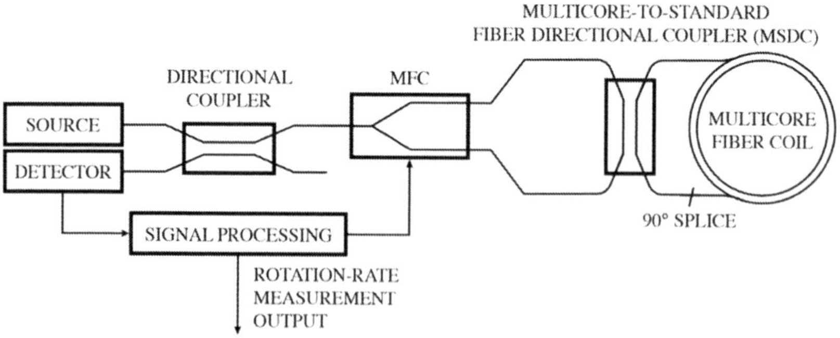

Figure 8.19 Multicore FOG including a multicore fiber ring, a single multicore-to-standard directional coupler to couple 100% of the light into and out of the ring, and a 90-deg splice to transfer light from one waveguide to another so that the light passes multiple times through the multicore fiber coil.

also be a multicore-to-multicore fiber coupler, where one core from one fiber couples to one core of the other fiber. The MFC must then be aligned with the relevant core of the multicore fiber. In this design, all of the light from the MFC would couple into the multicore fiber ring in one of the waveguides. The light would pass through the fiber coil once and be spliced to a second waveguide in the fiber. Then it would pass through the fiber coil again and be coupled to a third waveguide, and this process would continue until the light is coupled back into the original waveguide and then removed by the directional coupler. As a result, it would look like a normal fiber coil with a path length N times the length of the fiber, where N is the number of waveguides within the fiber.

In the two-coupler design, Fig. 8.20, light in the MFC pigtail is coupled through a multicore-to-standard directional coupler (MSDC 1) into waveguide #1 of the multicore fiber. This light propagates through the multicore fiber coil and is transferred to waveguide #2 in the 90-deg splice. The light then propagates again through the fiber coil and is then transferred to waveguide #3 at the splice. Then it propagates again through the coil and on to waveguide #4, where it transits the coil a fourth time and is

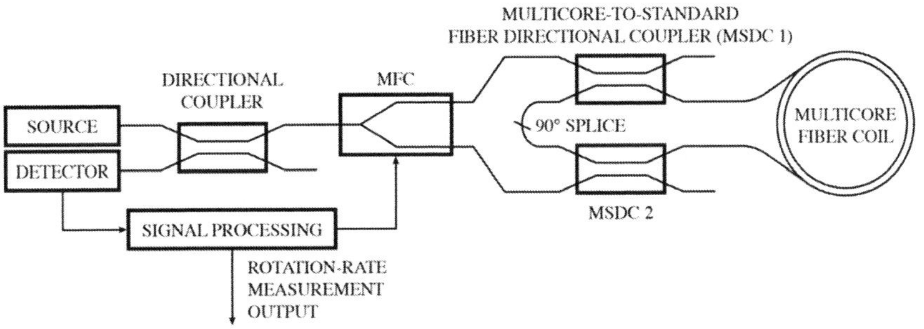

Figure 8.20 Multicore FOG including a multicore fiber ring and two multicore-to-standard directional couplers to couple 100% of the light into and out of the ring and a 90-deg splice to transfer light from one waveguide to another so that the light passes multiple times through the multicore fiber coil.

removed by MSDC 2 and sent to the MFC. Any light remaining in waveguide #4 does not have to be transferred back to waveguide #1, but it can be removed by a loss element that only affects the connection between waveguide #4 and waveguide #1. In the other direction, MSDC 2 couples light from the other MFC pigtail into waveguide #4, and this light propagates through waveguides #3 and then #2 and then #1, where it is removed after traveling through the multicore fiber by MSDC 1. Thus, as in a normal FOG, light travels a single optical path in opposite directions. Reciprocity applies as necessary for good performance. The two-directional-coupler approach has the advantage that no unwanted light will reach the photodetector even if the directional couplers do not transfer 100% of the light from the MFC pigtail fiber to the multicore fiber in the coil. Another advantage to using the two-directional-coupler design is that a loss element can be inserted between the directional couplers for the one waveguide into which the light is being coupled to avoid light passing through the fiber loop more than N times.

The hope is that any time-varying strain of the fiber would affect all of the waveguides within the fiber by the same amount. If this is the case, then the Shupe error could be drastically reduced,[17] which would serve as a compelling reason to pursue this approach. The barriers to this approach are many. Most notably, the fiber is not easy to produce despite now being part of the regular product line at Fibercore Ltd. Furthermore, a polarization-maintaining (PM) multicore fiber is not available. Despite these challenges, a PM multicore FOG has the potential to make a small, robust, and accurate sensor.

8.7 Summary

This chapter provided an update on our progress and described a modulation scheme for closed-loop fiber optic gyroscopes. The transfer function of digital values to optical phase steps through the modulator varies slowly and needs to be updated slowly. We presented our scheme for accomplishing this without significantly contributing to bias offset if glitch gating is incomplete. We also presented a novel source-sharing design that provides more efficient use of the available source light. We also discussed an idea for future research that could result in a multicore FOG that is less prone to errors due to a changing environment.

References

[1] R. A. Bergh, G. Kotler, and H. J. Shaw, "Single-mode fiber-optic directional coupler," *Electron. Lett.*, vol. **16**, pp. 260–261, (1980).

[2] R. A. Bergh, H. C. Lefevre, and H. J. Shaw, "Single-mode fiber-optic polarizer," *Opt. Lett.*, vol. **5**, pp. 479–481, (1980).

[3] H. C. Lefevre, "Single-mode fiber fractional wave devices and polarization controllers," *Electron. Lett.*, vol. **16**, pp. 778–779, (1980).

[4] H. C. Lefevre, Ph. Graindorge, H. J. Arditty, S. Vatoux, and M. Papuchon, "Double Closed-Loop Hybrid Fiber Gyroscope Using Digital Phase Ramp," *Prooceedings of OFS 3I'85, San Diego, OSA/IEEE*, Postdeadline Paper 7, (1985).

[5] R. A. Bergh, "Method and Apparatus for obtaining a digital measure of absolute rotation," US Patent # 4869592, (1989).

[6] R. A. Bergh, "Occasional calibration phase-difference modulation for Sagnac interferometer," US Patent # 8610995, (2013).

[7] V. Vali and R. W. Shorthill, "Fiber Ring Interferometer," *Applied Optics*, vol. **15**, pp. 1099–1100; (1976).

[8] R. Ulrich, "Fiber-optic rotation sensing with low drift," *Opt. Lett.*, vol. **5**, 173–175 (1980).

[9] H. Arditty, M. Papuchon, C. Puech, and K. Thyagarapan, "Recent developments in guided wave optical rotation sensors," in *Digest of Topical Meeting on Integrated and guided wave optics*, (Washington DC), paper TuC2, OSA (1980).

[10] R. A. Bergh, H. C. Lefevre, and H. J. Shaw, "An overview of fiber optic gyroscopes," *J. Lightwave Technol.*, vol. **LT-2**, pp. 91–107, (1984).

[11] H. C. Lefevre, *[The Fiber-Optic Gyroscope Second Edition]*, Artech House, London, (2014).

[12] M. J. Digonnet, *Private communication*, (2018).

[13] R. A. Bergh, "Dual-ramp closed-loop fiber gyroscope," *Proceedings of SPIE - The International Society for Optical Engineering 1169 January* (1989).

[14] H. C. Lefevre and P. Martin, "Optical-Fiber Measuring Device, Gyrometer," US Patent # 5141316, (1992).

[15] R. A. Bergh, "Method and apparatus for multiplexing multiple Sagnac interferometers with single input for source light," US Patent # 9518827, (2016).

[16] R. A. Bergh, "Interferometer employing a multi-waveguide optical loop path and fiber optic rotation rate sensor employing same," US Patent # 9103676, (2015).
[17] D. M. Shupe, "Thermally induced nonreciprocity in the fiber-optic interferometer," *Appl. Opt.*, vol. **19**, 654–655 (1980).

Chapter 9
Recent Developments in Laser-Driven and Hollow-Core Fiber Optic Gyroscopes

M. J. F. Digonnet and J. N. Chamoun
Stanford University

9.1 Introduction

The interferometric fiber optic gyroscope (FOG) has been a successful commercial product for over 25 years, and it is used in a variety of applications, including ship and sub-sea inertial navigation, along with stabilization and positioning.[1] Despite these significant achievements, improvements to the FOG in terms of performance and cost are required to broaden its applicability to other markets, including navigation of aircraft. One area of weakness is the scale-factor stability of conventional FOGs, which is typically limited to 10–100 ppm,[1,2] compared to the 1–5-ppm stability required for the navigation of an aircraft or a submarine. The scale factor of a FOG is proportional to the reciprocal of the wavelength of the Er-doped superfluorescent fiber source (SFS) used to interrogate it. An SFS emits light in a broad bandwidth of typically a few tens of nanometers. Historically, it has proven difficult to stabilize the mean wavelength of this radiation to better than 10 ppm, leading to relatively poor scale-factor stability. In addition, the noise of an SFS-driven FOG is limited by the relatively large excess noise, or relative intensity noise (RIN), of the SFS, which originates from amplified spontaneous emission.[3] Techniques have been developed both to stabilize the mean wavelength of an SFS to the sub-ppm level[4] and to subtract the excess noise.[5] However, these solutions add significant complexity and cost, making the FOG less competitive with other technologies, such as the ring laser gyro (RLG). The two primary limitations of commercial FOGs therefore originate solely from the use of an SFS. A somewhat less significant issue is thermal transients in the FOG sensing coil, which induce a phase error known as the Shupe effect that can limit the bias stability of the FOG.[2] This error has been substantially overcome through quadrupolar coil winding,[2] a special winding procedure designed to ensure that sections of fiber that are symmetrically located with respect to the mid-point of the sensing loop are physically in close proximity on the fiber spool. Despite its successes, this solution is not always reproducible, and it does not reduce the Shupe error sufficiently for high-end applications. Finally, another minor problem is the optical nonreciprocity introduced by exposure of the fiber coil to the earth's varying magnetic field, which produces a phase bias error in the fiber coil via the Faraday effect.[2] This issue can be effectively mitigated by shielding the coil in a mu-metal enclosure, though this solution increases the weight, size, and cost of the sensor.

Two classes of solutions have been proposed and investigated as remedies to these issues. The first one is to replace the broadband source with a semiconductor laser around 1.5 μm, which addresses the first two, more serious limitations described above. The most significant benefit of this solution is that the wavelength of such a laser can be easily stabilized to below 1 ppm, which should enable a FOG with a scale-factor stability that satisfies the navigation-grade requirement. Second, semiconductor lasers do not exhibit the excess intensity noise intrinsic to broadband light sources. This translates to a lower noise in a laser-driven FOG. A laser-driven FOG with a noise lower than the same FOG driven by an SFS has indeed been demonstrated.[6] A semiconductor laser operated around 1.5 μm offers several additional advantages over a broadband source. First, these light sources are ubiquitous and inexpensive owing to their widespread use in the telecom industry, whereas broadband sources are a niche item and thus more expensive. Second, lasers have a better wall-plug efficiency than broadband sources, which would lower the FOG's overall energy consumption. These additional advantages have the potential to make the FOG more competitive in the market for high-performance inertial-navigation systems.

Broadband light sources were incorporated into the conventional FOG design early on because the use of broadband light essentially eliminated several error mechanisms that depend on the temporal coherence of the light used to interrogate the FOG, namely, the nonlinear Kerr effect,[7] backscattering,[8,9] and polarization coupling.[10] The Kerr effect is a third-order nonlinear optical effect taking place in the FOG sensing fiber coil that generates a phase error between the counter-propagating beams traveling in the coil when they do not carry exactly the same power. Coherent backscattering noise and drift occur as a result of parasitic interference between the primary fields circulating in the fiber coil and spurious fields backscattered by randomly distributed defects along the fiber. Similarly, polarization-coupling errors arise from interference between these same primary fields, which are linearly polarized, and spurious fields that are coupled by fiber defects from the residual orthogonal polarization mode into the primary polarization. Replacing the broadband source with a highly coherent laser reintroduces these three error sources, so any practical implementation of a laser-driven FOG requires methods that mitigate these three deleterious effects. Sections 9.2–9.4 show that it is possible to reduce these errors to very low levels by reducing the temporal coherence of the laser emission.

The second class of solutions addresses the residual errors due to the Shupe effect and the Faraday effect, as well as the Kerr effect. It involves replacing the conventional solid-core fiber in the sensing coil with a hollow-core fiber (HCF). In a conventional fiber, the optical modes are guided by total internal reflection at the interface between a core and a cladding of lower refractive index. In contrast, in an HCF the cladding region consists of a 2D photonic crystal formed by a periodic array of holes separated by nanometrically thin silica membranes. The photonic crystal is designed such that light of the target wavelength cannot propagate in the cladding; light is then guided by a defect (a hollow core) introduced in the middle of the photonic crystal. In an HCF, the fundamental mode is therefore confined almost entirely in the air contained in the hollow core. Compared to silica, air has a much lower Kerr constant, Faraday constant, and index dependence on temperature. Thus, the residual Shupe effect and the Faraday effect are both significantly reduced in a FOG using a coil of hollow-core fiber. In addition, while a laser-driven FOG utilizing a conventional solid-core fiber may be susceptible to drift due to the Kerr effect at sufficiently high circulating power, the low nonlinearity of an HCF ensures that the Kerr effect is greatly reduced in a FOG utilizing HCF, even at high powers. The backscattering coefficient is likewise far lower in air than in silica, suggesting that a laser-driven FOG based on HCF could have a much lower backscattering noise and drift than a comparable FOG based on solid-core fiber. However, in at least some HCFs the frozen capillary waves that exist on the surface of the core membrane induce backscattering at a level that exceeds that of scattering in a conventional fiber.[11]

This chapter presents a review of recent research in both laser-driven FOGs and hollow-core FOGs. It begins with a review of the physics of the noise and drift due to coherent backscattering and polarization coupling in laser-driven FOGs, and presents an analysis of the theoretical dependence of the magnitude of these two errors on the laser linewidth. Analytical models and simulations predict that the laser must have a linewidth of about 40 GHz or greater for these two effects to be suppressed enough in a FOG with a ~1-km coil to meet the stringent drift requirement for inertial navigation of an aircraft. Four techniques to broaden the linewidth of a laser to such levels are then described, namely: (1) reducing the laser drive current and modulating the laser phase with an external phase modulator driven by (2) a sinusoidal waveform, (3) a pseudo-random bit sequence (PRBS), or (4) a broadband

Gaussian white noise (GWN). The latter technique (GWN phase modulation) produced a laser-driven FOG with very low error that satisfies the aircraft-navigation-grade requirements simultaneously for noise, drift, and mean-wavelength stability. The previously reported performance of an experimental FOG utilizing 235 m of HCF and interrogated with a broadband light source is also described in order to emphasize the greatly reduced Kerr-induced drift, thermal drift (Shupe effect), and Faraday-induced drift achievable in a FOG made with a hollow-core fiber.

9.2 Backscattering Errors in a Laser-Driven FOG

A laser-driven FOG consists of a conventional FOG that is interrogated by a laser instead of an Er-doped SFS (Fig. 9.1). The laser output is coupled by means of an optical circulator into the same type of multifunction integrated optic circuit (MIOC) as in a conventional FOG. The MIOC contains an integrated polarizer, a Y-junction, and phase-modulation electrodes for dynamic biasing of the FOG at the loop proper frequency.[2] Unless otherwise stated, the biasing modulation consists of a square wave with a 0.84π over-modulation amplitude. The input light is split by the Y-junction into two signals of nominally equal power that propagate clockwise (cw) and counter-clockwise (ccw) around the sensing coil. The signals returning from the coil are recombined at the Y-junction and directed by the circulator to an InGaAs photodetector. The rotation-rate signal is demodulated from the photodetector output voltage using a lock-in amplifier. Most FOG systems are operated in a single spatial and polarization mode to ensure reciprocity of the counter-propagating light fields, thereby reducing the impact of environmental fluctuations on the Sagnac phase measurement. Single-spatial-mode operation is achieved by using single-mode optical fiber throughout the system, while stable single-polarization operation is enforced by a high-extinction-ratio

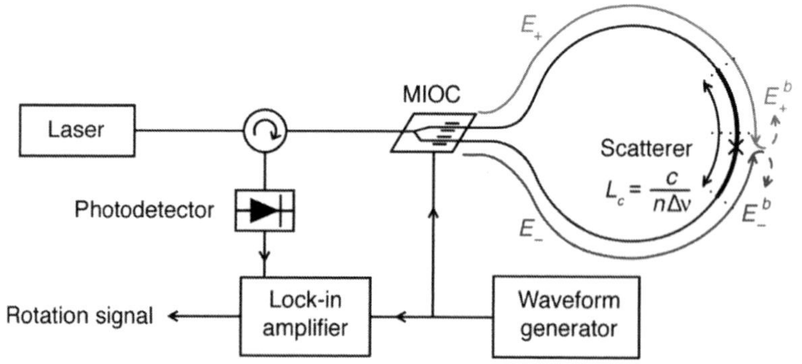

Figure 9.1 Schematic depiction of a laser-driven FOG indicating the important features of coherent backscattering.

polarizer, which is typically integrated into the MIOC, and by using polarization-maintaining (PM) fiber.

Coherent backscattering was recognized in the early days of FOG research as an important error source.[8] The two primary sources of backscattered fields in the FOG are distributed scattering in the sensing fiber and discrete Fresnel reflections at connections within the FOG optical circuit. The latter have been successfully minimized by angling the interface between the MIOC and the fibers. Distributed scattering originates from microscopic material inhomogeneities that are randomly distributed along the fiber.[12] Distributed scattering is an inextricable property of light propagation in a medium that cannot be avoided in conventional silica optical fibers. When the cw propagating field E_+ encounters a scatterer or a reflecting surface (see Fig. 9.1) at a point z along the fiber loop, a small portion E_-^b of it is coupled to the ccw propagating field. This scattered field travels back to the loop coupler, where it adds to the two primary fields E_+ and E_- coherently or incoherently, depending on z. When they interfere at the coupler, the fields $E_\pm$ and $E_\pm^b$ have traveled optical paths that differ by a length $\pm(L - 2z)$, where L is the loop length. If this path difference is smaller than the coherence length L_c of the interrogating light, then $E_\pm$ and $E_\pm^b$ interfere coherently. This is true for all scatterers located within $\pm L_c/2$ of the fiber coil's midpoint (see Fig. 9.1). This process leads to two deleterious effects. First, because there is a path imbalance between the backscattered (or reflected) and primary fields, interference between them converts the phase noise in the light into random intensity fluctuations (noise) in the output signal. Second, because the optical paths traversed by the fields backscattered in the cw and ccw directions are different, backscattering produces a non-zero mean signal in a non-rotating FOG, which constitutes a bias error. Environmental perturbations of the coil—in particular temperature variations and acoustic noise—cause this bias error to fluctuate randomly at a slow rate, which gives rise to an output drift that is indistinguishable from a rotation-induced change and thus constitutes an error in the gyro-output measurement. The rest of the scatterers in the coil produce backscattered fields with a path difference greater than L_c. Thus, backscattered fields from these regions are incoherent with the primary fields and do not contribute to the drift error, though they likely contribute to the noise.[13]

Early models of backscattering errors in a laser-driven FOG[8,9] vastly overestimated the noise and drift because they did not consider the impact of the nonzero laser linewidth. The study of backscattering errors was not pushed any further until fairly recently because broadband sources were demonstrated to be effective at essentially eliminating backscattering errors, so there was no need to study the magnitude of these errors. To revise these predictions, a quantitative model of noise and drift in a Sagnac fiber interferometer was developed.[12] The fiber loop is divided into N segments of equal length (N is typically ~100–1000). Each segment represents a scatterer, characterized by a complex scattering coefficient. Its amplitude varies from segment to segment along the fiber as a random variable with a Gaussian distribution, normalized such that the integrated signal backscattered by the fiber loop is consistent with the fiber's known backscattering coefficient α_B. To account for the laser coherence, the input electric field from the laser is sampled in the time domain. The source phase noise is modeled by adding a random phase to each successive time sample of the input light field. The distribution of the phase increments has a Gaussian distribution, with a standard deviation calculated from the laser linewidth.[12]

Figures 9.2(a) and 9.2(b) plot the dependence on linewidth of the noise and drift, respectively, due to backscattering (blue curves), simulated for the experimental FOG reported further on and described in Ref. 14. The noise is expressed as an angular random walk (ARW).[15] The simulation modeled a gyroscope with a PM-fiber coil of length $L = 1085$ m and a coil diameter $D = 8$ cm, and a backscattering coefficient $\alpha_B = 1.5 \times 10^{-7}$ m^{-1}, which was interrogated by a laser of wavelength $\lambda = 1.55$ µm. In this simulation, the gyro phase was dynamically biased using sinusoidal push–pull modulation. From Figs. 9.2(a) and 9.2(b), as the source linewidth is decreased from a high value ($L_c \ll L$), the noise increases. It has been argued that unlike previously thought, the noise does not originate mainly from scatterers within approximately a coherence length of the coil's mid-point (region indicated by the black double-ended arrow in Fig. 9.1)[12] but from scatterers distributed throughout the coil.[13] The reason the noise increases with decreasing linewidth in the broad linewidth regime ($L_c \ll L$) is because the noise spectrum is increasingly concentrated around the loop's proper frequency (the frequency range in the detection bandwidth of the lock-in amplifier). This

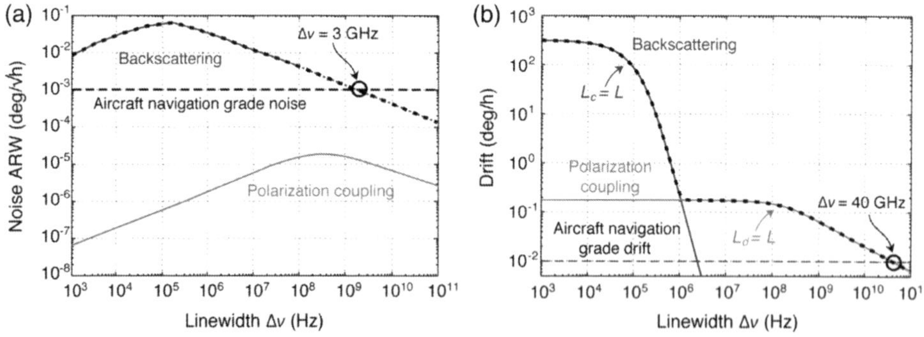

Figure 9.2 Simulated dependences on laser linewidth of (a) the noise and (b) the drift (upper bound value) due to backscattering and polarization coupling for an experimental FOG with a 1085-m coil along with the quadrature sum of both mechanisms (dashed curves). The dotted portion of the backscattering noise curve for linewidths above 100 MHz indicates an extrapolation from simulations.

view explains the $L_c^{-1/2}$ dependence of the noise on linewidth predicted in Fig. 9.2(a), a dependence that was confirmed by measurements (see Fig. 6 in Ref. 14). In contrast, if the noise originated mainly from coherent scatterers within L_c of the mid-point, then the noise would scale proportionally to the number of coherent scatterers, i.e., as L_c^{-1}, which is not borne out by either experiments or modeling.

This trend continues until the linewidth is such that $L_c \approx L$, at which point further decreasing the linewidth reduces the source phase noise, and the noise decreases (see Fig. 9.2(a)). The trend is similar for the phase bias drift when $L_c \leq L$, but when L_c exceeds L the offset remains constant because the drift arises from environmental fluctuations (in temperature in particular) and is, unlike the noise, independent of the laser phase noise. Another key conclusion from Figs. 9.2(a) and 9.2(b) is that both the noise and the drift, even at their respective maximum, are considerably smaller than they were believed to be until this work was published.

9.3 Polarization-Coupling Errors in a Laser-Driven FOG

As described in Section 9.2, a polarizer is employed in most FOGs to ensure single-polarization operation and the corresponding reciprocity of the counter-propagating primary fields. The effectiveness of a polarizer in blocking optical energy in the undesired polarization is characterized by its extinction ratio ε^2. While the extinction ratio of the best integrated $LiNbO_3$ polarizer can be very high (more than 80 dB in some cases[16]), it is still finite, meaning that some portion of light in the unwanted polarization is launched into the FOG's sensing coil. Polarization-coupling errors can then arise when light in the unwanted fiber polarization mode is coupled along the coil fiber (generally by fiber defects) into the main polarization mode. More specifically, the input light field typically contains a strong primary component E_x along the desired polarization axis, and a weak spurious component E_y along the orthogonal axis (see Fig. 9.3). The returning field has reciprocal components E_{xx} and E_{yy}, which entered and exit the FOG in the same polarization, but also nonreciprocal components E_{xy} and E_{yx}, which exit the FOG in a different polarization.[10] Of particular concern is the spurious

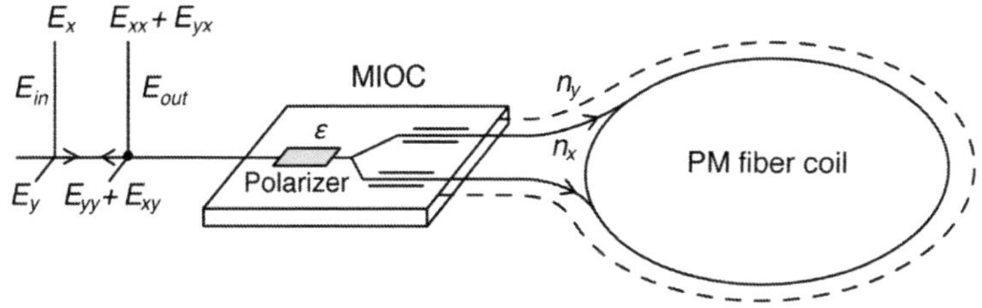

Figure 9.3 Illustration of polarization coupling in a FOG.

component E_{yx}, which interferes with the primary returning field E_{xx}. This parasitic interferometer has the effect of (1) converting laser phase noise into FOG output noise and (2) converting slow environmental fluctuations of quantities such as the fiber birefringence $\Delta n = n_x - n_y$ into FOG output drift.

Polarization coupling occurs at discrete optical interfaces such as a misalignment between the eigen-axes of spliced PM fibers between the light source and the input to the MIOC (see Fig. 9.1), or at the fiber–MIOC junctions. Polarization coupling also takes place throughout the sensing fiber due to statistically distributed micro-twists. In spite of the large number of potential coupling points, some simplifying assumptions can be made to formulate a straightforward analysis of the noise and drift resulting from polarization coupling.[10] First, distributed polarization coupling in the sensing fiber is the dominant coupling mechanism for all but the shortest FOG coils, so it is the only mechanism that needs to be considered for the long coils used in high-sensitivity FOGs. Second, the phase error due to polarization coupling can be divided into two terms: an *amplitude*-type error proportional to ε, and an *intensity*-type error proportional to ε^2.[17] Integrated-optic polarizers currently in use in FOG MIOCs achieve extinction ratios of $\varepsilon^2 = 80$ dB or better,[16] so the intensity-type error can be neglected. With these and a few other justified assumptions,[14] this model computes estimates of the noise σ_{pc} and drift φ_{pc} from polarization coupling as a function of the coil length L, the input polarizer extinction ratio ε^2, the PM fiber's polarization-holding parameter h, the laser linewidth $\Delta\nu$, and other parameters of potential interest, such as fiber polarization-dependent loss (PDL). In the absence of PDL, the noise and drift are given by

$$\sigma_{pc} = \varepsilon\sqrt{h}\rho_{in}\mathrm{Re}[N(0)]^{1/2}R_\phi\left[\frac{1}{2}L\tau_c(1 + e^{-2L\Delta n/c\tau_c}) - \frac{c\tau_c^2}{2\Delta n}(1 - e^{-2L\Delta n/c\tau_c})\right]^{1/2}, \tag{9.1}$$

$$\varphi_{pc} = \varepsilon\sqrt{h}\rho_{in}n_{\omega_m}R_\phi\left[\frac{c\tau_c}{2\Delta n}(1 - e^{-2L\Delta n/c\tau_c})\right]^{1/2}, \tag{9.2}$$

where $N(0)$, n_{ω_m}, and R_φ are constant factors, all of order unity, relating to the specifics of the FOG biasing modulation. Equations (9.1) and (9.2) are plotted in Fig. 9.2 as a function of linewidth $\Delta\nu = 1/\tau_c$ for $\varepsilon^2 = -66$ dB, $\rho_{in}{}^2 = -26$ dB, $h = 10^{-5}$ m^{-1}, $\Delta n = 0.001$, and $L = 1085$ m (polarization-coupling curves). For narrow linewidths (less than 10 GHz), the polarization-coupling noise increases with increasing linewidth due to increasing laser phase fluctuations. As the linewidth continues to increase past 10 GHz, the polarization-coupling noise reaches a maximum and then decreases with increasing linewidth. This behavior arises in the broad-linewidth regime because the phase fluctuations measured in the FOG are increasingly spread out in the frequency space, and less noise falls within the FOG's detection bandwidth.[14] The significance of the turn-over point is that it occurs at the linewidth where the depolarization $L_d = c/\Delta n \Delta \nu$ equals the fiber loop length L. The depolarization length is the distance in fiber over which two fields generated by the same source of linewidth $\Delta \nu$ propagating in orthogonal polarizations lose their temporal coherence due to the fiber birefringence. Importantly, Fig. 9.2(a) shows that for this set of parameters, the polarization-coupling noise is insignificant compared to the backscattering noise. This trend persists for any realistic set of FOG parameters, such that the polarization-coupling noise is always around four orders of magnitude lower than the backscattering noise.

Figure 9.2(a) shows that the qualitative behavior of the polarization-coupling noise and backscattering noise is similar. In particular, the scaling of the noise with linewidth is the same for both noise mechanisms in the narrow/broad regimes. This similarity should not come as a surprise given that both the backscattering and polarization-coupling noise originate from statistically distributed coupling centers that give rise to parasitic interferometers. More precisely, the noise and drift in polarization coupling is analogous to the noise and drift of the coherent backscattering error, except that for polarization coupling the relevant characteristic length scale is L_d instead of L_c. Because $L_d/L_c = n/\Delta n \approx 10^3$, the linewidth for which the polarization-coupling noise and polarization-coupling drift begin to roll off is shifted to larger linewidths by a factor of 10^3 relative to coherent backscattering (see Fig. 9.2). Another difference between the calculated drift dependence of these two mechanisms is that while at narrow linewidths the polarization-coupling drift is far weaker than the backscattering drift (see Fig. 9.2(b)), the latter rolls off much more rapidly with increasing $\Delta \nu$. Thus, coherent backscattering dominates the drift for narrow linewidths (below a few hundred kHz), and polarization-coupling drift dominates for broad linewidths (greater than $\sim$3 MHz). For linewidths up to $\sim$100 MHz, the polarization-coupling drift is constant at a level of 0.2 deg/h. For broader linewidths, it decreases as $\Delta \nu^{-1/2}$ due to the reduced source coherence.[18]

The total FOG noise and drift is the quadratic (uncorrelated) sum of contributions from backscattering and polarization coupling. They are plotted as dashed curves in Figs. 9.2(a) and 9.2(b), respectively. From Fig. 9.2(a) it is clear that a linewidth greater than $\sim$3 GHz is required for backscattering noise to be below the aircraft-navigation-grade noise limit of 0.001 deg/$\sqrt{\text{h}}$. For the drift, Fig. 9.2(b) shows that a linewidth around 40 GHz is needed to reduce the polarization-coupling drift below the aircraft-navigation-grade level of 0.01 deg/h. Thus, achieving aircraft-navigation-grade noise and drift with a laser-driven FOG requires source linewidths significantly broader than the $\sim$10-MHz linewidths available from off-the-shelf DFB semiconductor lasers.

Besides using a very broad linewidth, an independent method to suppress the polarization-coupling drift is birefringence modulation.[17] This technique applies a deterministic modulation to the birefringence-induced phase between the two polarization components E_{xx} and E_{xy}. This time-dependent phase adds algebraically to the fluctuating environmental phase and removes the amplitude-effect term through time averaging. This technique was experimentally demonstrated to reduce the polarization-coupling drift by a factor of 60 or more.[17]

9.4 Kerr-Induced Drift in a Laser-Driven FOG

The Kerr-induced phase error in a FOG has been comprehensively studied and is well understood.[2,19,20] This error arises when different light intensities propagate in the two directions through the fiber loop. The small but finite Kerr nonlinearity of the fiber then causes the two signals to accumulate a different nonlinear phase shift. The difference between these two phase shifts constitutes the Kerr phase error, which depends on the splitting ratio K of the MIOC Y-junction, the average propagating intensity I, the nonlinear Kerr coefficient n_2, and the coherence length and photon statistics of the light.[21] When either the splitting ratio or the output power of the light source varies, this error varies too, leading to a bias drift. When $K = 0.5$, the counter-propagating powers are equal, and there is no Kerr phase error. Improvements over the last few decades in the manufacturing tolerances on the Y-junction's splitting ratio therefore contribute to mitigating the influence of the Kerr effect in a FOG driven by coherent light. The Kerr error also decreases with decreasing coherence length,[2,22] so a coherent source with a broad linewidth will have a reduced Kerr error. Because FOGs now use low-loss fibers around 1.55 μm (as opposed to visible and 800-nm light in the early days of FOG research), the power circulating in the coil fiber is now lower, which also contributes to reducing the Kerr-induced drift in a laser-driven FOG. For example, for a FOG of length $L = 1085$ m with $n_2 = 3 \times 10^{-14}$ μm^2/μW for a silica fiber, $K = 0.49$ (a likely pessimistic value), and light with a coherence length $L_c = 1$ m ($\Delta\nu = 100$ MHz), the rotation error predicted using Ref. 21 is 4×10^{-3} deg/h. This is below the navigation-grade requirement (0.01 deg/h). Subsequent linewidth broadening will further reduce this already small value. In summary, the Kerr-induced drift in a 1-km FOG using modern components and a light source with a very broad linewidth (tens of GHz) is essentially negligible.

9.5 Techniques for Broadening the Laser Linewidth

Sections 9.2 and 9.33 make it clear that to achieve low overall noise and drift in a laser-driven FOG, the laser linewidth needs to be as broad as possible. Specifically, to meet the requirements for aircraft inertial navigation it must be larger than 40 GHz, where the drift is limited by polarization coupling.

Several schemes have been proposed and demonstrated to broaden the linewidth of a laser. These methods include deterministically sweeping the laser frequency[23] or applying feedback to a multimode laser to induce coherence collapse.[24] Frequency sweeping can broaden the linewidth to 1 GHz or more, but it may also degrade the mean-wavelength stability of the laser through coupling of the laser frequency to the output power. A coherence-collapsed multimode laser has large intensity noise and poor mean-wavelength stability, so it is believed to be unsuitable for this application. Three other promising techniques are discussed below, namely, operating a laser closer to its threshold or applying either an external or an internal phase modulation.

9.5.1 Linewidth broadening through optimization of the laser drive current

It is well understood theoretically that as the output power P of a laser approaches threshold from above, its linewidth increases[20] according to a function $\Delta\nu(P)$. Physically, this linewidth broadening comes from the increasing influence of spontaneous emission, which perturbs the phase of the primary lasing mode. This linewidth dependence on power was measured for four DFB lasers: Mitsubishi FU-68PDF-520 PDL, Lucent D2525P, QPhotonics QDFBLD, and Anritsu GB5A016. These measurements were done using a self-heterodyne method.[25] These dependencies, plotted in Fig. 9.4(a), show that even though all four lasers had similar linewidths specified around 10 MHz at their nominal operating power, the linewidths at lower optical powers, close to threshold, were one order of magnitude larger. Linewidths larger than 100 MHz can be obtained by operating the laser in this region. It is therefore possible to lower the backscattering noise by decreasing P.

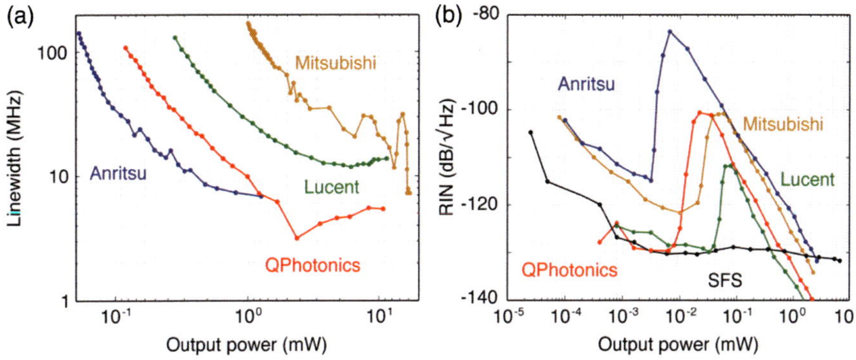

Figure 9.4 (a) Measured RIN at 600 kHz as a function of output power for DFB lasers from four manufacturers, along with the RIN of an SFS as a function of output power. (b) Linewidth as a function of output power for the same four lasers.

The extent to which operating a laser at low power results in an overall reduction in the overall FOG noise depends on the contribution of other noise sources, including relative intensity noise (RIN) and detector noise, which may become significant at low power. To model the balance of these noise sources, the total noise σ_{tot} (in radians/$\sqrt{\text{Hz}}$) in a FOG can be expressed as the quadrature sum of the three main noise contributions, namely the backscattering noise $\sigma_{bs}(\Delta\nu)$, the laser RIN σ_{RIN}, and the detector noise σ_{det}:

$$\sigma_{tot}^2 = \frac{\sigma_{bs}{}^2(\Delta\nu(P)) + \sigma_{RIN}{}^2(P) + \sigma_{det}^2}{S^2}, \tag{9.3}$$

where S (proportional to P) is the rotation sensitivity of the FOG, expressed in V/rad. The RIN is also a function of laser power. In addition to having a broad linewidth, a laser should therefore also have a low RIN for a given output power in order to obtain the lowest possible σ_{tot}.

To determine which laser had the lowest RIN, the RIN was measured as a function of output power for the same four lasers, as well as for an Er-doped SFS (see Fig. 9.4(b)). At low power (below threshold), the RIN decreases weakly with increasing output power. At threshold, it increases abruptly as the laser turns on. Further increases in output power are accompanied by a decrease in RIN with a power law dependence of P^{-1}.[20] It is clear that a lower RIN can be achieved with a laser than with the broadband source when the laser is operated at high power. This is because at higher power, lasers typically have a much lower excess noise than an SFS. However, at low power, where one would want to operate these lasers because their linewidths are broad and help reduce backscattering noise, the RIN is generally larger than the RIN of the SFS. That said, it is clear from Fig. 9.4(b) that the Lucent laser has the lowest RIN of all the candidate lasers examined (with the QPhotonics laser a close second), while also having a relatively broad linewidth for a given output power (only the Mitsubishi is broader, and it has a much larger RIN).

Figure 9.5 shows the measured noise in the 1085-m laser-driven FOG as a function of received power for two different biasing modulation amplitudes: $\varphi_0 = \pi/2$, where the FOG sensitivity is maximized, and $\varphi_0 = 0.84\pi$. In these measurements, the received power was adjusted by changing the laser output power. Using Eq. (9.3) and Figs. 9.2(a) and 9.4, it is now possible to describe how the various error mechanisms contribute to the overall noise in this FOG as the laser power is increased from a low value. Focusing on the curve for $\varphi_0 = \pi/2$, for the lowest powers the laser is below threshold, there is virtually no output light, and the FOG phase noise is limited by detector noise, which varies as P^{-1}. Above threshold, the RIN increases rapidly, and the FOG noise becomes limited by the RIN. As the current is increased further, the RIN

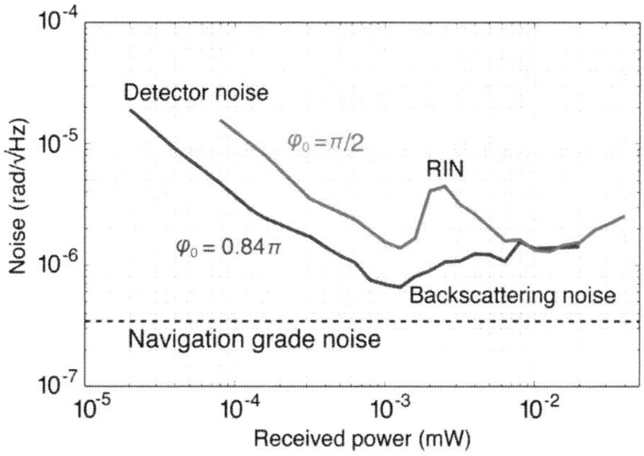

Figure 9.5 Measured 1085-m FOG noise vs. received power for two different biasing phase modulation set points.

decreases, but the laser linewidth narrows,[20] and as a result the coherent backscattering noise grows (see Fig. 9.2(a)). Eventually, the coherent backscattering noise exceeds the RIN, and the FOG noise increases with increasing laser power. For $\varphi_0 = \pi/2$, the power dependence of all three noise sources in Eq. (9.3) results in two identifiable local noise minima in Fig. 9.5: one at a received power close to 0.001 mW, where the detector noise and the RIN are balanced, and one at a received power around 0.01 mW, where the RIN and the backscattering noise are balanced. The value of the noise at either of these minima is around 1.4 μrad/$\sqrt{}$Hz.

Besides optimizing over the laser output power P, an additional noise reduction is possible by optimizing the noise over the biasing modulation amplitude φ_0. It is well understood that in the presence of intensity noise such as RIN and backscattering noise, the optimum signal-to-noise ratio is not achieved for maximum sensitivity biasing ($\varphi_0 = \pi/2$) but is shifted towards the dark fringe ($\varphi_0 = \pi$).[2] For this FOG, the global minimum noise was achieved for $\varphi_0 = 0.84\pi$, where $P \approx 0.001$ mW. Under these conditions, the noise was around 0.7 μrad/$\sqrt{}$Hz—two times lower than for $\varphi_0 = \pi/2$. In angular random walk units, this noise is equal to 0.002 deg/$\sqrt{}$h, which is within a factor of 2 of the aircraft-navigation-grade noise. Thus very low noise in a laser-driven FOG is possible by optimizing the laser current and the biasing modulation amplitude.

A disadvantage of this technique is that the ultimate outcome, which is the FOG noise, depends sensitively on the choice of laser, and the optimized laser biasing condition (drive current) that minimizes the noise does not necessarily minimize the drift. Additionally, and as was expected at the outset, for none of the DFB lasers that we studied was this technique able to achieve the tens-of-GHz linewidths required for navigation-grade noise and drift.

9.5.2 Linewidth broadening through external phase modulation

9.5.2.1 Principle and advantages

External phase modulation of the laser field is an effective alternative means of broadening the laser linewidth. The laser field is directed into an electro-optic phase modulator (EOM) driven by an RF electrical signal, which modulates the phase of the light (see Fig. 9.6). This technique is widely used in long-haul telecommunications systems where information is encoded using the phase and/or amplitude of the modulated light field. Because external phase modulation decouples the light generation and the linewidth-broadening functions, it offers several critical advantages over other schemes. First, because the light is modulated using linear optics, the phase modulation has no effect on the mean wavelength of the light, and the excellent intrinsic mean-wavelength stability of the source laser is preserved. Second, the maximum attainable linewidth is limited only by the EOM, which may have an electro-optic bandwidth of 10 GHz or more, far exceeding the intrinsic linewidth of any laser. Thus, very large effective linewidths are attainable irrespective of the laser properties. Integrated phase modulators based on $LiNbO_3$ technology are already in use in commercial FOGs for the purposes of dynamic phase biasing, so adding an input modulator for linewidth broadening to the MIOC does not in principle present a significant increase in cost or complexity.

9.5.2.2 Linewidth broadening using sinusoidal modulation

The conceptually simplest waveform for linewidth broadening is a sinusoid. Sinusoidal phase modulation at frequency ν_m adds discrete phase-modulation sidebands shifted by ν_m to the laser spectrum. For $\nu_m \geq 100$ MHz, the beats between sidebands are shifted far outside the electrical detection bandwidth of around 1 kHz, so the sidebands are effectively decorrelated. The reduction in backscattering noise with sinusoidal phase modulation then depends on the number of sidebands, where a larger number M of sidebands reduces the noise

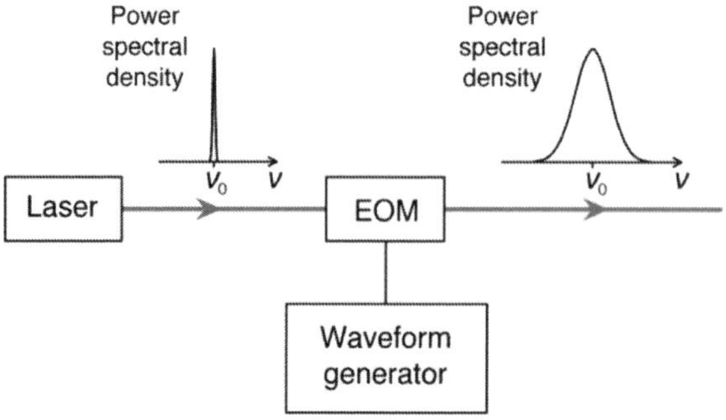

Figure 9.6 Linewidth broadening based on an external electro-optic modulator (EOM) driven by an RF waveform.

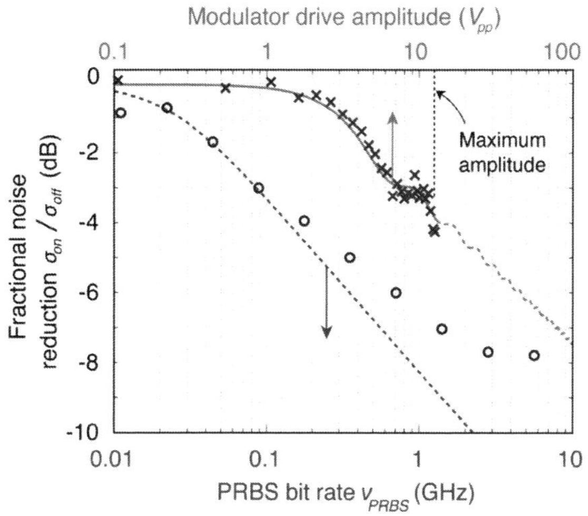

Figure 9.7 Noise reduction achieved with linewidth broadening of the laser using an external electro-optic modulator driven by either a RF sinusoidal waveform (upper theoretical curve and data experimental crosses) or a pseudo-random bit sequence (lower theoretical curve and experimental open circles).

by approximately $\sigma_{off}/\sqrt{M}$, where σ_{off} is the noise without modulation. The upper solid curve in Fig. 9.7 plots the calculated fractional reduction in noise using sinusoidal modulation at $\nu_m = 300$ MHz, along with measurements of this noise reduction in the 1085-m FOG (crosses). It is clear that the noise decreases monotonically in response to an increase in EOM drive voltage. The noise roll-off is slow for low voltages ($V \leq V_\pi$) and it increases to around –5 dB/decade at higher voltages ($V > V_\pi$). The maximum achievable noise reduction is constrained by the maximum drive amplitude permissible in the EOM (12.6 V, indicated in Fig. 9.7 by the vertical dotted line). For the particular EOM used in this work, the maximum measured noise reduction was a modest 4.2 dB. Thus, despite its simplicity, sinusoidal phase modulation presents two disadvantages. First, because the laser spectrum is split into frequency-shifted copies of the laser carrier, the reduced noise is still tied to the intrinsic linewidth of the laser source through its dependence on σ_{off}. Second, because the phase shift in high-speed phase modulators is usually limited to a few radians ($M \leq 5$), it is not possible to easily achieve more significant reductions in the backscattering noise by increasing the modulation amplitude.

Noise reduction in a laser-driven FOG using sinusoidal frequency modulation was also demonstrated in Ref. 26. In this work, direct sinusoidal current modulation of a DFB laser at a frequency of 300 MHz was shown to broaden the self-heterodyne linewidth of the laser from 3.28 MHz to several GHz. Adding a second sinusoidal tone at 99 MHz reduced the impact of carrier harmonics or, equivalently, reduced the appearance of coherence revivals at long path delays. Using this modulated laser to drive a FOG with a coil of length 186 m and a diameter of 20 cm, an ARW of 0.033 deg/$\sqrt{}$h and a drift of 1.43 deg/h were measured.[26] This technique reduced the noise by a factor of 2 and the drift by a factor of 6 compared to the same FOG driven by an unmodulated laser. However, this noise and drift are still worse than when the same FOG was interrogated with a broadband light source, and not low enough for aircraft navigation. Ultimately, that work showed that the choice of waveform can have a significant impact on the effectiveness of linewidth broadening through modulation and that higher-frequency content (two tones in that study) can generate improved broadening.

9.5.2.3 Linewidth broadening using pseudo-random bit sequence modulation

Pseudo-random bit sequence modulation is among several alternative waveforms that overcome the limitations of sinusoidal phase modulation. A random bit sequence (RBS) is a series of bits where the bit value of zero or one is chosen with some probability p, and in this case $p = 0.5$. In a pseudo-random bit sequence (PRBS), the pattern of zeroes and ones repeats after some very large number N of bits, e.g., $N = 2^{31} - 1 \approx 2 \times 10^9$. As a phase-modulation waveform, the PRBS zeroes and ones are represented by phase shifts $\pm\varphi_m$, and the sequence is characterized by a bit rate ν_{PRBS}. It can be shown that if the PRBS parameters are chosen such that $\varphi_m = \pi/2$, then the modulated laser spectrum is a sinc^2 function with a width that equals ν_{PRBS} and is *independent of the original laser linewidth.*[27] Unlike sinusoidal modulation, this property of a suppressed optical carrier satisfies one of the promises of external phase modulation: a broad linewidth output that is decoupled from the properties of the laser. The open circles in Fig. 9.7 show the fractional reduction in backscattering noise measured in the 1085-m FOG using PRBS modulation at bit rates from 10 MHz (the approximate linewidth of the unmodulated laser) to 6 GHz, along with a theoretical fit based on a laser linewidth of 20 MHz (dashed curve). This theoretical curve was generated using the $\Delta\nu^{-1/2}$ noise scaling from Ref. 12 and assuming a completely suppressed optical carrier. The effective linewidth was taken to equal the quadrature sum of 20 MHz and ν_{PRBS}. A noise reduction of nearly 8 dB was observed using PRBS modulation, which is significantly more than that with sinusoidal modulation. One interpretation of this result is that even though both sinusoidal and PRBS modulations have comparable RMS amplitudes, the PRBS modulation fills in the available electro-optic frequency space more uniformly, leading to more effective broadening of the optical linewidth and correspondingly lower backscattering noise. Using PRBS modulation at 2.8 GHz, near the minimum of the experimental data points in Fig. 7, the measured FOG angular random walk (ARW) was 0.00073 deg/$\sqrt{}$h, which is 30% *below* the noise specification for aircraft navigation.[28] The corresponding drift, which is as low as 0.03 deg/h, is only a factor of 3 larger than the navigation-grade drift specification of 0.01 deg/h.

For frequencies larger than ~3 GHz, the measured noise reduction deviates from the predictions of theory (see Fig. 9.7). This is likely due to the effects of a finite-bandwidth system on a PRBS. In an ideal PRBS, the square pulses representing each bit have zero rise time, but such a waveform requires an infinite bandwidth. In a real finite-bandwidth system, the square pulses are low-pass filtered, which leads to modulation sidebands at multiples of the bit rate that look like frequency-shifted versions of the optical carrier.[27] These carrier harmonics have a narrow linewidth, and consequently they increase the backscattering noise above the predictions based on an ideal PRBS (dashed curve in Fig. 9.7). It is generally understood that a system with a bandwidth BW can transmit square pulses at a rate of about $BW/3$, so in the system of Fig. 9.7 with $BW = 10$ GHz, finite bandwidth effects become important around 3 GHz, as seen in the figure. In essence, PRBS phase modulation can only use around 30% of the available electro-optic bandwidth for linewidth broadening.

Significant improvements in the noise and drift of a FOG were also reported when broadening the laser by direct modulation of its drive current with a PRBS sequence.[29] The incentive of this work is that direct instead of external phase modulation may reduce the cost and power consumption of the gyroscope. The test gyroscope used 235 m of PM fiber quadrupolar wound in a coil about 8 cm in diameter. The measured coherence length of the unmodulated, temperature-stabilized DFB laser was 3.8 m. When modulated with a PRBS signal at 500 MHz, it decreased to below 80 mm (a FWHM linewidth of ~1 GHz). The measured ARW of the FOG decreased from 0.172 deg/$\sqrt{h}$ without modulation to 0.076 deg/$\sqrt{h}$ with modulation, or a factor of ~3.6 dB. The measured drift decreased by a similar factor, from 5.862 deg/h to 2.298 deg/h.[29] The noise and drift of the same FOG interrogated with a broadband light source were still significantly lower, namely 0.026 deg/$\sqrt{h}$ and 0.177 deg/h, respectively. The reduction in noise and drift were not as large as observed with external phase modulation with PRBS,[28] likely in part because the linewidth of the broadened laser was not as large (~1 GHz in Ref. 29 versus more than 10 GHz in Ref. 28). The two techniques would need to be applied to the same gyroscope and laser in order to determine which one produces the greatest absolute noise and/or drift.

9.5.2.4 Linewidth broadening using a Gaussian white noise modulation

A third option for broadening the linewidth through external phase modulation is using a Gaussian white noise (GWN) phase-modulation waveform.[30,31] GWN phase modulation substantially overcomes the two problems of PRBS modulation, namely the limited bandwidth and the presence of carrier harmonics. An experimental FOG with a GWN phase-modulated laser source

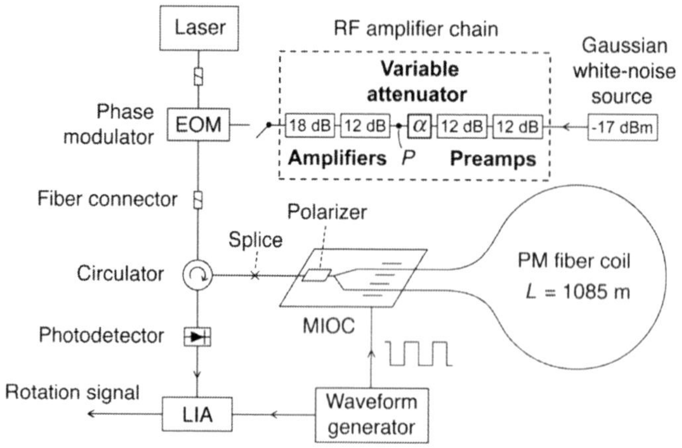

Figure 9.8 FOG interrogated by a laser whose linewidth is broadened through Gaussian white noise (GWN) phase modulation.

from Ref. 31 is depicted in Fig. 9.8. The light source was either a broad DFB laser (Lucent D2525P, 10-MHz linewidth) or a narrow-linewidth external-cavity laser (RIO ORION, 2-kHz linewidth). The laser was connectorized to a 12-GHz EOM from Photline (MPZ-LN10) with a V_π of 4.7 V. The EOM was driven by a broadband Gaussian white-noise source (Noisewave NW10-G) with a 3-dB bandwidth of 11 GHz. This noise was amplified by three 14-GHz preamplifiers (Minicircuits ZX60-14012L) and a 12-GHz amplifier (SHF 100 CP). A variable attenuator was used to adjust the GWN signal level in the RF amplifier chain. The attenuation was fine-tuned by adjusting the supply voltage to the first preamplifier. This broadened laser was evaluated in the same 1085-m FOG mentioned in earlier sections. The phase modulators in the FOG were driven by a square wave at the loop proper frequency for dynamic biasing of the interferometer, with 0.84π over-modulation.[2] The MIOC and coil were placed in a thermally isolated housing. All fibers were PM.

The spectrum of the laser with and without GWN phase modulation was characterized using an optical spectrum analyzer (OSA). Without phase modulation, the narrow laser lineshape (see the notional dotted line in Fig. 9.9) was much narrower than the OSA resolution, so it appeared as a peak corresponding to the measured OSA impulse response (dashed curve in Fig. 9.8). With phase modulation (solid curve in Fig. 9.9) the broadened laser spectrum was considerably broader than the original laser linewidth, and it had the Gaussian shape expected from theory.[31] Figure 9.9 illustrates the two primary advantages of GWN phase modulation over PRBS modulation. First, there are no carrier harmonics evident in the phase-modulated laser spectrum—the only narrow-linewidth feature is the residual optical carrier (the small bump at the top of the broadened spectrum), which is greatly reduced relative to the unmodulated case, although not fully suppressed.

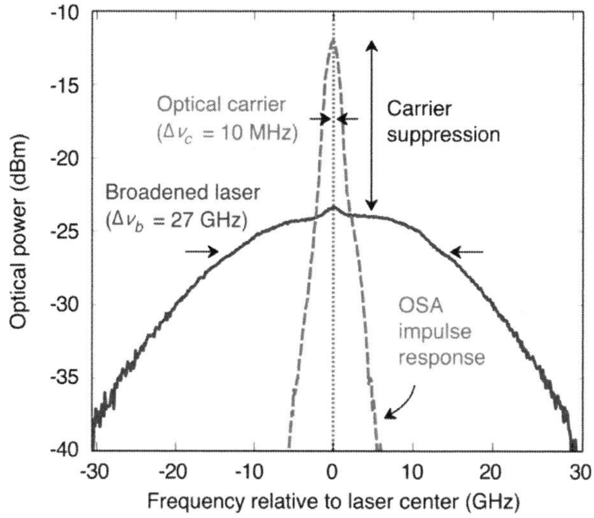

Figure 9.9 Example measured spectrum of a laser with and without GWN phase modulation.

Second, the width of the broadened spectrum was measured in this case to be 27 GHz, which is far larger than the 10-MHz intrinsic laser linewidth.

Even larger linewidths are attainable by adjusting the RF attenuation. Tuning the RF attenuation in the amplification chain (see Fig. 9.8) modifies both the carrier suppression f_c and the FWHM $\Delta\nu_b$ of the broadened component. These measured values of $\Delta\nu_b$ and f_c are plotted in Fig. 9.10 as a function of the propagating RF power at point P (defined in Fig. 9.8). The reason for selecting this location is that none of the amplifiers preceding point P in the chain are saturated, whereas the amplifiers following point P are saturated for low attenuations, and the carrier suppression therefore depends most naturally on the power at that point. The width $\Delta\nu_b$ was extracted from measured spectra like the one shown in Fig. 9.9 by fitting them to a Gaussian. The carrier suppression f_c was measured from both a fit to the OSA data and a method based on measurements of the drift reduction obtained with a given broadened laser, as described in Ref. 31. The dashed curve in Fig. 9.10 shows that as the RF power at P is increased, the width of the broad component increases monotonically from 17 GHz to around 44 GHz. At the same time, the carrier suppression (solid curve, measured by fitting OSA spectra) decreases from close to 0 dB (no suppression) to as low as –44 dB, then it starts to increase, approaching an asymptotic value around –8 dB. This behavior originates from the interplay between the V_π of the EOM and the saturation properties of the last two amplifiers in the chain. For noise powers at P below –1 dBm, the noise output from the last amplifier was well below its saturated output level of $V_{sat} = 12.6$ V. Decreasing the attenuation increased the noise power driving the EOM and therefore increased the corresponding

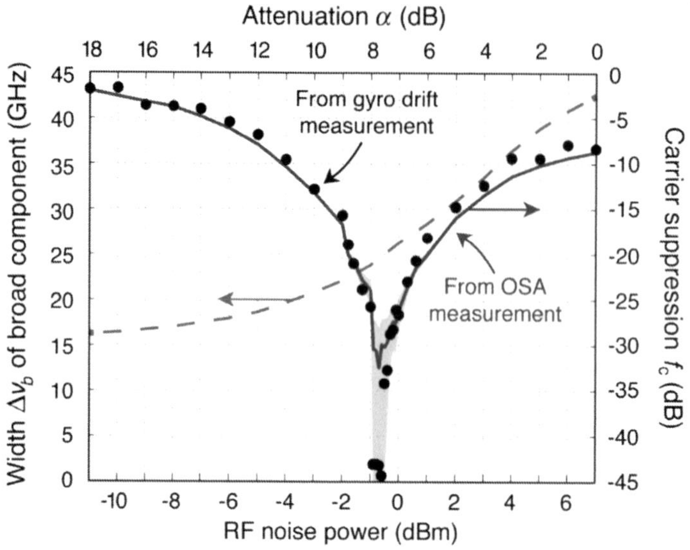

Figure 9.10 Full width at half maximum $\Delta\nu_b$ of the broad spectral component (dashed curve) and carrier suppression f_c measured directly with an OSA (solid curve). The shaded band represents the 95% confidence interval associated with the fit parameter for the carrier fraction in the vicinity of the largest carrier suppression. The carrier suppression inferred from a measurement of the gyro drift suppression (black filled circles) is also shown.

phase fluctuations added onto the laser field, which suppressed the optical carrier to levels as low as –32 dB (lowest point in the solid curve of Fig. 9.10). In the opposite limit of noise powers larger than –1 dBm, the last amplifier was saturated. In that case, further decreasing the attenuation (or equivalently increasing the RF power at P) did not significantly increase the noise power driving the EOM. Instead, it clipped the amplified noise waveform at measured levels of $\pm V_{sat}/2$, and this nonlinear amplifier response caused the carrier component to increase. To understand why, consider the case of an EOM driven by a strongly saturated amplifier whose output voltage jumps instantaneously between $\pm V_{sat}/2$ randomly. In the worst case of $V_{sat} = 2\ V_\pi$, the phase shift imparted to the optical field jumps between $\pm\pi$, which has no effect on its instantaneous frequency, and thus produces no carrier suppression and no broadening. For general values of V_{sat}, the carrier suppression $f_c|_{sat}$ in the saturated regime (i.e., when the amplifier driving the EOM is strongly saturated) is given by[31]

$$f_c|_{sat} = \frac{1}{2}(1 + \cos(\pi V_{sat}/V_\pi)). \tag{9.4}$$

In the system of Fig. 9.8, $V_{sat} = 12.6$ V, and $V_\pi = 4.7$ V, for which Eq. (9.4) gives $f_c = -6.4$ dB, which is close to the observed carrier suppression of –8 dB in the saturated regime (rightmost point of the solid curve in

Fig. 9.10). To reach this limit after an initial rapid decrease, f_c must start to increase, achieving a minimum for an intermediate value of the RF attenuation, as observed close to $\alpha = 7.8$ dB in Fig. 9.10.

For carrier suppressions smaller than about –20 dB, which is the region of most interest, it was difficult to measure the very small residual carrier accurately with this fitting method. This led to a large uncertainty in the estimate of the suppression, shown by the 95% confidence intervals in Fig. 9.10 (solid shaded band). To overcome this limitation, the drift in the FOG driven by the RIO laser broadened with GWN modulation was measured, and the carrier suppression was inferred from the measured reduction in the drift compared to the unmodulated case using the model in Ref. 14. For all points outside the critical power range of –1 to –0.5 dBm, the drift was calculated as the peak in the Allan variance.[15] For the points inside this range, it was calculated as the endpoint in the Allan variance because no Allan variance peak was observed. The inferred values of the carrier suppression are plotted as filled black circles in Fig. 9.10. They agree well with the OSA measurement (solid curve), except in the critical power range where, as expected, this second method provides a much more accurate determination of f_c, with a minimum as low as –44 dB. This carrier suppression is remarkably good: less than 0.004% of the optical power is contained in the narrow-linewidth carrier.

Figure 9.10 shows that carrier suppression as strong as –44 dB has been obtained with GWN modulation, compared to only –8 dB for PRBS modulation.[28] In addition, the linewidth of the broad component was measured to be as large as 44 GHz, a factor of 4 larger than the EOM bandwidth. These measurements demonstrate how GWN phase modulation overcomes the two primary issues of PRBS modulation and suggest that it can produce a laser source with a linewidth broad enough and a carrier suppression large enough for navigation-grade drift.

To characterize the noise and drift of the FOG in Fig. 9.8, the Allan deviation was measured when the FOG was driven by either the broadened Lucent laser, the broadened RIO laser, or a broadband Er-doped SFS with a spectral FWHM of 17 nm (see Fig. 9.11). For the phase-modulated lasers, the RF attenuation was chosen for maximum carrier suppression, where the broadened linewidth was 24 GHz (see Fig. 9.10). The maximum detected optical power was –17 dBm. The ARW with either modulated laser is 5.5×10^{-4} deg/$\sqrt{\text{h}}$. This is the lowest noise yet reported in a laser-driven FOG, and it is almost 50% below the aircraft navigation requirement of 0.001 deg/$\sqrt{\text{h}}$. Figure 9.11 also shows the first demonstration of a modulation-broadened laser-driven FOG with noise below that of the same FOG driven by a conventional SFS. This noise reduction is a direct consequence of the lower RIN of the lasers compared to the SFS. The measured drift is 0.0068 deg/h with either the Lucent or the RIO laser (Fig. 9.11). This is below the aircraft-navigation drift

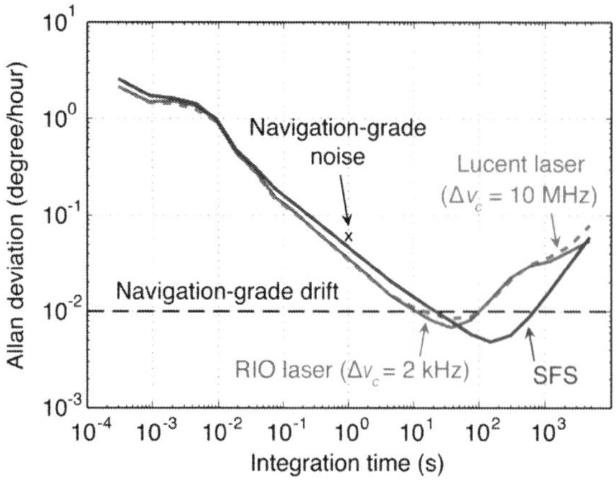

Figure 9.11 Measured Allan deviation of the FOG signal driven by either one of three light sources: a RIO or a Lucent laser broadened using Gaussian-white-noise phase modulation, or a broadband Er-doped superfluorescent fiber source.

requirement of 0.01 deg/h by ~30%, and more than 3 times lower than the lowest drift previously reported in a laser-driven FOG.[28]

In addition to this demonstration of unprecedentedly low noise and drift in a laser-driven FOG, the confluence of the low noise and drift with all three light sources points to several important conclusions. First, the close match between the drift measured with the two lasers under conditions of maximum carrier suppression is remarkable given that in the absence of phase modulation, the drifts in the FOG driven by these two lasers differ by more than three orders of magnitude (as a result of their very different linewidths, and therefore backscattering drift). This result demonstrates that this broadening technique fulfills the promise of low noise and low drift independently of the intrinsic linewidth of the unbroadened laser. Second, the similarity in the noise and drift measured with the broadened lasers and the SFS (see Fig. 9.11) indicates that these two errors are likely in part limited by coherence-independent mechanisms. Noise sources such as mechanical vibration or electronic noise associated with the open-loop demodulation scheme do not depend on the light source, so it is reasonable to expect that a lower noise could be obtained through better vibration isolation and elimination of all technical noise. Likewise, the drift may be limited by the Shupe effect, which is caused by thermal transients in the coil and does not depend on source coherence. Lower drift could then be achieved through careful thermal design of the sensor along with temperature modeling of the output, as is commonly done in commercial FOGs.

With this broadening technique, because the light field is phase modulated after it has been emitted by the laser, the good mean-wavelength stability of the laser is expected to be preserved, thereby leading to a superior scale-factor stability compared to an SFS. This point was confirmed by measuring the mean wavelength of the Lucent laser broadened with GWN phase modulation.[31] This measurement was carried out at the power level typically used in a FOG. It involved measuring the laser with an OSA every 1.7 s for 16 h. The Allan deviation of the resulting time-series data then provided the long-term stability of the mean wavelength of the broadened laser. This measurement shows that the mean-wavelength drift is less than 0.15 ppm, which is significantly lower than the 1–5-ppm requirement for aircraft navigation. This mean-wavelength drift is the same with and without phase modulation. The measured drift includes any drift from the OSA, so the true wavelength stability could be even lower than this value.

9.5.3 Measured dependence of noise and drift on laser linewidth

A summary of the noise and drift measured in the 1085-m FOG driven by lasers of different linewidths is shown in Figs. 9.12(a) and 9.12(b), respectively. These experimental results are superimposed on the predictions of the theoretical models presented in Refs. 12 and 14. Measurements were taken for a variety of laser sources, including a PRBS-broadened and a GWN-broadened laser, which together cover more than seven orders of magnitude of effective linewidth, from 2.2 kHz to 24 GHz. While the experiments were carried out using square-wave dynamic biasing modulation, the models used to compute the noise and drift from backscattering assumed sinusoidal dynamic biasing.[12] This was because it was difficult to obtain convergence of the models with square-wave modulation, owing in part to the inherent discontinuities in a square waveform. The polarization-coupling model, on the other hand, correctly assumed square-wave dynamic biasing[14] and presented no convergence issue. Despite this minor discrepancy, the measured noise and drift agree well with the predictions of theory and confirm the main predictions of Refs. 12 and 14. The main observations are that (1) the backscattering noise dominates at all but the broadest linewidths where technical noise may be dominant, (2) the drift is dominated by backscattering below 1 MHz and by polarization coupling above 1 MHz, and (3) the polarization-coupling drift decreases for linewidths broader than ~100 MHz. With the GWN-modulated laser, the noise is almost 50% below, and the drift about 30% below, their respective requirement for aircraft navigation.

9.6 Hollow-Core Fiber Optic Gyroscope

The studies reported above show that a laser can be used in a FOG to greatly improve the scale-factor stability while also keeping the coherent noise and drift to levels consistent with aircraft-navigation requirements. However, these studies do not address the issues of drift due to the Shupe and Faraday effects, which are independent of coherence. The solutions described in Ref. 2

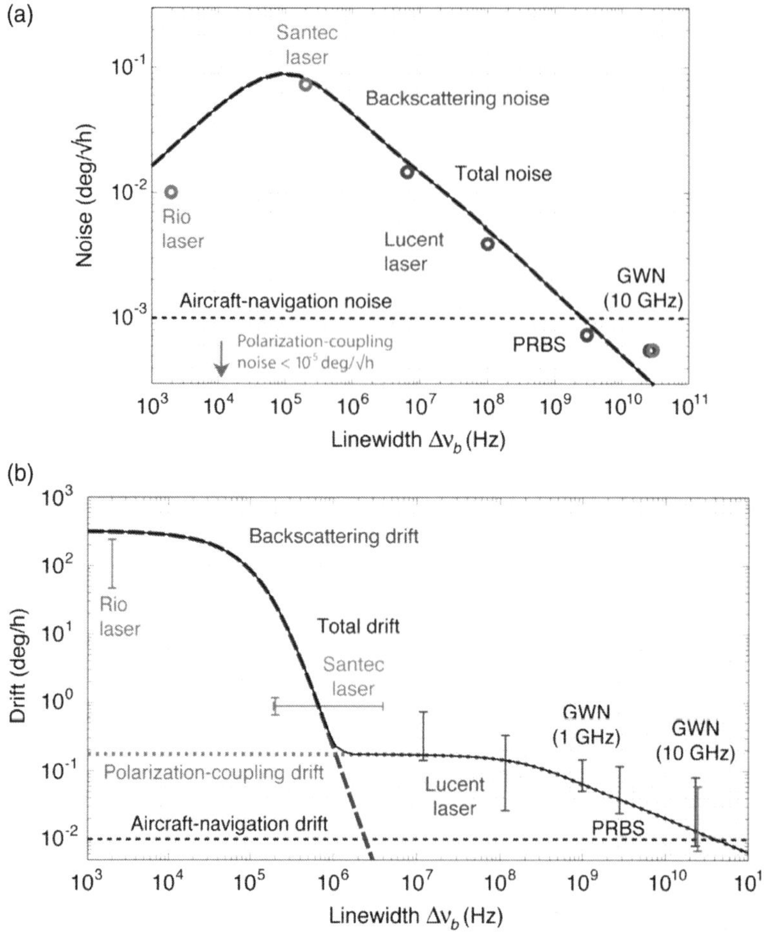

Figure 9.12 (a) Noise and (b) drift as a function of source linewidth in a 1085-m FOG: theory and experiments.

and mentioned in the introduction to reduce these two sources of drift work reasonably well, but they are not perfect—they increase cost and complexity, and they may reduce manufacturing yield. One promising method for reducing both of these sources of error is to replace the solid-core fiber in the FOG's sensing coil with a hollow-core fiber.[32,33] As mentioned earlier, in an HCF light is confined mostly in air, which leads to a substantial reduction in the Kerr constant,[34] Shupe constant, and Faraday constant experienced by light in the fiber, and therefore leads to a concomitant reduction in the drift associated with these three effects. In addition, a significant reduction in radiation hardening should be expected, a feature of importance for space applications.

9.6.1 Kerr-induced drift

To quantify the reduction in effective Kerr constant in an HCF FOG, a gyroscope with a configuration similar to Fig. 9.1 was tested,[34] except that the fiber coil was made of 235 m of hollow-core fiber (HC-1550-02 from NKT Photonics) quadrupolar wound on a mandrel 8 cm in diameter, and fiber components were used instead of an MIOC. Because the Kerr-induced drift was expected to be immeasurably small under normal operating conditions, its magnitude was magnified by increasing the signal power (up to 50 mW) and strongly unbalancing the loop coupler to a 90:10 splitting ratio. Had the Kerr effect been sizable, as would be the case with a conventional sensing fiber, such drastic modifications would have induced a large offset in the FOG output. Instead, no change in the output signal drift was observed when the input power was increased from 4 mW to 50 mW.[34] Thus, in spite of using high power, coherent light, *and* strongly unbalanced circulating powers, the Kerr phase shift was essentially eliminated. Analysis of the return signal led to the conclusion that the effective Kerr constant of the fundamental mode in this HCF was suppressed by a factor of up to ~170 compared to an SMF-28 fiber, in agreement with a model of the Kerr effect in HCFs.[34] In a laser-driven HCF FOG operated under normal conditions (counter-propagating signals with low and nominally equal power), a similar reduction in Kerr-induced drift is expected: in short, in a FOG using an HCF, the drift due to the Kerr effect is negligible.

9.6.2 Shupe effect

When the small temperature change dT is applied to a short length of HCF, the optical phase delay of the fiber mode changes by $d\phi$ as a result of (1) thermal expansion of the fiber and jacket, (2) thermal expansion of the fiber cross-section, (3) the temperature dependence of the refractive indices of the fiber materials (silica and air), and (4) stresses across the fiber due to thermal expansion of the fiber, which modify the materials' refractive indices through the elasto-optic effect.[35] These four effects are jointly described by the Shupe constant $S = (1/\phi)d\phi/dT$. This constant was computed numerically by modeling all four terms separately using the basic laws of thermal expansion (to determine all dimensional changes) and Hooke's law (to determine all stresses), and then using a numerical code to determine their combined effects on the phase of the fundamental mode.[35] These investigations showed that the dominant contribution for an HCF is linear expansion of the fiber length; the other three terms are considerably reduced compared to a standard fiber. For the HC-1550-02 HCF, the predicted value for S is consequently smaller (1.4 ppm/°C) than for a solid-core fiber such as the SMF-28 fiber (8.2 ppm/°C). These conclusions were quantitatively confirmed experimentally.[35]

The thermal drift due to the Shupe effect in a FOG is proportional to n^2S, where n is the mode effective index.[32] Because n is smaller in the HC-1550-02 fiber than in a solid-core silica fiber (~1 versus ~1.44), the thermal sensitivity should be ~7.6 times lower in a FOG made with this HCF than in a conventional FOG. This prediction was verified experimentally by applying a known thermal gradient to one side of the 235 m quadrupolar-wound coil of HC-1550-02 fiber and recording the resulting change in the FOG output signal.[32] This FOG had a similar configuration as the laser-driven FOGs reported in previous sections, i.e., an MIOC was used instead of all-fiber components, and the sensor was interrogated with an Er-doped SFS. The time-dependent output signal of the FOG during the application of the thermal gradient was found to be proportional to the time derivative of the applied temperature gradient, as expected from theory. The thermal sensitivity, expressed in terms of rotation error (in deg/s) per unit of applied temperature gradient (in °C/s), extracted from these measurements was 2.9×10^{-4} deg/°C. The same measurement repeated after substituting the HCF coil with a coil of SMF-28 fiber of similar length (300 m) yielded a much higher slope of 2.4×10^{-3} deg/°C. The thermal sensitivity of the hollow-core FOG is clearly much lower than that of a conventional FOG of comparable length. After normalizing this second measured thermal sensitivity value to what it would be if the conventional fiber had the same length as the HCF, the thermal sensitivity of the HCF FOG was 6.5 times lower than that of the conventional FOG, in good agreement with the value of 7.6 obtained from independent measurements of the Shupe constant of the two fibers. These measurements indicate that the Shupe effect is significantly reduced in an HCF FOG. Even lower Shupe constant are predicted with straightforward HCF design modifications.[35]

9.6.3 Faraday-induced drift

The Verdet constant of the HC-1550-02 HCF around 1.5 μm was measured using a polarimetry setup that determined the amount of polarization rotation induced in the fiber's fundamental mode when the fiber was subjected to a longitudinal magnetic field of known magnitude.[33] The Verdet constant was found to be 6.1 ± 0.3 mrad/T/m. A numerical model of this Verdet constant was also developed. This model calculated the spatial overlap of the mode with the fiber's silica membranes and air holes, taking into account the very different Verdet constants of these two materials. The prediction was in broad agreement with the measured value, confirming that in this fiber the Verdet constant is dominated by the contribution of the silica membranes. This measured Verdet constant was found to be 90 times weaker than the Verdet constant of an SMF-28 fiber measured by the same method (0.55 ± 0.01 rad/T/m).[33] The drift resulting from exposure to the earth's magnetic field is therefore expected to be greatly reduced in a FOG utilizing a hollow-core fiber in its sensing coil, thus reducing the amount of mu-metal shielding, or possibly eliminating it completely.

9.6.4 Noise and drift performance of HCF FOGs

To evaluate the noise performance of this experimental HCF FOG, the two MIOC pigtails were terminated at an angle to eliminate backreflections, then butt-coupled to the two ends of the HCF coil.[34] Each of the HCF fiber ends was cleaved at normal incidence, which produced very little backreflection since the mode effective index is very close to the index of the surrounding air. A fiber polarization controller placed on a solid-core fiber segment of the Sagnac loop controlled the polarization inside the coil (the HCF did not maintain polarization) and maximized the return signal transmitted by the polarizer. The modulator was operated at the loop proper frequency (638 kHz). The measured ARW was found to decrease monotonically with increasing source power, and to asymptotically reach a minimum independent of power at high power (~10 μW and above). This behavior was quantitatively consistent with a noise dominated by detector noise at low power, and SFS excess noise at higher power, as in a conventional SFS-driven FOG. The minimum ARW was 0.013 deg/$\sqrt{}$h, a value limited by the source excess noise. The drift, measured during a 60-h run, was 0.94 deg/h.

9.7 Conclusions

The use of a sufficiently broad laser (multiple-GHz linewidths) to interrogate a FOG is an effective way to mitigate the two main sources of coherent errors (noise and bias drift) in a laser-driven FOG, namely backscattering and polarization coupling, while greatly improving the scale-factor stability. The measured dependence of the noise and drift on laser linewidth in an experimental laser-driven FOG utilizing a 1085-m coil of PM fiber is in excellent agreement with the theoretical predictions of these two contributions. The main observations are (1) the noise is dominated by backscattering at all linewidths, (2) the drift is dominated by backscattering below ~1 MHz and polarization coupling above ~1 MHz, and (3) the polarization-coupling drift decreases for linewidths above ~100 MHz. By broadening a laser from its original linewidth of ~10 MHz to ~2.8 GHz using an external electro-optic modulator driven with a pseudo-random bit sequence, a FOG with a noise of only 0.00073 deg/$\sqrt{}$h and a drift of 0.03 deg/h is demonstrated, the lowest values reported for a FOG interrogated with coherent light. This noise value is lower than the requirement for inertial navigation of an aircraft by ~30%, and the drift is only a factor of 3 above the requirement. When the conventional solid-core fiber in the sensing coil is replaced by a hollow-core fiber (HC-1550-02 fiber), measurements show that the thermal drift is reduced by a factor of ~6.5, the Kerr-induced drift by a factor of up to ~170, and the Faraday-induced drift by a factor of 90. These benefits all result from the fact that the HCF mode is largely guided in air, in which each of these three effects is reduced compared to in silica. A FOG utilizing a 235-m length of this HCF and probed with broadband light has a measured noise of 0.013 deg/$\sqrt{\text{h}}$ limited by the source excess noise, and a drift of 0.94 deg/h. Improved performance is expected with better optical connections within the FOG, as well as reduced loss and thus backscattering in the fiber.

References

[1] G. A. Pavlath, "Fiber optic gyros: The vision realized," in *Proc. SPIE*, Aug. 2006, vol. **6314**, Photorefractive Fiber and Crystal Devices: Materials, Optical Properties, and Applications XII, 63140G.

[2] H. Lefèvre, [The Fiber Optic Gyroscope], 2nd Edition, Artech House, Boston & London, **26**, 95-116, 25, 38–47 (2014).

[3] P. R. Morkel, R. I. Laming, and D. N. Payne, "Noise characteristics of high-power doped-fibre superluminescent sources," *Electron. Lett.* **26**(2), 96–98 (1990).

[4] H. G. Park, M. J. F. Digonnet, and G. S. Kino, "Er-doped superfluorescent fiber source with a ±0.5 ppm long-term mean-wavelength stability," *J. of Lightwave Technol.* **21**(12), 3427–3433 (2003).

[5] H. Joachin, J. Bonnefois, E. Ducloux, and H. Lefèvre, "Interferometric filtering of the excess relative intensity noise of the broadband source of a fiber optic gyroscope," *Proc. SPIE 9157, 23rd International Conference on Optical Fibre Sensors*, 91572D (June 2, 2014).

[6] S. Lloyd, S. Fan, and M. J. F. Digonnet, "Experimental observation of low noise and low drift in a laser-driven fiber optic gyroscope," *J. of Lightwave Technol.* **31**(13), 2079–2085 (2013).

[7] K. Petermann, "Intensity-dependent nonreciprocal phase shift in fiber-optic gyroscopes for light sources with low coherence," *Opt. Lett.* **7**(12), 623–625 (1982).

[8] C. C. Cutler, S. A. Newton, and H. J. Shaw, "Limitation of rotation sensing by scattering," *Opt. Lett.* **5**(11), 488–490 (1980).

[9] J. Mackintosh and B. Culshaw, "Analysis and observation of coupling ratio dependence of Rayleigh backscattering noise in a fiber optic gyroscope," *J. of Lightwave Technol.* **7**(9), 1323–1328 (1989).

[10] W. Burns and R. Moeller, "Polarizer requirements for fiber gyroscopes with high-birefringence fiber and broad-band sources," *J. of Lightwave Technol.* **2**(4), 430–435 (1984).

[11] K. Z. Aghaie, M. J. F. Digonnet, and S. Fan, "Modeling loss and backscattering in a photonic-bandgap fiber using strong perturbation," *Proc. SPIE 8632, Photonic and Phononic Properties of Engineered Nanostructures III*, 86320K (February 21, 2013).

[12] S. W. Lloyd, M. J. F. Digonnet, and S. Fan, "Modeling coherent backscattering errors in fiber optic gyroscopes for sources of arbitrary line width," *J. of Lightwave Technol.* **31**(13), 2070–2078 (2013).
[13] J. Chamoun, [A laser-driven fiber optic gyroscope for inertial navigation of aircraft], Doctoral dissertation, 2017, Stanford University, Ch 4.
[14] J. N. Chamoun and M. J. F. Digonnet, "Noise and bias error due to polarization coupling in a fiber optic gyroscope," *J. of Lightwave Technol.* **33**(13), 2839–2847 (2015).
[15] "IEEE standard specification format guide and test procedure for single-axis interferometric fiber optic gyros," *IEEE Standard* 952–1997 (1998).
[16] T. Buret, D. Ramecourt, and F. Napolitano, "From space qualified fiber optic gyroscope to generic fiber optic solutions available for space applications," *International Conf. on Space Optics*, 14–17 (2008).
[17] S. L. A. Carrara, B. Y. Kim, and H. J. Shaw, "Bias drift reduction in polarization-maintaining fiber gyroscope," *Opt. Lett.* **12**(3), 214–216 (1987).
[18] W. Burns, "Phase error bounds of fiber gyro with polarization-holding fiber," *J. of Lightwave Technol.* **4**(1), 8–14 (1986).
[19] R. Bergh, B. Culshaw, C. Cutler, H. Lefevre, and H. Shaw, "Source statistics and the Kerr effect in fiber-optic gyroscopes," *Opt. Lett.* **7**(11), 563–565 (1982).
[20] G. P. Agrawal, [Semiconductor Lasers], Van Nostrand Reinhold, New York, 269–275 (1993).
[21] N. J. Frigo, H. F. Taylor, L. Goldberg, J. F. Weller, and S. C. Rashleigh, "Optical Kerr effect in fiber gyroscopes: effects of nonmonochromatic sources," *Opt. Lett.* **8**(2), 119–121, (1983).
[22] A. E. Kaplan and P. Meystre, "Large enhancement of the Sagnac effect in a nonlinear ring resonator and related effects," *Fiber-Optic Rotation Sensors and Related Technologies: Proceedings of the First International Conference MIT, Cambridge, Mass., USA*, (1981).
[23] S. Blin, M. J. F. Digonnet, and G. S. Kino, "Fiber-optic gyroscope operated with a frequency-modulated laser," *19th International Conference on Optical Fibre Sensors*, 7004 (2008).
[24] I. S. Kim, P. Tantaswadi, and J. Blake, "Coherence-collapsed 1.3-μm multimode laser diode for the fiber-optic gyroscope," *Opt. Lett.* **20**(7), 731–733 (1995).
[25] T. Okoshi, K. Kikuchi, and A. Nakayama, "Novel method for high resolution measurement of laser output spectrum," *Electron. Lett.* **16**(16), 630–631 (1980).
[26] T. Komljenovic, A. T. Minh, M. Belt, S. Gundavarapu, D. J. Blumenthal, and J. E. Bowers, "Frequency modulated lasers for interferometric optical gyroscopes," *Opt. Lett.* **41**(8), 1773–1776 (2016).
[27] K. P. Ho and J. M. Kahn, "Spectrum of externally modulated optical signals," *J. of Lightwave Technol.* **22**(2), 658–663 (2004).

[28] J. N. Chamoun and M. J. F. Digonnet, "Pseudo-random-bit-sequence phase modulation for reduced errors in a fiber optic gyroscope," *Opt. Lett.* **41**(24), 5664–5667 (2016).
[29] S. Gundavarapu, T. Komljenovic, M. A. Tran, M. Belt, J. E. Bowers, and D. J. Blumenthal, "Effect of direct PRBS modulation on laser driven fiber optic gyroscope," *4th IEEE International Symposium on Inertial Sensors and Systems, Kauai, Hawaii* (2017).
[30] B. Anderson, C. Robin, A. Flores, and I. Dajani, "Experimental study of SBS suppression via white noise phase modulation," *Proc. SPIE*, 2014, vol. **8691**, Fiber Lasers XI: Technology, Systems, and Applications, 86911W-1–7.
[31] J. N. Chamoun and M. J. F. Digonnet, "Aircraft-navigation-grade laser-driven FOG with Gaussian-noise phase modulation," *Opt. Lett.* **42**(8), 1600–1603 (2017).
[32] S. Blin, H. K. Kim, M. J. F. Digonnet, and G. S. Kino, "Reduced thermal sensitivity of a fiber-optic gyroscope using an air-core photonic-bandgap fiber," *J. of Lightwave Technol.* **25**(3), 861–865 (2007).
[33] H. Wen, M. A. Terrel, H. K. Kim, M. J. F. Digonnet, S. Fan, and G. S. Kino, "Measurements of the birefringence and Verdet constant in an air-core fiber," *J. of Lightwave Technol.* **27**(15), 3194–3201 (2009).
[34] V. Dangui, M. J. F. Digonnet, and G. S. Kino, "Laser-driven photonic-bandgap fiber optic gyroscope with negligible Kerr-induced drift," *Opt. Lett.* **34**(7), 875–877 (2009).
[35] V. Dangui, H. K. Kim, M. J. F. Digonnet, and G. S. Kino, "Phase sensitivity to temperature of the fundamental mode in air-guiding photonic-bandgap fibers," *Opt. Express* **13**(18), 6669–6684 (2005).

Chapter 10
Optical Fibers for Fiber Optic Gyroscopes

Chris Emslie
Fibercore

10.1 Introduction

With the possible exception of nonsilica glasses, each time a new fiber technology has emerged, someone has applied it to FOGs and reported beneficial results.

The development of the FOG has essentially tracked that of commercial optical fibers for more than 40 years, running parallel since 1976, when Vali and Shorthill[1] first demonstrated the concepts originally proposed by Pircher and Hepner[2] in early 1967. The nine-year gap between first proposal and experimental demonstration was undoubtedly caused because, in 1967 practical, single-mode fibers simply did not exist, and even by 1976 they were in their infancy and confined to a handful of research laboratories around the globe. FOG technology received a further boost at the beginning of the 1980s when stress-birefringent polarization-maintaining (PM) fibers were developed almost simultaneously by multiple commercial and academic institutions, including NTT,[3] Corning,[4] Hitachi,[5] Bell Laboratories,[6] and the University of Southampton.[7] It was the advent of PM fibers that paved the way for commercial FOG development by delivering practical immunity to environmental variations such as vibration and temperature. To this day, precision-wound coils of PM fiber dominate the FOG industry. Nevertheless, with FOG technology challenging the ring laser gyro (RLG)[8,9] in navigation-grade sensors and even making in-roads into strategic levels of precision, virtually every new development in fiber technology has been investigated extensively to discover if theoretical performance advantages in FOGs may be realized.

The development of photonic bandgap fibers in the 1990s[10,11] were quickly seized upon by FOG researchers who recognized that using air as the guiding medium could free FOGs from the fundamental limitations of doped silica, particularly in terms of Rayleigh scattering and the Kerr, Faraday, and Shupe effects. Similarly, when multicore fibers[12] were first proposed for use in differential strain measurement and ultra-high-bandwidth telecommunications[13] around 1980, their potential to reduce the physical dimensions of the sensor coil and enhance its performance was eventually realized[14] (although to be fair, multicore technology did not really begin to gather mainstream momentum until the mid-2000s). Neither of these two examples of the use of "new" fiber technology has entered the mainstream because, whilst their potential beneficial impact has been confirmed through experimentation, they remain dogged by practical problems. These problems may doubtless be solved but will require a FOG project of strategic importance in which conventional PM fibers simply cannot do the job to drive the technology forward.

The fiber in a FOG cannot be treated in isolation as a standalone component—the fiber is part of a highly complex and still not fully-understood "composite" structure comprising the "glass," polymer coating(s), potting adhesives, the coil hub/armature/former, and winding scheme. As with the majority of fiber sensors, FOGs work because the guiding properties of an optical fiber enable a very long path-length capable of magnifying what is essentially a relatively weak phenomenon (the Sagnac effect) within a compact package. This attribute is probably best illustrated by comparison–today, it is possible to measure the rotation rate of the earth by using a FOG that would easily fit within a teacup, whereas Michelson and Gale's ground-breaking 1925 experiment[15] required 287,846 m^2 of the Illinois plains to achieve the same result using a conventional, optical circuit comprising straight optical paths deviated by mirrors. The fundamental attribute also introduces a massive degree of randomness from the myriad interactions that may occur within a multi-layer coil that may be several km in length. This fundamental randomness and complexity is the prime reason why the area is yet to be analyzed fully and is still beset with misunderstanding and folklore. The ability to wind effective coils in commercial quantities is probably the most significant technical barrier to entry into the FOG market.

In FOGs, "specialty" or application-specific fiber is used not just in the sensor coil but also in ASE sources, polarizers, depolarizers, and fused splitters/couplers. As with arguably every technological discovery, the parallel development of system-infrastructure components is essential and in reality will determine whether the technology succeeds or fails. So whilst these component fibers may be used in far smaller volumes than the coil fibers, they are arguably of equal importance and as such have earned their place in this chapter.

10.2 Coil Fibers

The vast majority of current FOGs use PM fiber because it delivers the optimal combination of simplicity of design, ease of packaging, and stability.

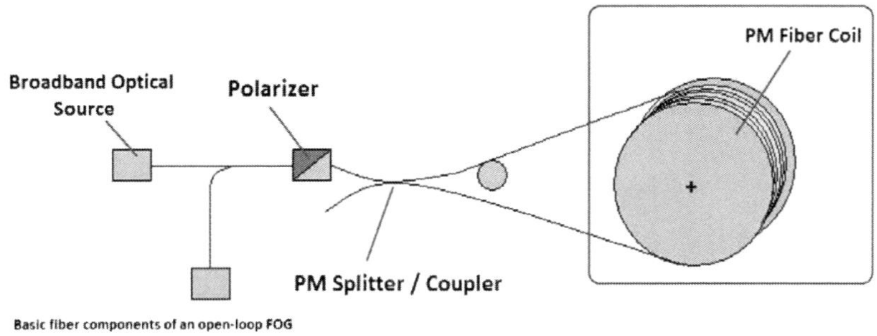

Figure 10.1 Schematic of a single-axis, open-loop FOG identifying the location of the sensor coil.

Nevertheless, depolarized FOGs are still encountered, primarily in highly specialized applications for example, in deep-space vehicles, where very high levels of radiation tolerance are easier to engineer in the simpler construction and chemistry of a more conventional, single-mode fiber.

The more recent advent of micro-structured or bandgap fibers in the 1990s has also generated interest in FOG communities with the prospect of a fiber formed from a single type of glass in which the absence of thermal stress promises temperature stability that no conventional, stress-birefringent PM fiber could deliver. Neither did the advent of multicore fibers in the mid-2000s escape the attention of the FOG-development community, and despite the technical challenges of actually getting light into and out of the coil, the promise of multiplying the path length without increasing the volume of the coil and enhancing mechanical stability continues to intrigue.

Under laboratory conditions, polarization maintenance may be demonstrated in virtually any single-mode fiber—provided that it is kept short enough, straight enough, and isolated from any form of environmental perturbation. Problems tend to occur when it becomes necessary to use the fiber in more practical situations.

The fundamental (TEM_{00}) mode that propagates within a single-mode fiber is actually a degenerate combination of two, orthogonally polarized modes (Fig. 10.2). In a conventional, telecommunications-type fiber, these two components have the same propagation constant, i.e., they travel at the same velocity. This property makes it very easy for optical energy to transfer, or "cross-couple," from one of these modes to the other if it encounters any sort of perturbation within the fiber. These perturbations may be intrinsic, i.e., caused by microscopic geometric variations within the core or residual thermal stress, locked-in by the fiber fabrication process, or extrinsic, i.e., induced by the environment in which the fiber has been deployed. Extrinsic perturbations are usually mechanical-stress-related phenomena caused by micro- and/or macro-bending, typically in combination with the effects of the thermal behavior of the fiber-coating (buffer) material. In

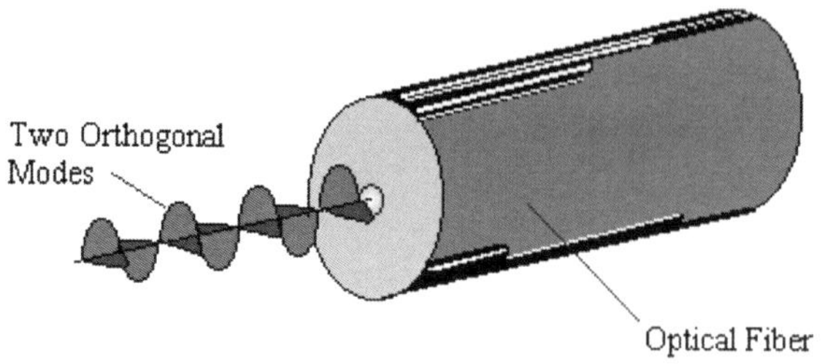

Figure 10.2 The two modes of a single-mode optical fiber.

essence, these are the same phenomena that generate polarization mode dispersion (PMD) for the simple reason that birefringence and PMD are essentially the same thing.[1]

In the real world, where fibers cannot reasonably be protected from environmental stress, temperature fluctuations, etc., it is necessary to use a purpose-designed PM fiber. PM fibers are engineered in such a way that the two orthogonally polarized modes are forced to travel at different velocities, i.e., with different propagation constants. This difference in velocities makes it very difficult for optical energy to cross-couple, with the result that the polarization state of the transmitted light is preserved.

This difference is created through the introduction of anisotropy within the core of the fiber, either geometric (by making the core elliptical) or, more typically, through the application of a controlled, uniaxial stress. These two designs are described as form birefringent and stress birefringent, respectively.

10.2.1 Stress- and form-birefringent fiber types

The vast majority of PM fibers used today have one of the three basic stress-birefringent geometries: bow-tie, PANDA, and elliptical jacket (Fig. 10.2). In all three designs, the core is flanked by areas of high-expansion glass that shrink more than the surrounding silica as the fiber is drawn and then freeze the core in tension. This tension induces birefringence, i.e., it creates two different indices of refraction, a higher index parallel to and a lower index perpendicular to the direction of the applied stress. In essence, the phenomenon is very similar to that which creates visible interference fringes when transparent plastics are stressed, except that in a PM fiber the effect is highly controlled and its magnitude is at least an order of magnitude lower than may be achieved in an organic glass (polymer/plastic).

1. This is not to suggest that modern telecoms fibers with very low PMD values are highly birefringent or polarization maintaining—in fact, they are the very opposite. They are designed with very low values of intrinsic birefringence to make them effectively transparent to the polarization state of the transmitted light. These fibers are still adversely affected by environmentally induced birefringence.

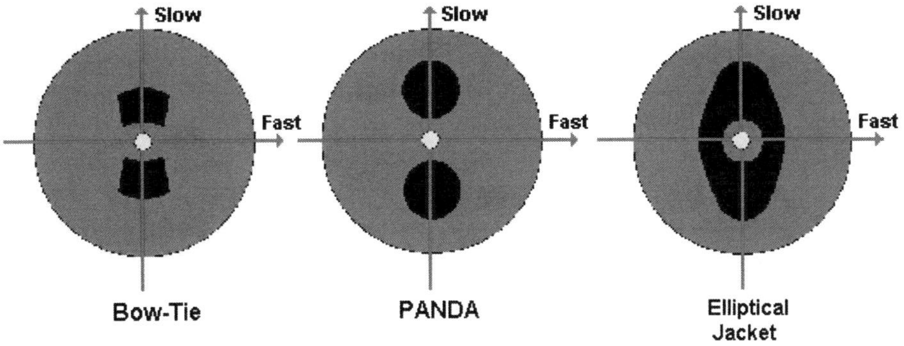

Figure 10.3 Cross-sections of bow-tie, PANDA, and elliptical-jacket geometries.

In any of these designs, when polarized light is launched along the "slow axis," it is forced to travel at a lower velocity than if it had been launched along the "fast axis," and vice versa. The cross-coupling of light from one axis to the other therefore becomes very difficult because it would require a perturbation capable of making a significant change in the velocity of the transmitted light. The greater the applied stress is, the greater the difference in propagation constant (light velocity) between the two axes and the higher the birefringence. The PM ability is enhanced because a larger perturbation is therefore needed to generate cross-coupling from one axis to the other. Note that polarization maintenance is not a loss mechanism, as is sometimes believed—under the vast majority of practical circumstances, there is no measurable difference in attenuation between the fast and slow axes. However, to this day, most applications use the slow axis exclusively. Although the origins of this practice are thought to lie in the theoretically superior resistance to bend-induced loss provided by the slight increase in numerical aperture on this axis, the principal benefit has been to provide a useful degree of standardization.

Each of the three designs is capable of generating sufficient birefringence for even the most demanding of applications, so the precise choice of fiber is typically determined by other criteria, ranging from handling characteristics to history. In telecommunications applications, the fiber of choice is usually the PANDA design, invented by Nippon Telegraph and Telephone in the early 1980s[3] and then developed and commercialized in the US and Europe throughout the 1990s. In essence, PANDA is a telecommunications fiber that has been modified by the insertion of stress rods (usually referred to as stress-applying parts, or SAPs) to provide PM properties. The fiber was conceived to support the large volume of Japanese work in the area of coherent communications that had been driven by that country's unique geography: a series of islands interlinked by stretches of ocean that could be 100 km or more in width. This topography challenged the direct-detection technologies of the day but provided an ideal test case for the improved receiver sensitivity of Cocomms. Fiber attenuation and mode-field diameter (MFD) were well matched to those of single-mode telecommunications fibers so that when Cocomms was overtaken by the EDFA in the late 1980s, it was a natural extension to continue using the fiber in conjunction with the IO modulators that successfully made the transition between the two technologies.

To this day, bow-tie fibers are most typically encountered in sensor applications; indeed, the majority of FOGs worldwide use a fiber of this design. It should therefore come as no surprise that the fiber was conceived in the early 1980s as a sensor fiber, developed by the University of Southampton Optical Fiber Group[7] in support of the FOG program at British Aerospace. Without the constraints of a telecommunications fiber design, the bow-tie was introduced with a high numerical aperture (NA) to provide increased resistance to the bend-induced loss that could arise in small-diameter sensor coils, and the elliptical core generated by its fabrication process was accepted readily for sensor use. In extremis, the bow-tie design can be shown capable of creating more birefringence than any other stressed design[16,17] for the simple reason that it is based on two opposing wedges, the simplest and most efficient means of applying stress to a point. However, in all but the most exotic of applications, the fundamental design is implemented in a sub-optimum condition in the interest of manufacturing yield and consequent cost. It is interesting to note that although re-designed variants of both PANDA and bow-tie have subsequently been introduced to address the Sensor and Telecommunications markets, respectively, more than 20 years later, both designs continue to dominate the applications that they were originally conceived for. The bow-tie fiber has been developed and commercialized by Fibercore Limited, a spin-out company from the University of Southampton Optical Fiber Group since the fiber's invention in 1983.

The elliptical-jacket fiber also had its origins in Japan, but this time with Hitachi. In common with PANDA, the original interest was in a fiber suitable for coherent communications use. However, the fiber was manufactured in the US initially by a company called EoTech that was subsequently acquired by 3M, who are probably best-known for the commercialization of this product as a sensor fiber throughout the USA and Europe. Whilst the elliptical jacket fiber is capable of similar levels of birefringence to the PANDA and bow-tie designs, this performance is achieved at the expense of handling characteristics. As can be seen from the above figure, this fiber is unique in that the SAP extends all around the core, generating a significant amount of parasitic stress that serves to reduce the fiber's birefringence. The oversized stress-member necessitated by this characteristic can compromise performance in the reduced-diameter fibers used in many sensor applications, for the simple reason that there can be insufficient room to locate a SAP of the necessary dimensions. Furthermore, levels of parasitic stress can be considerable, leading to uneven fracture when cleaved and reducing fusion splice yields.

10.2.1.1 Elliptical-core form-birefringent fiber

In conventional, single-mode fibers, core ellipticity is undesirable for the simple reason that it creates birefringence (i.e., PMD) that reduces performance in high-data-rate systems. In a form-birefringent fiber (Fig. 10.4), this effect is taken to the extreme.

Form-birefringent fibers, with their simple design of a highly elliptical core, combined with a very high NA and small mode, predate stress-birefringent fibers and may be traced back to Hitachi in the late 1970s. However, the high levels of attenuation and lack of compatibility, created by their highly-germania-doped cores and consequent small modes, made them unsuitable for telecommunications use. However, they did find some application in sensors as the "E-core" fiber,[18] manufactured by Andrew Antenna Corporation and sold throughout the 1980s, and the original Hitachi designs were developed by Corning as its PMF38 fiber throughout the latter part of the 1990s. However, despite efforts to demonstrate how the basic incompatibility brought about by the small mode and high loss could be overcome (and with very attractive pricing), the product failed to gain market acceptance. Today, form-birefringent fibers, tracing their lineage back to the original Andrew E-core, are used in the KVH range of FOGs.

Like the manufacturing process of more conventional fibers, the process for PM fibers comprises two parts: preform fabrication and fiber drawing. The fiber-drawing process is practically identical for both PM and non-PM fiber types. For this reason, only the preform-fabrication stage is addressed in this section. A more comprehensive treatment of conventional fiber-fabrication processes may be found in any one of a number of standard texts[19] and several websites.

10.2.1.2 Bow-tie fibers

Bow-tie preforms are fabricated on a lathe using inside vapor phase oxidation (IVPO) with stress members created by the process of gas-phase etching.[7] First

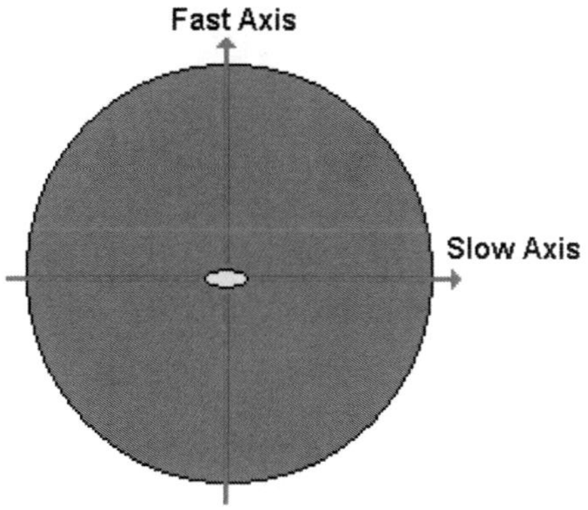

Figure 10.4 The cross-section of an elliptical-core, form-birefringent fiber.

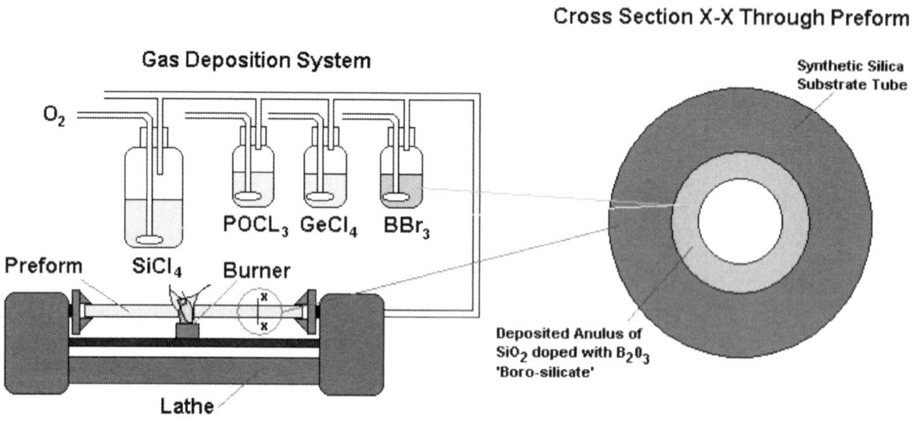

Figure 10.5 Deposition of the boron-doped ring.

of all, a ring of boron-doped silica (effectively boric oxide (B_2O_3) mixed with silica (SiO_2)) is deposited within a high-purity, synthetic-silica-substrate tube by the oxidation of boron tribromide (BBr_3) in combination with silicon tetrachloride ($SiCl_4$).

When a sufficiently thick layer has been created, the rotation of the lathe is stopped to allow two diametrically opposed sections to be etched away. The material is etched by passing a suitable etchant (typically sulphur hexafluoride (SF_6)) through the center of the tube and activating it by means of a narrow-zone etching burner (Fig. 10.6). The final shape of the bow-tie SAPs may be

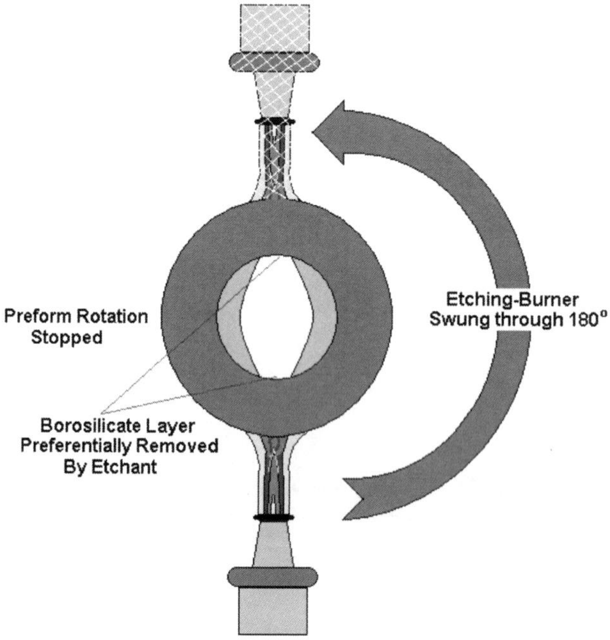

Figure 10.6 Gas-phase etching.

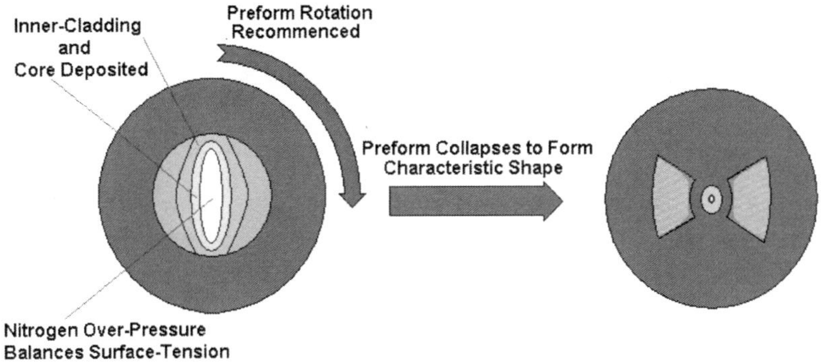

Figure 10.7 Core deposition and controlled collapse.

controlled and the stress levels optimized by varying the arc through which the etching burner is rotated.

After the completion of the etching stage, rotation is recommenced, and the inner cladding and core layers are deposited in sequence. The inner cladding is typically fused silica containing a suitable viscosity modifier (usually phosphorus pentoxide (P_2O_5), synthesized by the oxidation of phosphoryl chloride ($POCl_3$) and index-matched using fluorine). The core is germanosilicate (a mixture of germania (GeO_2) and SiO_2), created by the oxidation of a combination of $SiCl_4$ and $GeCl_4$.

Preform fabrication is completed with a controlled collapse process (Fig. 10.7). Overpressurized dry nitrogen is introduced to balance the surface-tension forces that gradually overpower the viscosity of the material as the process temperature is increased to almost 2,000°C. The characteristic bow-tie shape is created as the central hole is collapsed to form a solid, cylindrical preform. When this preform is drawn, surface-tension forces also ensure that the geometry of the preform is faithfully reproduced in the fiber.

10.2.1.3 PANDA fiber

As befits its origins as a modified telecommunications fiber, the starting point for a PANDA fiber is a circularly symmetric, "telecom-type" preform into which two diametrically opposed holes have been drilled ultrasonically. The PANDA preform is then completed by the insertion of a boron-doped stress rod into each hole (Fig. 10.8). These stress rods may be fabricated either by vapor deposition or by a sol-gel process. When drawn, the low melt viscosity of these stress rods relative to the surrounding silica ensures that the boron-doped material entirely fills the holes to form the characteristic PANDA shape.

Whilst the fabrication process for PANDA may appear relatively straightforward, the drawing process must be well controlled in order to prevent distortion of both the stress rods and the fiber itself and to achieve the exceptional geometric precision for which the design is renowned. Similarly,

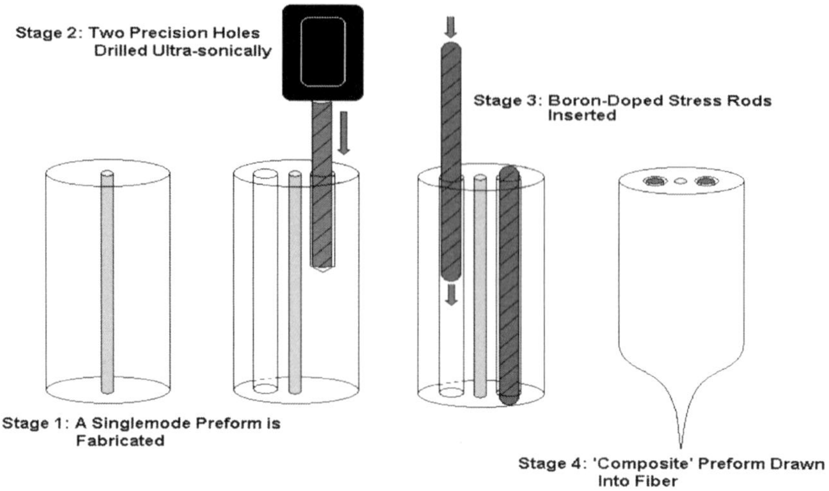

Figure 10.8 The fabrication of a PANDA preform by the machining and insertion of stress rods.

despite various proposals for alternative PANDA fabrication methods, the tight machining tolerances and tool stability necessary to drill side-holes of significant depth have continued to make the manufacture of very-high-yielding preforms technically challenging.

10.2.1.4 Elliptical-jacket fiber

The fabrication method for elliptical-jacket preforms combines processes for both bow-tie and PANDA manufacture: boric oxide deposition by MCVD and ultrasonic machining. The initial fabrication steps are identical to those for a bow-tie preform, except that the gas-phase etching is omitted, leaving an unbroken, circular annulus of boron-doped material surrounding the core.

After this circularly symmetric preform has been completed, its symmetry is broken by the machining of two flats, one on each side of the core (Fig. 10.9). Whilst it is possible to use conventional mechanical grinding techniques, the associated force and vibration would make the highly stressed preform vulnerable to fracture. For this reason, a more benign, ultrasonic machining process is typically used. When drawn, the high-surface-tension forces within the molten glass force the preform to circularize, thereby creating the characteristic elliptical-jacket shape (Fig. 10.10).

10.2.1.5 Elliptical-core, form-birefringent fiber

Some degree of ellipticity may be induced in the core of any single-mode fiber if process controls are inadequate. For example, in MCVD, if the surface tension forces within the collapsing preform are not accurately balanced by pressurizing the substrate during the final stages of the fabrication process, then the preform will "pull flat" to some degree and create an elliptical core.

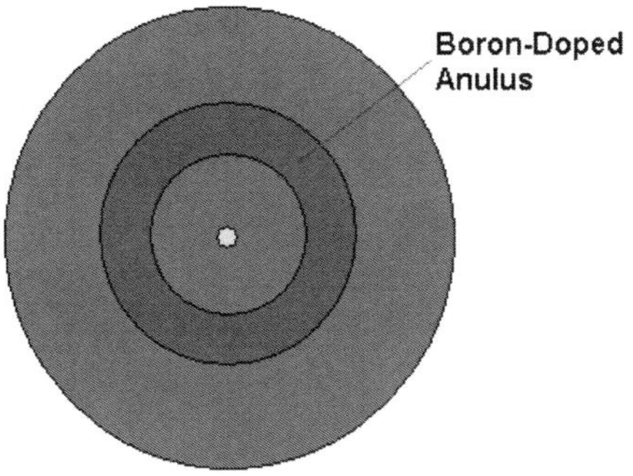

Figure 10.9 Stage 1 of elliptical jacket fabrication: MCVD preform with a boron-doped ring.

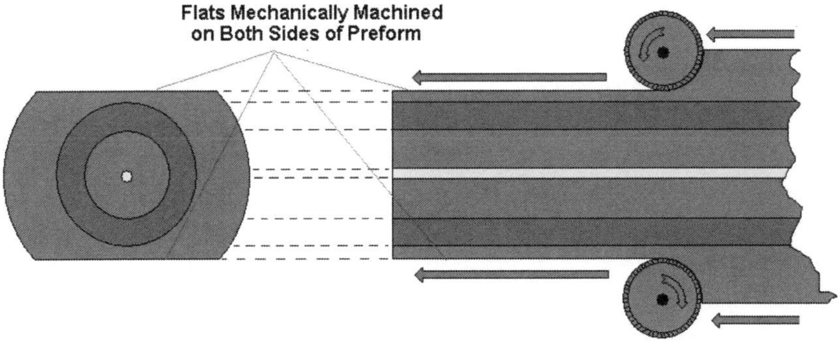

Figure 10.10 Stage 2 of elliptical-jacket fabrication: preform machining.

In order to fabricate a core that is deliberately elliptical, it is necessary to take this phenomenon to an extreme by removing the circularizing effect of the rotation of the lathe, i.e., depositing the core (by MCVD/IVPO) within a stationary substrate tube, and then collapsing it under vacuum to create the characteristic high-ratio ellipse. The resulting cladding ellipticity is low due to the very small amount of material movement actually taking place relative to the total volume of the preform, and any residual cladding ellipticity can either be removed by including an additional, high-temperature "rounding" step during preform fabrication or even during the fiber-drawing process itself.

10.2.2 Microstructures in hollow-core, photonic bandgap fibers

Hollow-core, photonic bandgap fibers (HC-PBGFs) comprise a central, hollow core surrounded by a periodic microstructure of air-filled silica cells of wavelength dimensions (Fig. 10.11). They propagate light in a way that is very different when compared with that of conventional, index-guiding fibers. The microstructure of silica cells creates a photonic bandgap that prevents light from propagating into this cladding region, thereby allowing tight optical

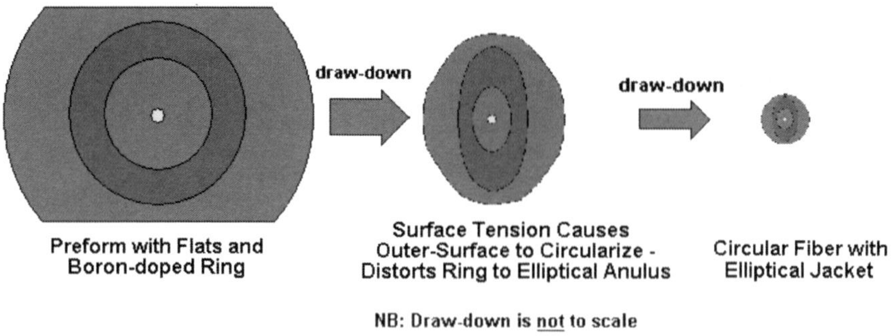

Figure 10.11 Stage 3 of elliptical-jacket fabrication: the flattened preform circularizes during fiber-drawing.

confinement and thus guidance within the air-core. As such, the guiding medium in a HC-PBGF is predominantly air, with only fractional penetration of the optical field into silica and consequent guidance.

As the development of FOG technology has addressed progressively higher grades of precision, the fundamental properties of silica that limit the performance of conventional fibers in FOGs have come under increasing scrutiny. In particular, Rayleigh scattering—together with the Shupe, Kerr, and Faraday effects—is between one and three orders of magnitude lower in air than in silica. The use of fibers of this type in FOGs would therefore be expected to reduce extraneous drift and noise associated with these phenomena significantly. Furthermore, if backscatter within an HC-PBGF could be made to approach theoretical limits, it should be possible to drive the FOG using a laser rather than a broadband ASE source, leading to a further reduction to noise and greater SF stability.

Practical experiments performed at the Ginzton Laboratory (Stanford University) have confirmed many of the anticipated theoretical advantages of bandgap fibers in FOGs, although not to the predicted degree.[20,21,22] Specifically, 170-fold, 6.5-fold, and 20-fold reductions in sensitivity to the Kerr effect, Shupe/temperature transients, and the Faraday effect, respectively, have been demonstrated. Nevertheless, the use of these fibers in FOGs remains dogged by three aspects: high attenuation, excessive scattering, and the fundamental lack of compatible, ancillary components that has restricted the impact of all "pre-commercial" optical fiber technologies. As will be illustrated in the following section, the effective fabrication of HB-PBGFs requires geometric perfection over many kilometers of fiber. Any perturbation to the shape or sequence of the cells or stress defect creates a scattering point that when summed over any significant distance gives rise to high levels of attenuation. Whereas in theory bandgap fibers could be effectively lossless, the current state-of-the-art lies in the 10–20-dB/km range. At this level, the path length that may be created within a FOG, and therefore the level of precision that may be achieved, is severely restricted. Furthermore as the root cause of this excessive loss is scattering, anticipated reductions to noise and improvements in SF stability may not currently be realized. The construction of practical FOGs using bandgap fibers that could demonstrate performance outside of a controlled laboratory environment is hindered for the simple reason that standard components (splitters, polarizers, pigtailed sources, etc.) that are fully compatible with these fibers are not yet available. Based on direct experience of the evolution of more conventional, specialty fiber technologies, it is likely to be difficult for optical bandgap fiber to progress to a level at which it becomes viable in FOGs unless its manifold attributes become indispensable in application areas that are better funded than sensors are currently—most likely telecommunications, but potentially also consumer products or very-high-power laser delivery.

10.2.2.1 Bandgap fiber fabrication

HC-PBGF is manufactured using a variation of a technique often described as 'Stack-and-draw' for which the process-flow is illustrated in Fig. 10.12. First of all, the primary preform is created by assembling several hundred high purity, low-hydroxyl silica capillaries into an array of the required geometry. This array is bound by a larger, silica tube and may also employ another tube to surround the area that will form the air core (although better results have been achieved by relying on the coalescence of the cells in the draw to for the air-core/microstructure cladding boundary.[23] The primary preform is then given to the draw tower (or specialized "cane-puller"), fused, and drawn down into a cane that is typically 1–3 mm in diameter. This cane is then inserted into

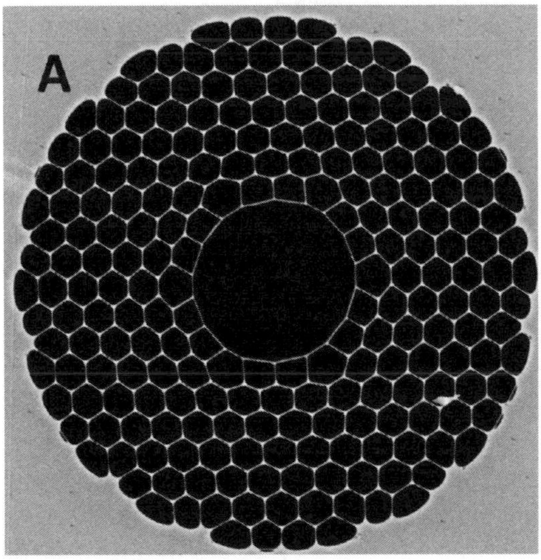

Figure 10.12 Scanning electron micrograph of a typical bandgap fiber. (Figure courtesy of the Southampton Opto-electronics Research Center.)

a synthetic silica sleeving tube and re-drawn using a conventional drawing tower to achieve the final desired dimensions of the fiber.

The creation of a highly consistent, geometrically precise, periodic structure of air holes over kilometric distances is essential to the effective fabrication of a HC-PBGF. For this reason, the drawing process itself is significantly more complex than is typical for more conventional, solid-core fibers. Draw temperatures are kept relatively low, e.g., 50°C below accepted norms, resulting in draw tensions of perhaps twice the usual level and a differential pressure maintained between the large central hole and the matrix of smaller peripheral holes. In this way, the many holes are discouraged from collapsing, and the desired final geometry is preserved.

10.2.3 Multicore fiber

Multicore fibers first started to attract serious attention in the late 1990s when interest in the measurement curvature through differential strain[24] in multicore fibers was first mooted. Interest in the technology has gained pace since around 2010, driven primarily by the telecommunications industry's search for a solution to the fundamental challenge posed by the world's apparently insatiable appetite for bandwidth,[25] combined with the reality of rapidly diminishing installed capacity. However one lesser-known application that remains intriguing is the use of multicore fibers in FOG coils.[14]

The fundamental argument for the use of multicore material in FOGs is three-fold:

1. Multiple optical paths within a single fiber enables a greater path length to be packaged within any given volume.
2. Bringing the fiber cores into closer proximity reduces the temperature gradient across the cores and should therefore enhance FOG performance by reducing the Shupe effect.
3. By effectively encapsulating multiple cores in silica, the volume fraction of (unstable) potting adhesive relative to (very stable) silica is reduced, thereby increasing the overall thermal stability of the coil.

In a multicore FOG, a continuous, helical optical path is generated by winding the coil and then indexing one of the fiber ends by a single core prior to fusion splicing. The same multipolar coil-winding schemes (quadrupole, octupole, etc.) used to mitigate the Shupe effect in conventional coils are equally valid when working with multicore fibers. The basic configuration of a multicore FOG is shown in Fig. 10.13.

Whilst the potential benefits of using multicore fibers in FOGs are clear and potentially material, uptake of the technology has been low to date. Strong interest in high-bandwidth/high-density communications from the telecommunications industry has meant that viable multicore fibers may now be sourced from several companies, but the components necessary to get light into and out

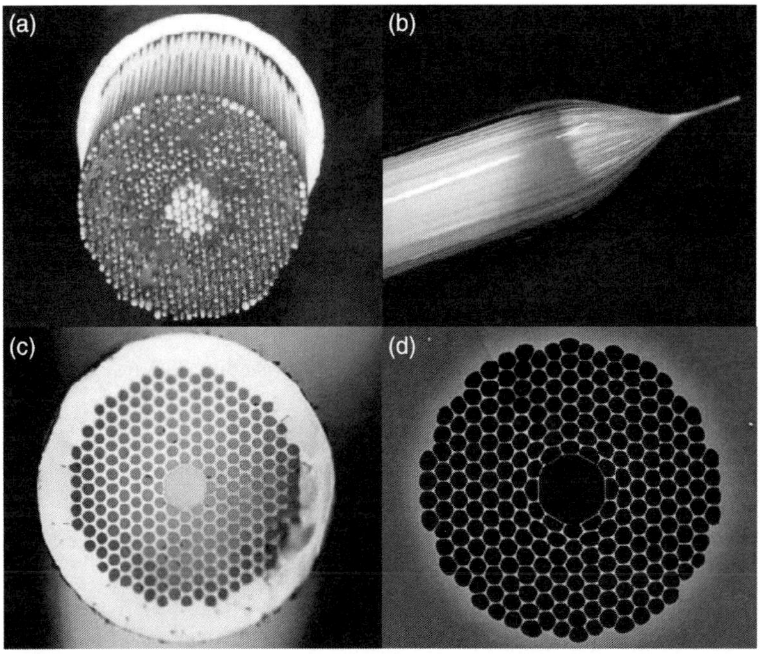

Figure 10.13 HC-PBGF fabrication process flow: (a) primary preform obtained by assembling a few hundred, high purity, low-hydroxyl silica glass capillaries into an array with the required geometry tightly-packed inside a glass tube; (b) first stage draw of the primary preform into a millimeter-sized cane; (c) optical image of the cross-section of the HC-PBGF cane showing elimination of the interstitial volumes and formation of a central core defect; (d) scanning electron microscope image of the microstructure region of the final fiber, showing a much increased diameter to pitch ratio of the cladding holes. (Figure courtesy of the University of Southampton ORC.)

of the coil are not readily available, and the challenges of realizing a practical, multicore FOG are therefore non-trivial. Once again, technologies developed for the telecommunications industry may yet come to the rescue in the form of the stacked and drawn or holographically inscribed "fan-out" components developed for use with multicore fiber in telecoms and datacoms.[26,27]

10.2.3.1 Fabrication

Multicore fibers are fabricated via at least two main methodologies: core insertion and the soot preform method (Fig. 10.14). The starting point for both procedures is a series of conventionally manufactured (MCVD, OVPO, PCVD, etc.) core rods stretched and/or etched to the desired dimensions.

In the case of core insertion, these core rods are then inserted into a substrate comprising a high-purity synthetic silica rod drilled with the required number of holes in the desired locations. This composite preform is then drawn under reduced pressure to ensure that the substrate collapses around the core rods and forms a void-free interface.

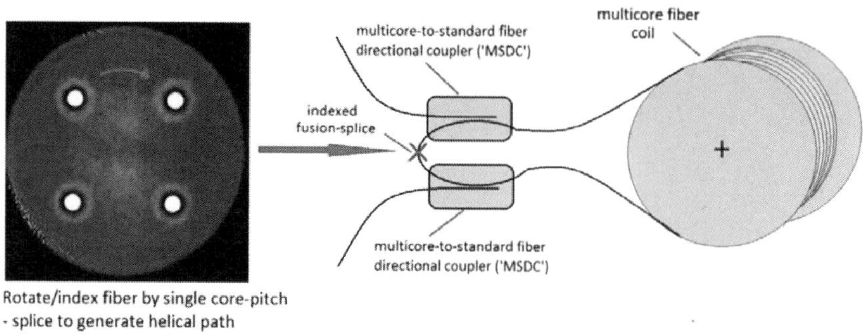

Figure 10.14 Basic configuration of a multicore FOG (four-core).

An alternative method involves mounting the core rods in a jig to locate them within an oversized synthetic-silica-substrate tube in the positions desired in the final fiber. This substrate tube is then filled with high-purity synthetic silica "soot" (micro-particles formed by burning silicon tetrachloride in oxygen) before being consolidated at a sufficiently high temperature to cause the soot particles to coalesce into a solid mass.[28] The aspect of forming a preform from a diffuse aggregate of particles has many parallels with both the outside (OVPO) and axial (VAD) vapor deposition processes used in the manufacture of conventional telecommunications fibers. As such, expedients for both drying and assisting void-free consolidation developed for OVPO and VAD, for example, a chlorine soak followed by helium purging to encourage out-diffusion, may also be used in the fabrication of multicore fibers. Both techniques are illustrated schematically in Fig. 10.14.

Although the soot process may ultimately prove to be more scalable for high-volume production, the relative immaturity of multicore FOG technology is such that neither fabrication method has established itself as optimal. Nevertheless, given the need to create a helical optical path by indexing the cores prior to achieving a low-loss fusion splice (Fig. 10.13), arguably the most important characteristic is geometric precision. A very high degree of precision is necessary over the entire length of the preform in order to minimize the potential for stacking tolerances and consequent progressive misalignment of the cores. Both processes probably offer a current best-case precision of about ±0.5 μm, which is sufficient for demonstrating feasibility, but significant development of either process is likely to be necessary if multicore FOG technology is ever to enter the mainstream.

10.3 Coil Fiber Design Considerations

10.3.1 Diameter

At first glance, diameter would seem to be one of the least complex design parameters for an optical fiber. However, in the context of FOGs it feeds into many key performance aspects, including coil form and packing factor, lifetime, handling characteristics in production, and even PM performance itself.

The fundamental purpose served by the fiber in any intrinsic optical fiber sensor is to confine a sufficiently long optical path inside the smallest practical volume. In this way, optical effects that are typically very weak may be magnified sufficiently to be of practical use.

The benefits of reduced-diameter fibers in intrinsic fiber sensors were quickly realized, leading to an almost universal adoption of an 80-μm ("low-profile") standard in the mid-1980s. When a primary coated 80-μm fiber occupies around half the volume of a comparable 125-μm fiber, it enables twice the optical path length to be confined within the same volume—and produces a corresponding improvement in sensitivity. The 50% reduction in mass was also a welcome improvement for those working on sensors intended for space vehicles. The impact of using a reduced-diameter fiber may be seen clearly in Fig. 10.15, which shows two 400-m lengths wound onto identical 15-mm-diameter spools, one with a glass diameter of 80 μm and a coating diameter of 165 μm, and the other with glass and coating diameters of 60 μm and 120 μm, respectively.

Universal interest in small-diameter sensor coils raised awareness of lifetime issues that, frankly, are still not generally (or fully) understood. Anybody who was working in fiber development up until the advent of practical UV-cured acrylate coating materials in the early 1980s will remember fibers that fractured frequently and spontaneously in storage. It is a simple fact that bends induce stress, and stress in a brittle material will

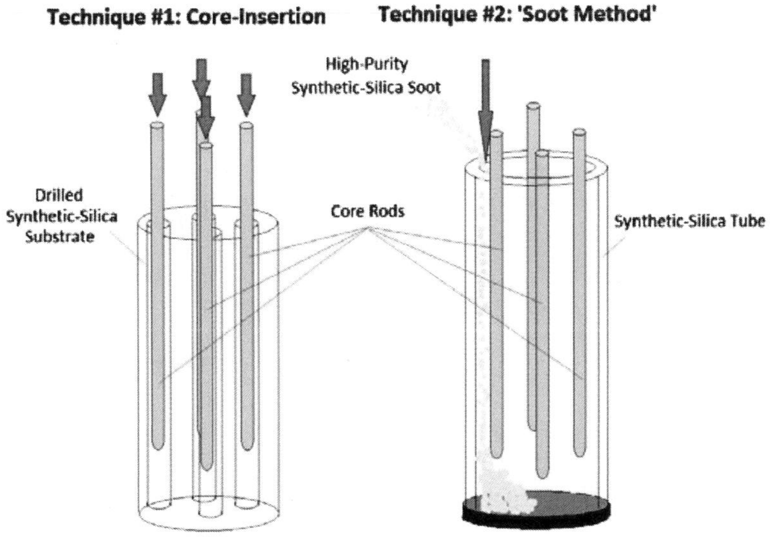

Figure 10.15 Multicore fiber fabrication.

ultimately lead to fracture—the smaller the bend diameter is, the greater the stress and the shorter the lifetime. The storage reels of the early 1980s were 1' / 30 cm in diameter; the potential problem would be far greater for a 1" / 25-mm gyro coil. Fortunately, the 40% reduction in coiling stress brought about by the introduction of 80 μm assisted by improving both fiber-coating and preform-substrate technology enabled reasonable lifetimes to be predicted. Figure 10.16 shows the impact of fiber glass diameter on lifetime.[29] Whilst an 80-μm fiber could be deployed with confidence in a 30-mm-diameter FOG coil intended to operate for 25 years or more, the diameter would need to be increased to 45 mm or more if a more conventional 125-μm fiber were used.

As indicated in Fig. 10.16, further reductions in fiber diameter would enable FOG designers to reap commensurate benefits in terms of reliability. It should not therefore be surprising that interest in ultra-low-profile fibers of 60 μm or even 40 μm has been strong, almost from the very beginning of commercial FOG technology. In 1987, York VSOP (UK) developed a 60-μm PM fiber for the FOG program of a leading, international aerospace company. This first commercial example of an ultra-low-profile fiber was a technical success but a commercial failure due to the lack of compatible components and customer resistance to the handling properties brought about by the fiber's extreme lack of stiffness. Stiffness (or flexibility) is determined by the cross-sectional area: 80-μm and 60-μm fibers have 40% and 23% of the stiffness of a comparable 125-μm fiber, respectively. Many production processes that employ optical fiber rely heavily on fiber stiffness to ensure that they function correctly. In simple terms, if a fiber is too flexible, it will simply not "point" in the intended direction. Similarly, a 60-μm fiber, even coated to 135 μm, can challenge the human senses of sight and touch sufficiently to frustrate attempts to load fixtures as well as demand re-calibration of precision coil-winding equipment. Nevertheless, the true barrier to the wholesale adoption of sub-80-μm fibers in FOGs remains the lack of commercial availability of compatible fiber components. The variation of fiber stiffness with glass diameter is illustrated in Fig. 10.17.

Figure 10.16 Comparison of 15-mm OD spools wound with 80/165 and 60/120 PM fibers.

The reduction in stiffness with diameter also makes ultra-low-profile fibers far more susceptible to the effects of microbending. Microbending manifests itself as an increase in optical attenuation that occurs when a fiber is subjected to partial bends of radii less than about one millimeter. Typically, these conditions occur when the fiber is tensioned over an uneven surface. The mechanism at work is a localized, stress-induced change in the refractive index difference occurring at the core–cladding interface and causing a reduction in the strength of guidance. These effects are additive, so when they occur over a significant length of fiber, the resultant increase in attenuation can be very large. In stress-birefringent PM fibers, microbending can also generate significant reductions in the h-parameter as bend-induced, extrinsic stress works against the fibers' intrinsic stress.

Fibers that have lower stiffness are, by definition, easier to bend (and also to microbend). A 60-μm fiber has little over half the intrinsic resistance to microbending when compared with an 80-μm fiber. Given that ultra-low-profile fibers enable the fabrication of smaller, more densely packed coils, it is clear that their use creates a more aggressive microbend environment that they are intrinsically less able to accommodate. For this reason, the birefringence of ultra-low-profile fibers must be increased substantially if they are to deliver comparable PM performance. The comparison of identical low-tension 200-m coils of 75-mm diameter wound with both 60-μm and 80-μm PM fibers has shown that a ~70% increase in birefringence is required to counteract the reduced microbend resistance of the 60-μm product[30] (Fig. 10.18).

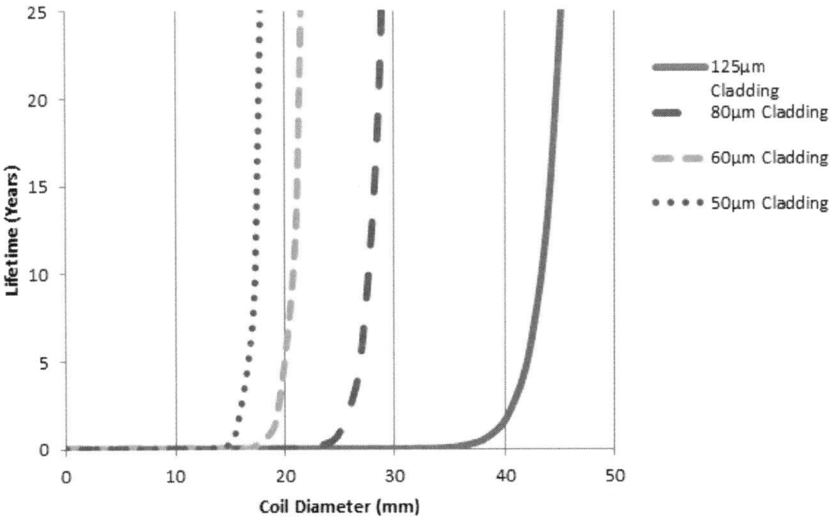

Figure 10.17 Comparison of predicted lifetimes for fibers of different diameter (100 kpsi / 1% strain proof-test).

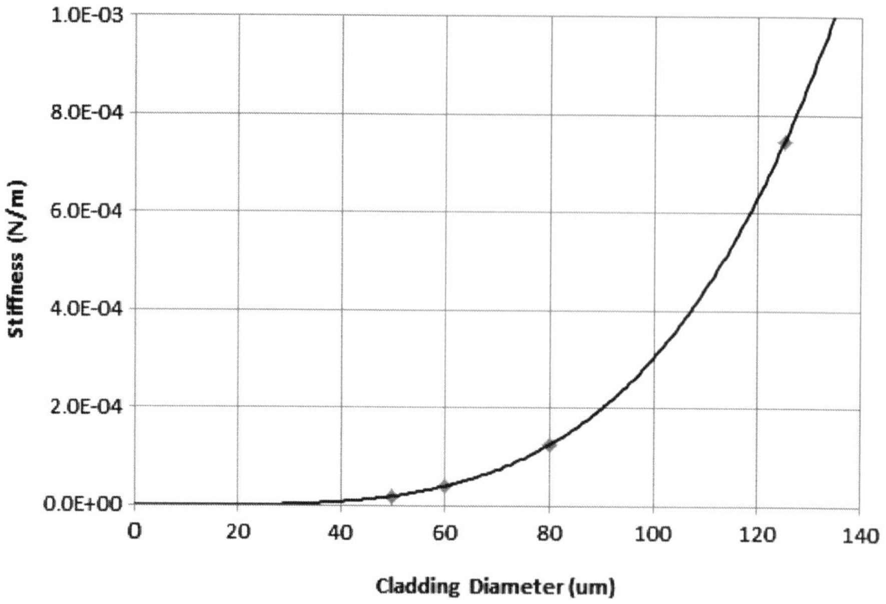

Figure 10.18 Variation of fiber stiffness with cladding (glass) diameter.

10.3.2 Wavelength

The choice of optimum sensor wavelength has taken different directions at different times dictated primarily by source technology beginning with 800–850 nm due to availability of ASE sources either from pioneers such as Superlum or through modification of diode lasers (by removal of the output facet). Shorter wavelengths still favored by European FOG manufacturers – likely combined historic, application, and material-cost related. Early investigations into 1310 nm (notably British Aerospace) to take advantage of telecoms synergies and reduced alignment tolerances enabled by larger core diameters overtaken by now almost universal migration to 1550 nm driven by the development of ASE sources based on EDF (temperature stability) and significant improvement in radiation-resistance. In PM FOGs, the wavelength has a proportionate impact on the length of fiber used in the coil and therefore the overall cost (fiber is typically one of the largest material costs on the BOM).

10.3.3 Attenuation

With typical coil lengths for tactical-grade FOGs in the 100-m (830 nm) to 400-m (1550 nm) range and around 1,000 m for navigation-grade FOGs, exceptionally low losses approaching the 0.2 dB/km found in telecommunications have simply been unnecessary. For many years, attenuation figures of <5 dB/km at 830 nm and <3 dB/km at 1550 nm were accepted as both normal and fit for purpose, particularly in light of the widespread adoption of high-power ASE sources based on erbium-doped fiber at 1550 nm. However, in recent years as the FOG has begun to position itself as a viable competitor to ultra-high-precision RLG technology (whilst beginning to lose out to MEMS in lower-performance tactical grades), more systems have been deployed in marine and submarine navigation as well as in space vehicles, satellites, and artillery positioning, demanding stabilities of fractions of a degree per hour. In order to achieve these levels of precision, coils of more than 5,000 m in length have become far more common. These longer coils naturally work better if the fiber attenuation is lower, e.g., high-precision applications that need attenuation figures substantially lower than 1.0 dB/km.

In conventional, telecommunication-type fibers, losses approaching 0.2 dB/km are routine. However, creating the essential combination of very low loss and very high birefringence presents a unique challenge because the boric oxide used to create the SAPs absorbs very heavily around 1550 nm. If the SAPs are brought closer to the core, as would typically be done to enhance birefringence, boron diffuses into the inner cladding and outer regions of the fiber core, pushing the attenuation up if birefringence of greater than $\sim 6 \times 10^{-4}$ (a beat length of ~1.0 mm at 633 nm) is required. This situation is exacerbated because exceptionally high levels of birefringence are necessary to provide acceptable levels of polarization extinction ratio (PER) due to the significant increase in fiber–fiber interaction and the consequent extrinsic stress in very long coils.

10.3.4 Polarized versus depolarized design

As in any interferometric fiber sensor, the use of PM fiber will guard against signal fade caused by environmental influences, e.g., stress, vibration, and temperature; for this reason, polarized FOG designs have been adopted almost universally in commercial applications. An alternative approach utilizes a fully de-polarized signal.[31] However, any anticipated reduction in fiber cost due to the less complex design of a non-PM fiber has typically been outweighed by the combination of the cost of the de-polarizer, greater length of fiber (relative to PM designs), and the greater complexity of signal processing required. In recent years, the trend towards small form factors with coil diameters of 30 mm or smaller will have removed any commercial advantage entirely by demanding the use of "specialty" bend-insensitive fibers. Nevertheless de-polarized FOGs are still encountered in highly specialized applications, e.g., systems deployed in deep-space probes, where the omission of the structures necessary to preserve polarization greatly simplifies the design of the truly radiation-hardened fiber demanded by that environment.

10.3.5 Birefringence

As in any interferometric fiber sensor, the ability of the fiber to preserve polarization, as indicated by its birefringence value, is of paramount importance. The very long length of sensor coils, which may range from

100 m to more than 5,000 m per axis, increases the degree of random interaction between the fiber layers and therefore increases the importance of this parameter—particularly considering the current trend towards higher precision, demanding long path lengths in increasingly smaller packages. For this reason, typical values of birefringence have been reduced from around 3.2×10^{-4} (beat length around 2.0 mm at 633 nm) to better than 2.5×10^{-4} (beat length < 1.0 mm at 633 nm) in order to cope with these increasing demands. Put simply, birefringence is created by a uniaxial, intrinsic stress introduced during the preform manufacturing process; maximizing this intrinsic stress (and with it birefringence) makes the PM ability more resilient to extrinsic stresses created by the fiber–fiber interactions within the coil.

Although at first glance it would appear that there should be no such thing as a FOG fiber having too much birefringence, in reality there are limits dictated by both optical and mechanical considerations. First of all, if the birefringence is too high (say, beyond $\sim 8 \times 10^{-4}$ or a beat length of 0.8 mm at 633 nm) the fast axis may be driven too close to the cutoff, thereby making the fiber likely to polarize, particularly if either the coil diameter is small or there are small-diameter fiber-management loops within the FOG enclosure. Secondly, the very high levels of intrinsic stress necessary to generate this birefringence are released, explosively when the fiber is cleaved. This stress "break out" causes significant end-face damage, opening up cracks around the SAPs which then "blow out" during the fusion-splicing process as the highly doped SAPs effectively boil. It is interesting to note that FOG architectures may still be encountered in which fusion splices are avoided through the expedient of pigtailing the fiber directly onto the integrated-optic chip. These designs tend to indicate that a FOG was originally designed using the elliptical-jacket fibers prevalent during the 1990s.

10.3.6 Numerical aperture

In order to cope with the small-diameter bends often encountered in sensor coils, many PM fibers are designed with higher numerical apertures than are generally found in telecommunications fibers. Typical NA values of 0.16 or 0.20 are achieved by incorporating higher levels of germania (GeO_2) within the fiber core. As a general rule, an NA of 0.16 would typically enable use in coils of several hundred meters in length down to 35 mm in internal diameter, with 0.20 providing sufficient guidance to tolerate coil diameters of 25 mm or smaller. In comparison, even index-trench-assisted "bend-insensitive" telecommunication-type fibers would be unable to guide over more than a few tens of meters under the same bending conditions.

Another unique aspect of PM fibers is that birefringence creates a lower and a higher value of NA for the fast and slow axes, respectively. In normal PM fiber designs, this difference is perhaps 2.5%, with the result that there is no practical difference between the two axes in terms of resistance to bend-induced loss.

However, if taken to extremes, e.g., in a polarizing fiber, a bending condition may be found for which the slow axis continues to guide yet the fast axis does not. In this case, the guidance of the slow axis is also assisted, to a small extent, by the presence of the low-refractive-index regions formed by the SAPs.

Fast and slow axes are clearly visible in the refractive-index profiles shown in Fig. 10.19, identified as sections through the relevant axes. The presence of the SAPs is indicated by the low-index regions that flank the core in the slow axis view. Numerical apertures for PM fibers are typically determined directly from the fiber refractive index profile using the equation

$$NA = \sqrt{(n_1^2 - n_2^2)} \approx \sqrt{2n_2 \delta n},$$

where n_1 is the core refractive index, n_2 is the cladding refractive index, and δn is the refractive-index difference between core and cladding.

Note that for the slow axis the value of n_2 is that of the inner cladding only, not that of the SAP. This method of measuring the NA directly from the fiber-refractive-index profile will typically generate a value higher than that derived from taking the sine of the half-angle of the output cone due to the influence of cutoff wavelength.

10.3.7 Coating package design

For many years, the opinion has been that the most important aspect of fiber design for FOG is birefringence—if the birefringence is maximized, then so

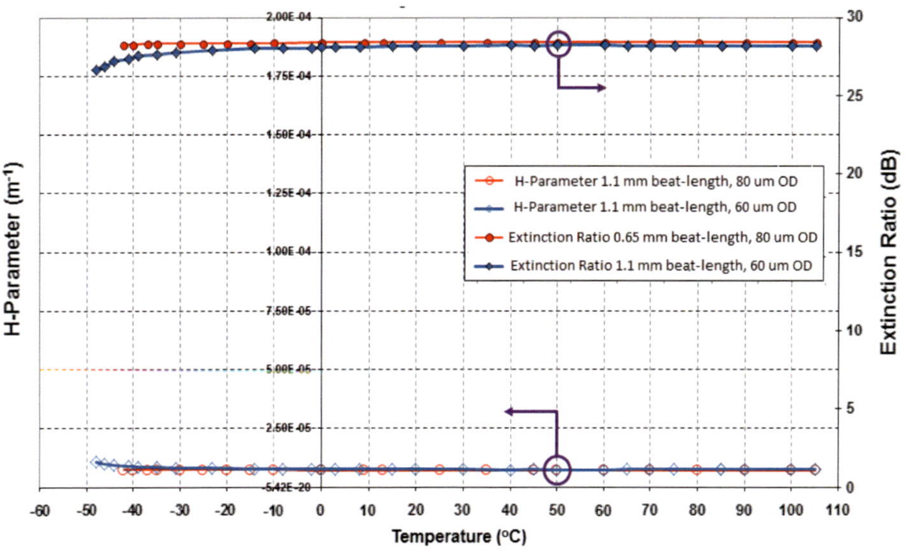

Figure 10.19 Comparison of h-parameter and polarization extinction ratio of an 80-mm PM fiber with those of a 60-mm PM fiber with enhanced birefringence.

too will the PM performance. However, in recent years it has become clearer that this perspective has been overly simplistic. In truth, the fiber itself represents only a single component within a highly complex system that also comprises the coating package, coiling adhesive or potting compound, and the coil design itself. In reality, even when working with fibers with exceptional birefringence (say, 0.8 mm Lp or better), a poorly designed coating package or incompatibility between the secondary coating and the coiling adhesive will all but destroy the underlying performance of the fiber.

The first coating packages designed for FOGs were, unsurprisingly, derived from those used for telecommunications fibers. This approach made sense because in both applications the coating package performs the task of protecting the fiber from the effects of microbending as well as imparting long-term strength and ease of handling. In reality, the fiber service condition in FOGs is far more severe than that encountered in telecommunications. In a typical sensor coil, more than 50 layers of fiber may be wound one on top of the other under small but significant tension to create an exceptionally severe microbending condition. The "microstress" that accompanies this microbe-nding effectively counteracts the intrinsic stress within a stress-birefringent fiber, reducing its ability to hold polarization and with it the fundamental performance of the sensor coil. Commonly encountered imperfections within the wind caused by fiber cross-overs or poor sequencing all contribute materially to the challenge. Furthermore, FOGs are often required to operate at –55°C, rather than the –40°C limit typical of telecoms. At these very low temperatures, all coating materials (typically urethane acrylates) become extremely rigid, thereby enabling the extrinsic coiling stress to be transferred directly into the fiber, effectively making an already bad situation even worse.

Early attempts to optimize the coating specifically for FOGs were aimed mostly at identifying the primary (inner) coating material with a glass-transition temperature of below –55°C, on the premise that only a material that stayed very soft throughout the operating temperature range would be capable of absorbing microbends and therefore prevent microstresses from transferring into the fiber itself. However, theoretical stress analysis (Fig. 10.20) shows that a dual-layer coating works by re-directing stress around the secondary (hard, outer) coating and away from the core of the fiber.

Viewed intuitively, this result simply confirms that fibers may be protected more effectively from the effects microbending by using the coating package to increase the overall stiffness of the fiber-coating composite and therefore make them more difficult to bend. Far from being the most important component in a dual-layer package, the purpose of the soft primary material is to act as a suspension medium with the correct emphasis shifting from the relative softness of the primary to the relative hardness and section of the secondary.[32]

Careful design of the coating package with the appropriate selection of primary and secondary resins and even optimization of cure conditions may

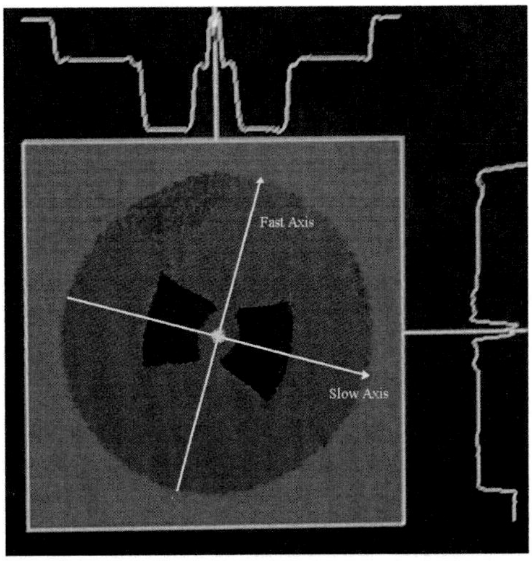

Figure 10.20 Refractive-index profiles of the two axes of a PM fiber.

be shown capable of reducing the coiling stress transferred to the core of the fiber by up to 50% below –40°C (Fig. 10.21).

Work reported at the Karlsruhe Symposium for Inertial Sensors and Systems in 2017 took this coating optimization concept a stage further to reduce the stress transferred into the core of the fiber by an additional 12% through matching the thermal expansion characteristics of the primary and secondary coatings[33] resulting in a 5-dB improvement in the polarization extinction ratio in small-diameter coils.

10.3.8 Radiation tolerance

With the increasing use of FOGs in satellites and space vehicles, the ability of the sensing fiber to resist the increase in attenuation induced by ionizing radiation (referred to as RIL or sometimes RIA) has become more important in recent years. In the vast majority of applications, changing the chemistry of the fiber, particularly within the light-guiding regions of the core and inner cladding, is considered preferable to shielding the sensor because it avoids any increase in mass or volume—both of which are of critical importance in space vehicles.

The nature of the mechanisms responsible for RIL is such that by far the simplest way of minimizing it is to operate at the longest possible wavelength, typically 1550 nm, where the impact of the intensified Urbach edge is negligible, and by avoiding the use of any dopants that are particularly susceptible to RIL within the guiding regions of the fiber. The influence of wavelength may be seen very clearly in Fig. 10.22, even for fibers with chemical compositions well optimized for use in high-radiation environments.

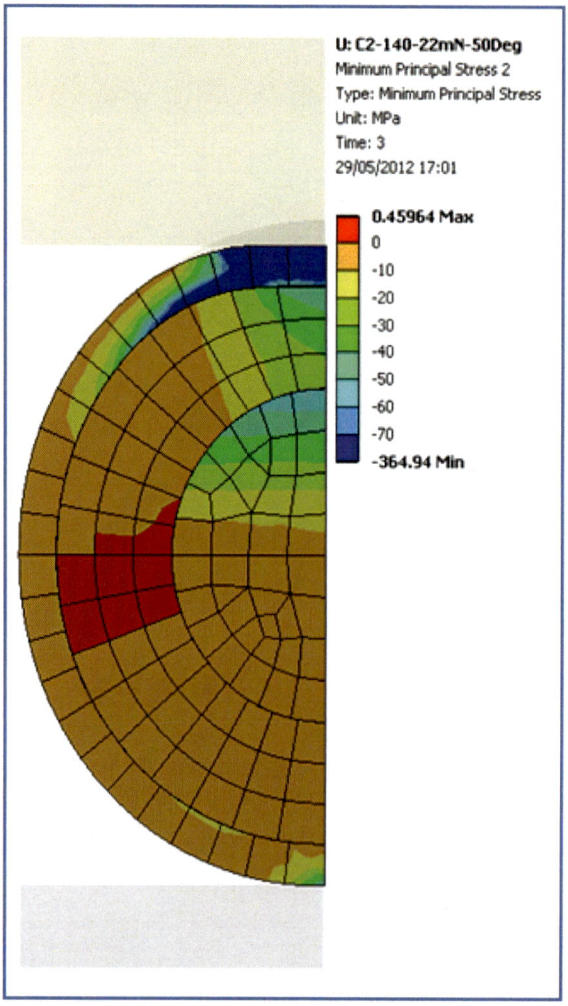

Figure 10.21 Stress distribution in a dual-coated fiber under load.

Tests performed at Fibercore to simulate the gamma exposure received by a satellite in geostationary orbit over a mission period of 20 years have indicated that RIA figures at 1550 nm may be reduced by 1–2 orders of magnitude, from ~100 dB/km to around 1–2 dB/km, simply with the judicious choice of dopants and the total avoidance of phosphorus in either the core or inner-cladding regions of the fiber (Fig. 10.23). The insensitivity of RIA to the fiber NA illustrates the relative radiation hardness of germanium, the element used to raise the refractive index in the vast majority of optical fibers. The removal of essentially all core and inner-cladding dopants to produce a silica-cored fiber may reduce the RIL by a further 1–2 orders of magnitude, but in practice it is unlikely to be necessary except in the longest, most precise FOG coils used in the most severe ionizing-radiation environments.

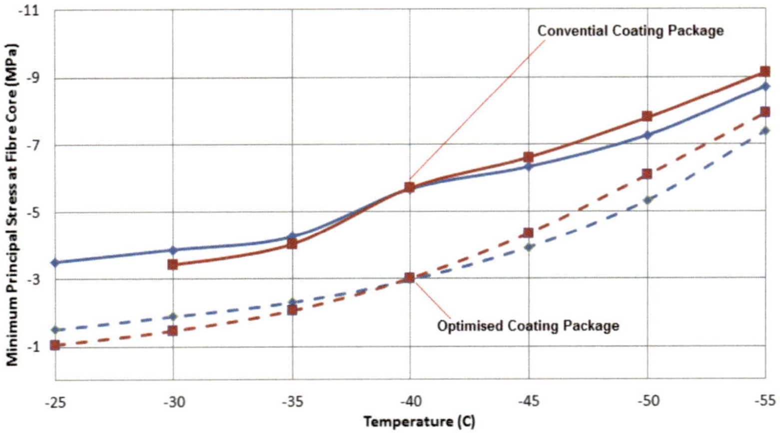

Figure 10.22 Reduction in stress transferred to the fiber core with an optimized coating package.

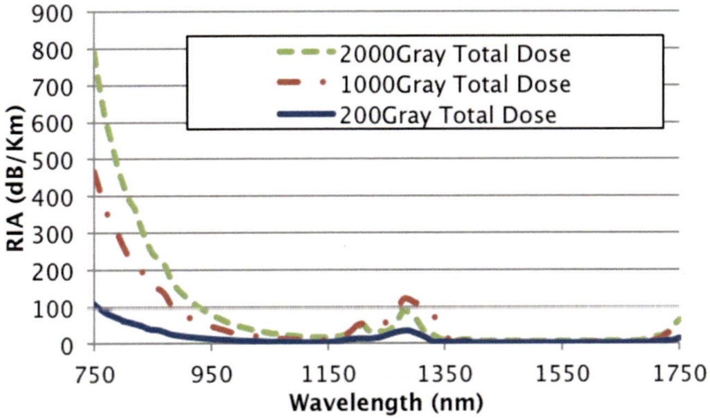

Figure 10.23 Radiation-induced attenuation (RIA) for an optimized core composition.

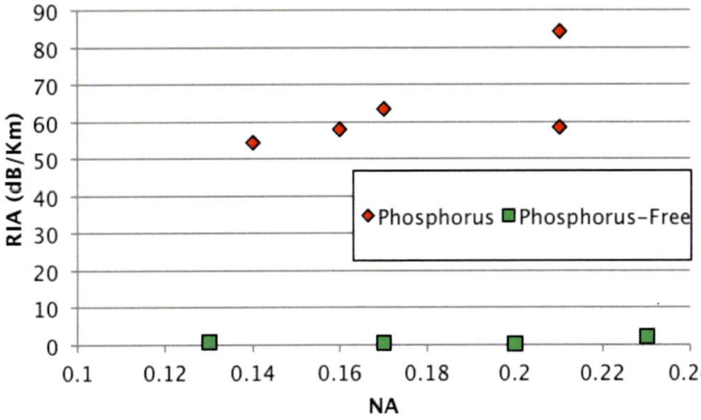

Figure 10.24 The impact of the presence of phosphorus on RIA.

10.4 Component Fibers

Whilst the sensor coil is arguably the most important component in any FOG and certainly the most complex, there are several others in which specialty fibers play an essential role, namely ASE sources, PM splitters, and polarizers.

10.4.1 ASE sources

In the early years of FOGs, most designs used super-luminescent diodes in the 800–850-nm window, and in fact the short coil lengths (half those of equivalent 1550-nm FOGs), consequent compact dimensions, and ease of packaging have meant that this wavelength range is still prevalent in Europe today. Nevertheless, there has been a widespread migration to 1550 nm over the past 20 years, driven by the relaxed alignment tolerances afforded by the larger MFD, significantly lower attenuation that allows a disproportionate increase in coil length, and greater radiation tolerance. The invention of the erbium-doped fiber amplifier (EDFA) in the mid 1980s[34] accelerated this migration by enabling the development of a broadband, incoherent light source powerful enough to drive all three axes simultaneously and providing wavelength stability two orders of magnitude better than that of typical semiconductor-based devices.

The design criteria for an erbium-doped fiber for an ASE source are somewhat differet from those encountered in the EDFAs used in telecoms and CATV. First, a substantial length of fiber (~20–30 m) is necessary to generate the levels of output power to drive all three FOG axes. In order to minimize this length, erbium doping levels tend to be significantly higher, more than ten times those of a C-band EDFA fiber. Very high NAs are also advantageous, both to raise the pump power density and to reduce the sensitivity to bend-induced attenuation—thereby enhancing the pump efficiency and also reducing the physical size of the finished component. For example, a typical ASE fiber would have a NA in the region of 0.3 (0.24) and a 1531-nm

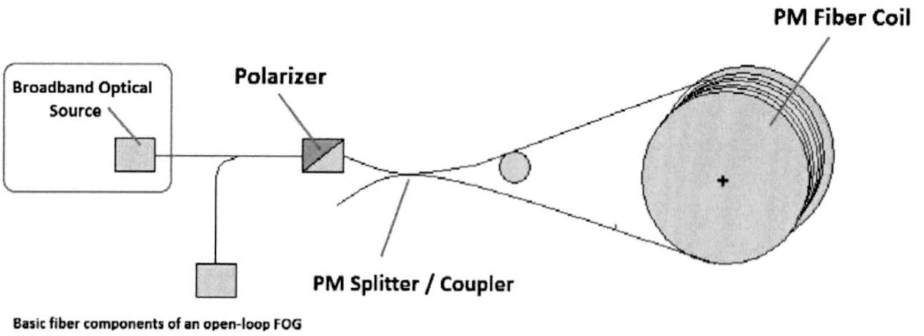

Figure 10.25 Schematic of a single-axis, open-loop FOG identifying the location of the ASE source.

absorption figure of up to 40 dB/m (8 dB/m); figures for a typical EDFA fiber are in parentheses.

One of the latest developments in ASE fibers for FOGs has been radiation tolerance.[35] Although ASE sources based on erbium-doped fiber have been used in space-qualified FOGs since at least the beginning of the 1990s, their performance was far from ideal. The spectra of these early ASE fibers was dominated by germania, making them narrow and biased towards 1530 nm but also (and accidentally) imparting degrees of both radiation tolerance and photosensitivity. As EDF designs were developed specifically for telecommunications and CATV, alumina replaced germania in the core chemistry, thus increasing the solubility of the Er^{3+} ions in the matrix and broadening the spectral emissions beyond 1550 nm. The new broadened, alumina-dominated spectrum also enhanced performance in an ASE source, but it was achieved at the expense of radiation tolerance. In reality, in terms of fiber-core dopants, only silica and germania offer any real resistance to ionizing radiation; if anything else is present (or at least anything that is currently known), RIL will climb very quickly indeed. As RIL builds, not only does the output of the ASE source decline but also the nature of the induced loss causes the spectrum to both narrow and distort, thus compromising the performance of the source massively. Whilst it has not been found possible to harden alumina-doped fibers against ionizing radiation, core chemistries have developed that effectively "self-bleach" under the influence of the pump, particularly if shorter wavelengths closer to the peak of the induced attenuation are used. ASE fibers developed by Fibercore and presented at SPIE OPTO, Optical Components and Materials in 2014 were shown capable of delivering more than 90% of their original output after 20 years in simulated geostationary orbit.

10.4.2 PM splitters and couplers

The need to split a polarized light input equally is fundamental to all FOG architectures. In the vast majority of current designs, this splitting is performed on a lithium niobate chip, in tandem with modulation and polarizing functions. However, in the early years of the technology, yields on polyfunctional integrated optical circuits (IOCs) were sufficiently poor to encourage a significant amount of investigation into PM fused-tapered couplers.

In a PM coupler, it is necessary to not only taper the fibers together in an essentially lossless manner but also align the birefringent axes of the two fibers throughout the tapering process (Fig. 10.26).

In terms of axis angular misalignment alone, the level of precision required does not appear particularly onerous, with 1 deg equating to approximately –35 dB and even 10 deg theoretically capable of delivering –15 dB. However, due to the complicated structure and composition of

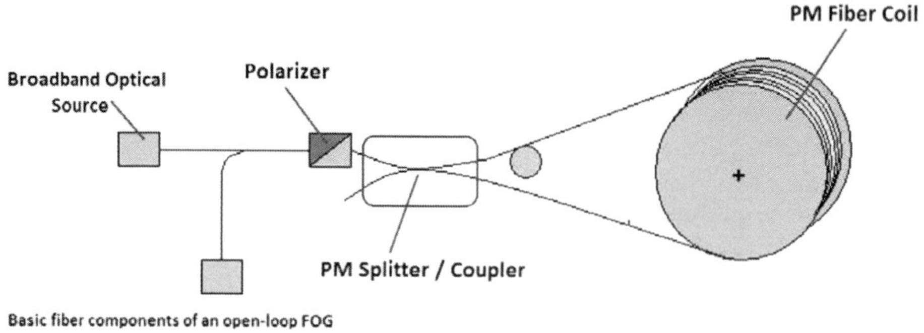

Figure 10.26 Schematic of a single-axis, open-loop FOG identifying the location of the PM splitter.

PM fibers, other factors must be considered. In a PM fiber, the stress members are heavily doped with boron (specifically B_2O_3) in order to generate the expansion mismatch necessary to generate birefringence. Boric oxide has a significantly lower refractive index than silica, which creates two index "pits" at either side of the core. These pits create a complex refractive-index profile that presents alternative guiding structures to the evanescent field as it attempts to couple from one core to the other, giving rise to exceptionally high excess losses, which may be of the order of several dBs. In order to bring these losses under control, it is necessary to re-design the profile of the PM fiber completely by moving the stress members further away from the core and by reducing their refractive-index contrast (Fig. 10.27). These modifications are represented for a bow-tie structure, but the same would hold true irrespective of the fiber cross-section.

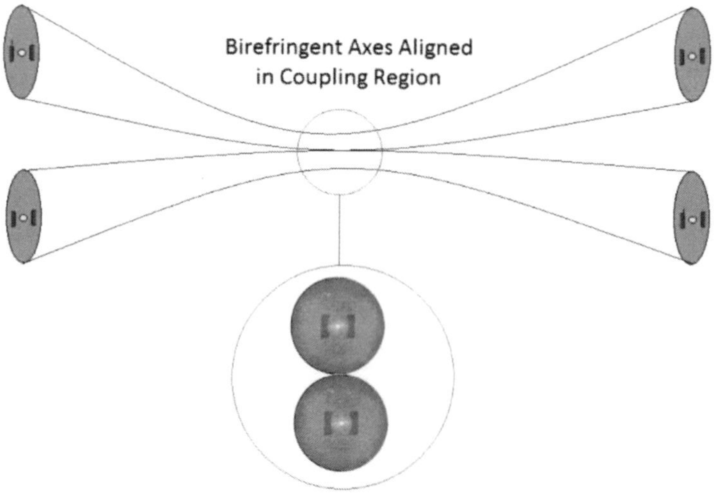

Figure 10.27 Axis alignment in a fused-tapered PM coupler.

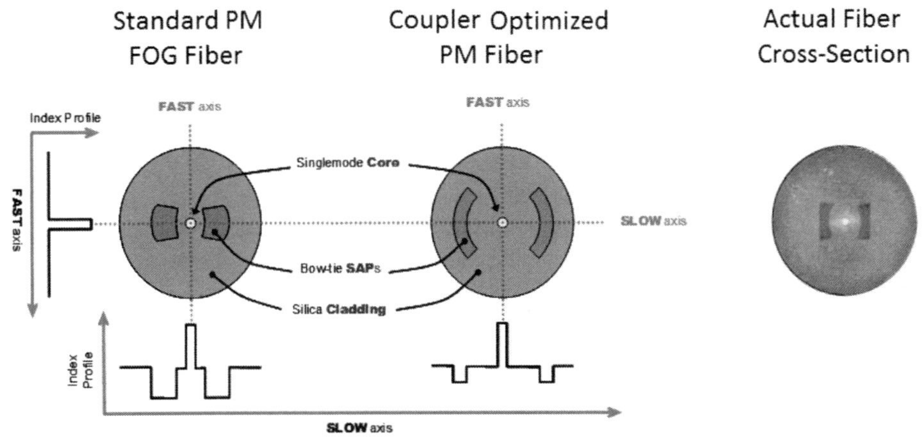

Figure 10.28 Coupler-optimized PM fiber.

The birefringence and therefore PM ability of these fibers is reduced significantly when compared with the typical fibers optimized for the sensor coil. In practice, this reduction has little impact since the coupler must only hold polarization for a few millimeters within the coupling region. In fact, experiments performed at Fibercore in which modified PM fibers were cleaved and examined at various stages throughout the tapering process indicated that the process of diffusion effectively causes the stress members to disappear—suggesting that "PM couplers" were actually non-PM but only at the center of the coupling region itself. With a fully optimized PM fiber, it is possible to fabricate PM couplers with extinction ratios and excess loss figures in the 18–25-dB and 0.2–0.5-dB ranges, respectively. These performance figures compare very well with those of couplers fabricated using the polished half-block method[36] whilst also providing the added robustness and environmental stability of a fused device.

Dramatic improvements in fabrication yields for poly-functional integrated optic circuits, fueled by the process-development funded by the telecoms boom of ~1996–2001, would have consigned fused PM couplers to the history books were it not for more recent interest in ultra-high-precision FOGs where the exceptionally strong spatial filtering and low insertion loss could still offer significant benefits.

10.4.3 Polarizing fibers

Polarizing, single-polarization, PZ, and Zing™ materials have been with us almost as long as PM fibers themselves[37] and were applied to a number of early FOG designs from the likes of British Aerospace (now BAE Systems) and Honeywell.

A single-polarization fiber is, in many ways, the ultimate manifestation of PM. It is a fiber in which the birefringence has been driven so high (typically

40% or more greater than typical coil-fiber values) that the fast axis reaches the threshold of guidance. At this point, the introduction of some physical perturbation, for example, by coiling several meters of the fiber or even by extruding a polymer jacket/buffer around the fiber, the fast axis may be encouraged to radiate and leave only the slow axis to propagate until the point at which it too will start to lose guidance and radiate (Fig. 10.29).

Due to the way a single-polarization fiber actually works, the characteristics of the polarizing window may be fine-tuned by varying the coil diameter and/or fiber length. Reducing the coil diameter will reduce the fiber cutoff wavelength and therefore both narrow the window and shift it to shorter wavelengths. Conversely, increasing the coil diameter will broaden the window and effect a long-wavelength shift. Increasing the length of fiber used within the polarizer will sharpen both the long- and the short-wavelength edges of the window, thereby creating a broader usable window that is far

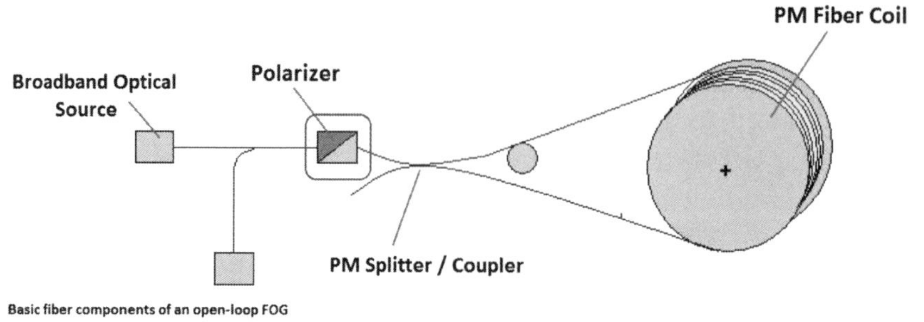

Figure 10.29 Schematic of a single-axis, open-loop FOG identifying the position of the polarizer.

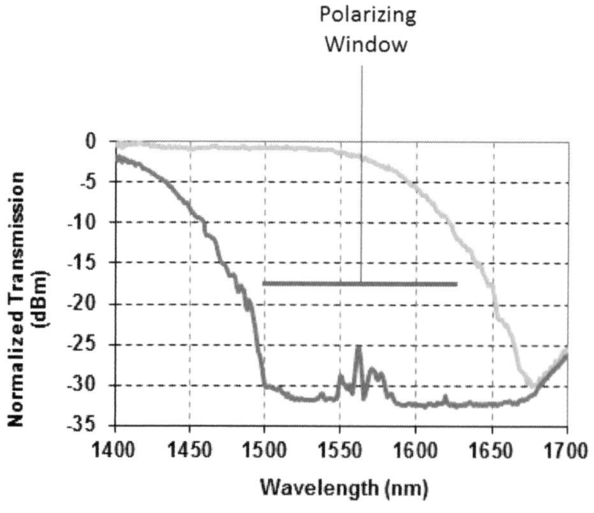

Figure 10.30 The polarizing window of a single-polarization fiber.

more compatible with the broad spectra of the ASE sources typically encountered in FOGs.

Opinions differ as to the optimum fiber structure necessary to generate an effective single-polarization fiber, with various publications referencing the use of a raised-index ring, sometimes referred to as a "wing," to create an anti-guide to encourage the fast mode to radiate more readily.[38] However, both theoretical and empirical work performed by Fibercore has shown birefringence to be by far the strongest driver of single-polarization performance. These exceptional levels of birefringence, approaching 1.3×10^{-3}, may most easily be realized by increasing the differential thermal stress generated by the SAPs, through increases in doping levels, locating the SAPs closer to the core, and, if the fiber design allows, optimizing their shape factor.

In the same way that interest in fused PM couplers for use in FOGs waned when manufacturing yields for poly-functional IO chips improved dramatically in the late 1990s (fueled by investment in process development for the telecommunications industry), polarizing fibers also fell from favor. Nevertheless, interest in these highly specialized variants of PM fiber has revived in recent years as FOGs have been driven into progressively higher grades of precision where the higher values of PER, lower insertion loss, and inherent spatial-filtering characteristic of all-fiber technology could deliver material performance benefits.

10.5 Epilogue

Since the very first days of practical FOGs in the early 1980s, researchers have discovered that virtually every new innovation in optical fiber design—from single polarization through multicore to photonic bandgap and beyond—is capable of demonstrating clear performance advantages under specific environmental conditions. Nevertheless, the development of optical fibers actually used in the series production of FOGs has followed a classical evolutionary path, dominated by stress-birefringent PM fibers. As FOG technology has become accepted within the mainstream, these fibers have been engineered progressively to address very specific, practical demands—both to increase the performance envelope (and therefore the range of gyroscope applications that may be addressed) and to improve and simplify the manufacturability of the FOG itself. As a result, the fundamental basis of the design of the fibers used today has actually changed very little and would be recognized immediately by anyone working within the field more than 35 years ago. Nevertheless, one should not interpret this apparent lack of progress as proof that current 80-μm stress-birefringent fibers represent some pinnacle of design from which there can be no further improvement. It is simply that these fibers, together with the essential suite of ancillary fibers and components that has developed around them, deliver the best compromise of performance and cost for the applications penetrated by FOGs thus far. For example, Fibercore delivered the very first 60-μm fiber into a commercial FOG program in 1987, and the engineers made it work very well indeed. Yet despite significant and almost continuous technical interest in the manifold benefits of ultra-low-profile fibers, the combination of increased susceptibility to microbending, lack of support from both compatible components and manufacturing equipment and manual handling characteristics has ensured that the commanding position of 80-μm fibers has never been challenged because no strategic application was dependent upon the new fiber for its success and because its benefits were outweighed by the difficulty and therefore cost of implementation.

In other words, whilst multicore and bandgap fiber technologies are little more than academic curiosities today, that does not necessarily mean that we will not be seeing more of these currently esoteric fibers in the future. All that is needed is new gyro applications that only FOGs can address and that contemporary stress-birefringent fibers cannot.

References

[1] V. Vali and R. W. Shorthill, 'Fiber Ring Interferometer', *Applied Optics*, Vol. **15**, 1976, pp. 1099–1100.

[2] G. Pircher and G. Hepner, 'Perfectionnnements aux depositifs du type gyrometer inetrferometrique a laser', French patent 1.563.720, 1967.

[3] T. Hosaka, K. Okamoto, T. Miya, Y. Sasaki, and T. Edahiro, 'Low-loss singl polarization fibers with asymmetrical strain birefringence', *Electron. Lett*, **17**, 530–531 (1981).

[4] US 4415230 A 'Polarization retaining single-mode optical waveguide', D. B. Keck, Corning Glass Works Nov 15. 1983.

[5] US 4426129 A 'Optical fiber and method of producing the same', H. Matsumura, T. Katsuyama and T. Suganuma, Hitachi, Ltd. Jan 17 1984.

[6] US 4354736 A 'Stress-induced birefringent single mode optical fiber and method of fabricating the same', M. S. Maklad, F. I. Akers and W. L. Thomas, International Telephone and Telegraph Corporation, 19 Oct 1982.

[7] R. D. Birch, D. N. Payne, and M. P. Varnham, 'Fabrication of polarization-maintaining fibers using gas-phase etching', *Electron. Lett*, **18**, 1036–1038 (1982).

[8] W. M. Macek and D. T. M. Davis, 'Rotation Rate Sensing with Travelling–Wave Ring Lasers', *Applied Physics Letters Letters*, Vol. **2**, 1963, pp. 67–68.

[9] W. W. Chow, et al., 'The Ring Laser Gyro', *Review of Modern Physics*, Vol. **57**, 1985, p. 61.

[10] P. Yeh, et al. 1978, 'Theory of Bragg Fiber', *J. Opt. Soc. Am.* Vol **68**, pp. 1196–1201.
[11] P St. John-Russell, 2003 'Photonic Crystal Fibers', *Science* Vol **299**, pp. 358–362.
[12] G. Meltz and E. Snitzer 'Thermal and Dispersive Characteristics of Muliticore Fibers', *Int. URSI Symp. on Electromagnetic Waves*, NBS Special Publications (1980), 314B/1-B/2.
[13] S. Inao, et al. 1979, 'High density multicore-fiber', *Conf. Proc. 28th International Wire & Cable Symposium, Cherry Hill, NJ.*
[14] WO 2009103015 A2, 'An interferometer employing a multi-waveguide optical loop path and fiber optic rotation rate sensor employing same', Bergh R.A. Aug. 20 2009.
[15] A. A. Michelson; Henry G. Gale (1925). "The Effect of the Earth's Rotation on the Velocity of Light, II". *Astrophysical Journal.* **61**: 140.
[16] M. P. Varnham, et al. 1983, 'Analytic solution for the birefringence produced by thermal stress in polarization-maintaining optical fiber', *J. Lightwave Technol.* **LT-1**(2).
[17] J. Noda, et al. 1986, 'Polarization-maintaining fibers and their applications', *J. Lightwave Technol.* **LT-4**(8).
[18] R. B. Dyott, et al. 1979, 'Preservation of polarization in optical fiber waveguides with elliptical cores', *Electr. Lett.* Vol. **15**, pp. 380–382.
[19] J. E. Midwinter, 1979, 'Optical Fibers for Transmission', pp. 178–183, John Wiley & Sons.
[20] M. Digonnet, et al., 'Sensitivity and stability of an air-core fiber-optic gyroscope', *Meas. Sci. Technol.* **18** (2007), 3089–3097.
[21] H. K. Kim, et al., 'Air-Core Photonic-Bandgap Fiber-Optic Gyroscope', *JLT*, vol. **24**, No. 8, August 2006.
[22] S. Blin, et al., 'Reduced Thermal Sensitivity of a Fiber-Optic Gyroscope Using an Air-Core Photonic-Bandgap Fiber', *JLT*, Vol. **25**, No. 3, March 2007.
[23] F. Poletti, et al., 'Hollow-core photonic bandgap fibers: technology and applications', *Nanophotonics* 2013; **2**(5-6):315–340.
[24] M. J. Gander, et al. 'Bend measurements using multicore optical fiber', *Proc. OFS* 12, 1997.
[25] B. Zhu, et al. 2011 'Space-, wavelength-, and polarization-division multiplexed transmission of 56-Tb/s over a 76.8 km seven-core fiber', Proc. OFC-NFOEC 2011 Post-deadline paper PDPB7.
[26] V. I. Kopp, et al. 'Pitch-reducing Optical Fiber Array and multicore fiber for space-division multiplexing', Photonics Society Summer Topical Meeting Series, 2013 IEEE, pp. 99–100.
[27] R. R. Thomson, Proc. SPIE 9774, 'Next-Generation Optical Communication: Components, Sub-systems and Systems V, 977400 Feb 13 2016.

[28] Rosendahl Nextrom Oy, Ensimmainen savu, PO Box 44, 01511 Vantaa, Finland www.rosendahlnextrom.com/fiber-optics/.

[29] W. W. Griffioen, et al., 'COST 218 evaluation of optical fiber lifetime models', Proc. Vol. 1791, Optical Meterials Reliability and Testing Benign and Adverse Environments (1993).

[30] A. Gillooly, and J. Hankey, 'Ultra-Low Profile, High Birefringence Fibers for Fiber Optic Gyroscopes', Inertial Sensors and Systems, Karlsruhe, September 2011.

[31] K. Bohm, et al., 'Low-drift fiber gyro using a superluminescent diode', *Elect. Lett.*, Vol. **17**, No. 10, pp. 352–353, 1981.

[32] C. Emslie, and J. Hankey, 'Optimization of Coating Package Design for PM Fibers Used in High Performance Fiber Optic Gyroscopes', Inertial Sensors and Systems, Karlsruhe, 2013.

[33] A. Gillooly, 'Next Generation Optical Fibers for Small Diamter Fiber Optic Gyroscope (FOG) Coils', Inertial Sensors and Systems, Karlsruhe, 2017.

[34] R. J. Mears, L. Reekie, I.M Jauncey, and D. N. Payne, 'Low-noise erbium-doped fiber amplifier operating at 1.54 mm', *Elect. Lett.* **1987** Vol.**23**(19) pp.1026–1028.

[35] M. Hill, et al., 'Fibercore AstroGain™ fiber: multichannel erbium doped fibers for optical space communications', Proc. Vol. 8982, SPIE OPTO, 2014, San Francisco CA, USA.

[36] US 4991922 A 'Optical fiber coupler and method', Dahlgren R. P., The Charles Stark Draper Laboratory Inc. Feb 12 1991.

[37] M. P. Varnham, et al., 'Single-polarization operation of highly birefringent, bow-tie optical fibers', *Elect. Lett.* **19**(7) pp. 246–247, 1983.

[38] M. J. Messerly, et al., 'A broad-band, single-polarization optical fiber', *J. Lightwave Technol.* Vo. **9**(7) 1991.

Chapter 11
Techniques to Ensure High-Quality Fiber Optic Gyro Coil Production

X. Steve Yao
General Photonics Corporation

A fiber coil is the most critical, yet least perfected component affecting the performance of a FOG. Coil winding is often considered an art, not a science, requiring technicians with magic skills to produce high-performance coils, mainly due to the lack of testing methods and capabilities. Numerous factors affect fiber-coil quality and performance, including the polarization crosstalk, coil asymmetry, fiber-winding tension, and properties of potting adhesives.

This chapter will first discuss all winding-induced imperfections and their relationship to FOG performance. Novel testing methods and equipment for characterizing the static performances of both PM fiber coils and single-mode fiber coils are then introduced. Techniques are presented for measuring the transient properties of coils, such as coil asymmetry, and the Shupe effect. Finally, 3D coil inspection is demonstrated by using optical coherence tomography (OCT).

With the aid of these test capabilities and associated manufacturing procedures, high-quality coils can be consistently produced for high-precision FOGs. It is now possible to claim that FOG coil winding is no longer an art but a science and can meet the demanding requirements for mass production and deployment of high-precision FOGs for various applications.

11.1 Introduction

A fiber coil is arguably the most critical optical component in a FOG because it is the sensing element of the FOG, and its quality determines the ultimate specifications of the critical parameters of the FOG, such as the bias

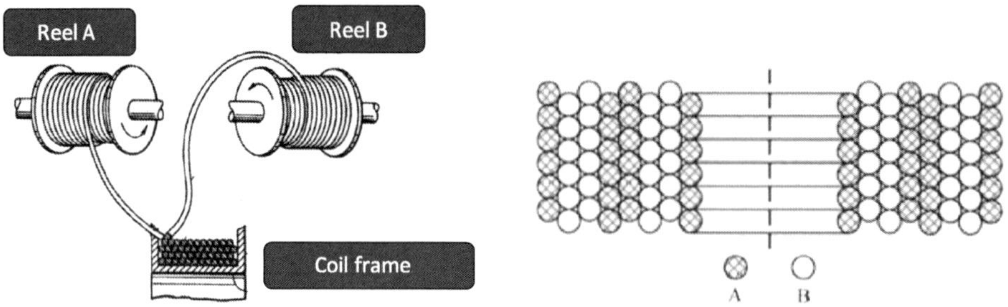

Figure 11.1 (a) Illustration of winding a quadrupole coil with two fiber supply reels A and B. (b) The relative positions of the fibers from the two reels.

instabilities at fixed as well as varying temperatures.[1–4] In order to minimize the rate errors induced by a temperature transient known as the Shupe effect,[5] the quadrupole winding pattern[6,7] was introduced, as illustrated in Fig. 11.1. In essence, the winding of the FOG coil starts from the center of the fiber. Before winding the coil, the fiber is first separated into two reels, A and B, and the center of the fiber is placed on one end of a coil frame as the starting point. First, a layer of fiber from reel A is wound on the frame from left to right, followed by two layers of fiber from reel B from left to right and back to left. Finally, the fiber from reel A is wound from right back to left, its home position, to complete a quad. The second quad starts from reel A again and follows the same routine, as shown in Fig. 11.1(b). More quads can be added for longer fiber coils. Note that the ends of reels A and B are connected to the two outputs of a multiple function integrated optical chip (MIOC) to form the Sagnac interferometer of the FOG. Because the fiber layers from reels A and B are always next to each other, they experience similar local temperature even under relatively fast temperature variations. Consequently, the rate errors induced by the temperature transients are minimized when a quadrupole coil is used in FOGs. To further minimize the Shupe effect, higher-order winding patterns, such as 8-pole, 16-pole, inter-leaved winding, and bifilar winding[8] have also been developed; however, all of them suffer more or less similar winding defects as quadrupole coils.

Although a large number of FOGs have been produced for various applications, the making of high-quality fiber coils still remains a work of art, with large variations in performance for coils made with the same process, sometimes requiring a super technician with magic hands to produce high-quality coils. Test standards are available for testing the performance of a whole fiber gyro during operation, including combined contributions from the light source, MIOC, passive components, fiber coil, mechanical structure and enclosure, electronics, and firmware; however, no published test standards or comprehensive methods are available for testing a fiber coil's performance.

Such tests are important for ensuring that high-quality fiber coils can be mass produced with good uniformity and efficiency. Without them, it would be extremely difficult and time consuming to identify and isolate problems in the FOG system, causing arguments among system integrators, electronics and mechanical designers, optical-component makers, assembly workers, and fiber-coil producers. In order to determine the quality of a coil alone, meaningful test parameters and associated test methods must be defined and developed. Such parameters can relate to winding imperfections for engineers to find the root causes of the problem coils and implement corrective measures or process improvements.

Up until now, fiber coil producers could only provide two specifications for the coils produced: the insertion loss (IL) and the polarization extinction ratio (PER). However, these parameters do not completely specify a FOG coil. The main reason for the lack of effective test methods and standards for fiber gyro coils is that the factors affecting the coil performance in a gyro are numerous, complicated, and subtle in many cases. As will be discussed next, different winding defects of a coil can propagate and sometimes combine to degrade the performance of the FOG using the coil.

In a perfect fiber coil, each fiber turn lies next to the adjacent turn, as shown with the bottom two layers in Fig. 11.2(a). In addition, each fiber layer is perfectly flat, with no bumps or humps. However, the following winding errors illustrated in Fig. 11.2(a) are commonly found in fiber coils:[9,10]

1. Fiber climbing: a fiber turn climbs above its current layer, caused by pressure from its neighboring fiber turns.
2. Fiber sinking: a fiber sinks down from its current layer, caused by inconsistent fiber winding tension.
3. Fiber gapping: two or more fiber turns settle incorrectly from their proper positions.
4. Concavo-convex defect of the fiber layer: mainly caused by the irregularity of the fiber diameter.
5. Fiber overlapping: often occurs at both ends of the coil when the fiber turn transitions from a lower layer to an upper layer during the winding process.

These winding defects create excess local stresses and asymmetries, which will degrade the coil performance under varying temperatures. As shown in Fig. 11.2(b), there are other winding defects, such as missing an A or B layer in a quad, or missing one or more fiber turns in a layer. All of the defects of the coil will propagate and cause different negative effects on the coil, including coil asymmetry, fiber stresses induced by bending, twisting, pressure, shrinkage of the potting adhesive, and coil-frame thermal expansion. These negative effects will then combine to cause problems, such as thermal non-reciprocity, polarization crosstalk, fiber strains (length and index variation), variations of Rayleigh scattering, and coil-size variation. Finally,

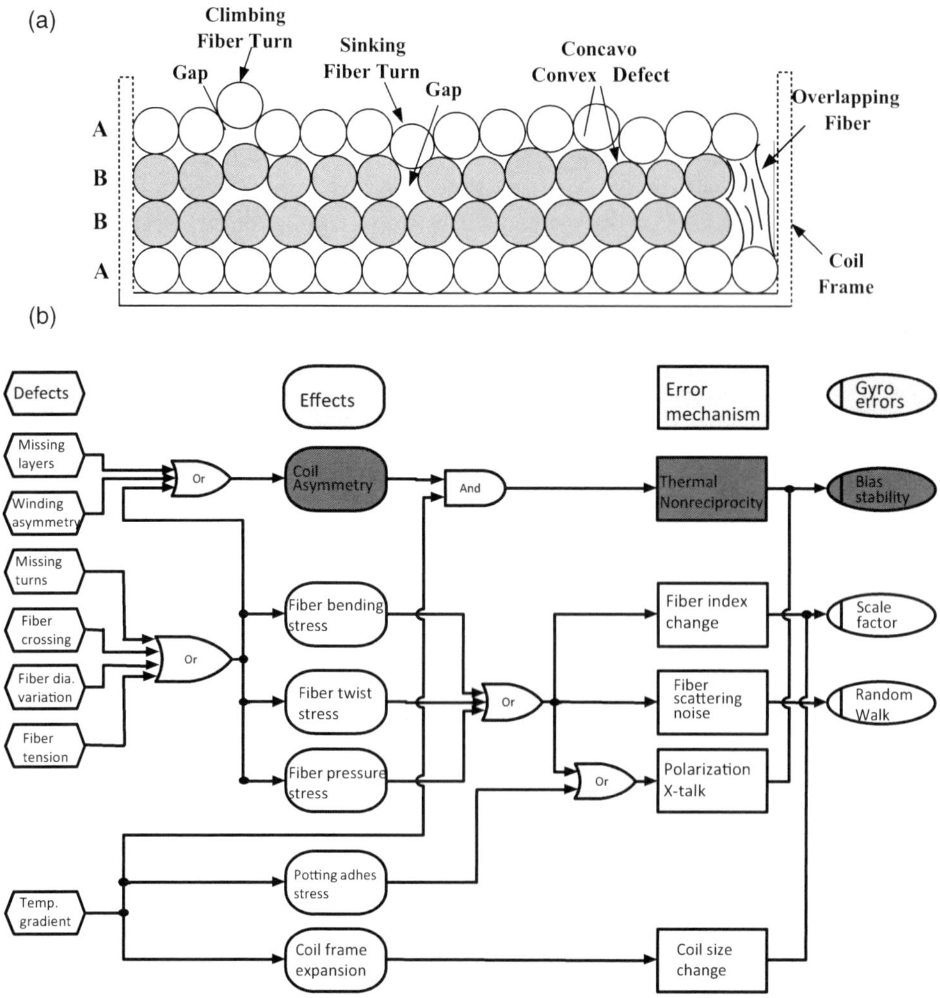

Figure 11.2 Illustration of (a) some different winding defects of a quadrupole coil and (b) how coil defects affect the performance of a FOG.

these problems combine to affect the FOG performance, such as bias stability, random walk, and scale factor under varying temperatures.

Insufficient work has been performed in this area to understand these defects and their impact on the FOG performance, despite limited early work.[6–10] In the past ten years, however, General Photonics has invested significant energy and capital to develop instruments and methods that fulfill the vacancies of almost every aspect of fiber coils for FOG applications.[11–16]

Sections 11.2 and 11.3 introduce two sets of parameters to describe the performance of a FOG: one is called the static parameters, and the other is called the transient parameters. The static parameters mostly reflect the FOG performance around a fixed temperature, such as the random walk and the bias instability at a stable temperature environment, whereas the transient parameters mostly reflect the performance of the coils in an environment with varying temperatures or vibration, such as the Shupe rate error when the environmental temperature of the FOG is changing. The sections further describe different instruments and methodologies developed to test parameters of different kinds of fiber coils and show how those parameters affect the coil performance in a fiber gyro. Finally, the excellent performance obtained for coils made with the assistance of our testing capabilities is presented.

11.2 Static Performance Parameters and Testing Methods

This section describes the parameters and testing methods to characterize the performance of fiber coils under relatively stable temperature or vibration environments. First, polarization-maintaining (PM) fiber coils and the parameters relating to their performances in FOGs are discussed, including the PER and polarization crosstalk, especially under different temperatures. Topics after that include the testing methods to characterize single-mode (SM) fiber coils and important parameters relating to their performance in depolarized FOGs, including polarization-dependent loss (PDL), polarization mode dispersion (PMD), and degree of polarization (DOP) in different environment temperatures.

11.2.1 Polarization-maintaining fiber coils

Most FOGs are made with PM fiber coils due to their superior performance compared with their SM fiber counterparts, although great performance has also been achieved with depolarized SM fiber coils.[17] The following subsection mainly discusses parameters important to PM fiber coils and the PM fiber itself, as well as methods and equipment to characterize these parameters.

11.2.1.1 Insertion loss and polarization extinction ratio

As mentioned previously, insertion loss is used to characterize the loss of light energy caused by fiber attenuation, as well as the winding-induced fiber bending loss, which may include macro-bending loss and micro-bending loss. The macro-bending loss is determined by the fiber itself and the tension during the winding process, whereas the micro-bending loss is mainly determined by the winding quality, including the suitability of the adhesives used to bond the fiber. Fiber crossovers can generally cause such micro-bending. Therefore, the difference between the IL of the coil and that of the unwound fiber itself is an indication of winding quality, especially under extreme temperatures. A large variation of IL under different temperatures may disqualify a coil for many applications. A power meter with a stable light source at the wavelength of interest can be used to accurately measure the IL of the coils.

PER is another commonly used parameter to characterize a PM fiber coil, and it directly affects the performance of the FOG incorporating the coil. In fact, it is the accumulation of all polarization cross-couplings or crosstalk along the PM fiber. Coils with a poor PER have a large number of polarization crosstalk events inside the coil, which may cause interference noise in the detection circuit, resulting in poor bias instability and random walk, even at fixed temperatures.[2] The PER of a coil includes the following five contributions: (1) the PER of the PM fiber itself, (2) the PER degradation caused by the winding process, (3) the PER degradation caused by the curing shrinkage of the fiber bonding adhesive, (4) the thermal expansion or contraction of the fiber bonding adhesive, and (5) the PER degradation caused by the coil frame or bobbin, especially the tension from the thermal expansion or contraction of the coil frame if the thermal expansion coefficients of the fiber and the frame are significantly different. Therefore, PER under different temperature is a good indication of the coil quality. A large PER variation implies large temperature-induced stresses inside the coil, which may cause bias instabilities, compounded with the Shupe effect.

The preferred equipment for measuring the PER is a PER meter with a high dynamic range of 50 dB or better. Such a PER meter is generally made with a rotating polarizer, such as ERM-202 from General Photonics Corporation. In practice, a stable broadband light source, such as a SLED or ASE source, should be used to get stable and repeatable PER readings. An alternative way to measure the PER of a fiber coil uses a polarimeter or a polarization analyzer with a tunable laser.[18]

11.2.1.2 Distributed polarization crosstalk analyzer

Unfortunately, PER itself cannot identify where and why a coil has problems. It is desirable to be able to see local polarization crosstalk along the fiber with a high spatial resolution, which are induced from the winding process as well as the bonding adhesives. Martin, Le Boudec, and Lefèvre[20] reported at the SPIE Fiber Optic Gyros 15th Anniversary Conference on the use of white-light interferometery to obtain such local crosstalk information in an optical fiber, and a resulting instrument named an optical coherence-domain polarimeter (OCDP) was once made commercially available by a company called Photonetics (the company was discontinued in the late 1990s in an industry consolidation). Another team at Northrop Grumman/Litef also reported using such a white-light interferometer to test FOG components at the same conference.[21] Unfortunately, such a powerful method was not widely adopted by the FOG industry. One possible explanation may be that the original white-light interferometer suffered from ghost polarization crosstalk peaks, which could cause confusion when interpreting the crosstalk data. Another possible reason is that the instrument was only commercially available for a short period in the marketplace and few people had the opportunity to practice with it in the coil-winding facilities.

Since 2011, General Photonics Corporation has offered a powerful instrument called the distributed polarization crosstalk analyzer (DPXA, model number PXA-1000),[11–14] shown in Fig. 11.3, specially designed to measure the local polarization crosstalk in fiber coils, as well as in a $LiNbO_3$ MIOC, as a function of distance. Similar to Ref. 20, the DPXA is essentially a white-light interferometer. However, it adopted a patented technique[21] to effectively remove the zeroth- and second-order ghost peaks and significantly reduce the third-order ghost peaks to below the noise level of –80 dB so that any third-order ghost polarization crosstalk peaks generated by crosstalk less than –27 dB are not visible by the DPXA. With such ghost-peak elimination and suppression, the DPXA can unambiguously identify polarization crosstalk along a PM fiber caused by external stresses and internal defects. At the General Photonics fiber-winding facilities (including its partner Suzhou Optoring's facility), the DPXA is the most powerful tool for winding process improvement and for coil-bonding-adhesive selection. In addition, as will be discussed later, it can be used to completely characterize all parameters of a PM fiber,[14] including its birefringence (or beat length), birefringence thermal coefficient, birefringence dispersion, and birefringence uniformity along the fiber. Finally, it can

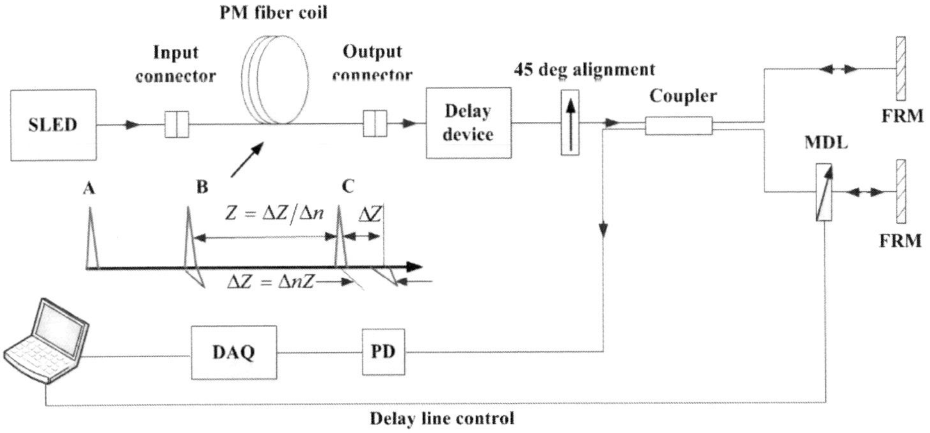

Figure 11.3 Illustration of a ghost-peak-free distributed polarization crosstalk analyzer using a scanning white-light Michelson interferometer. The inset shows the delay relation between the original and crosstalk components. Light with a short coherence length traveling in the fiber is polarized along its slow axis at input point A. Crosstalk is induced by a stress at point B, where a small portion of light is coupled into fiber's fast axis. A relative delay at the output point C between the two polarization components is ΔZ. The location Z of crosstalk point B can be obtained from a measurement of ΔZ. FRM, MDL, PD, and DAQ are Faraday rotation mirror, motorized delay line, photodetector, and data acquisition card, respectively.

accurately measure the PER of the fiber coil or a fiber jumper, without the errors from fiber interfaces, such as connectors or fusion splices.

As illustrated in Fig. 11.3, a polarized super-luminescent diode source (SLED) at a wavelength of interest with a very short coherence length (~25 μm, corresponding to a 3-dB Gaussian line width of 30 nm) is coupled into the slow axis of a PM fiber under test (point A of inset in Fig. 11.3). Assume that at point B a polarization crosstalk is induced by an external disturbance and some lights are then coupled into the fast axis of the PM fiber with a coupling coefficient parameter $h = I_1/I_2$, where I_1 and I_2 are the light intensities in the fast and slow axes of the PM fiber, respectively. Because the polarized light along the fast axis travels faster than that along the slow axis, at output of the fiber the faster light component will be ahead of the slow component by $\Delta Z = \Delta n Z$, where ΔZ is an optical path length difference, Δn is a group birefringence of the PM fiber, and Z is the fiber length between the point where the crosstalk occurs (point B) and the output end (point C). A polarizer oriented at 45 deg to the slow axis of the PM fiber was placed at the end of the fiber. Polarization components from both slow and fast axes were projected onto the same direction of the linear polarizer axis so as to produce an interference pattern between those two components in a scanning Michelson interferometer. When the relative optical path length is scanned, an interference peak appears when these two polarization components are overlapped in the space, but it disappears when they are separated by more than a coherence length of the light source (i.e., SLED). Then the *group birefringence* Δn of the PM fiber between points B and C can be calculated as

$$\Delta n = \Delta Z / Z. \tag{11.1}$$

It is evident from Eq. (11.1) that the accuracy of Δn depends on the measurement accuracies of both ΔZ and Z, and vice versa.

Note that the illustration in Fig. 11.3 assumes only one polarization crosstalk point along the fiber. If there are multiple polarization crosstalk points, second-order interference peaks will occur. That is, the light in the fast axis caused by the coupling at a crosstalk point will couple back to the slow axis at the subsequent crosstalk points down the fiber.

As shown in Fig. 11.4(a), consider a situation where there are three coupling points X_1, X_2, and X_3 along the PM fiber, and the light input to the PM fiber has no fast-axis component and is polarized along the slow axis of the PM fiber. At each coupling point, light is coupled not only from the polarization mode along the slow axis to the polarization mode along the fast axis but also from the polarization mode along the fast axis to the polarization mode along the slow axis. As a result of this coupling, the resulting wave-packet series output by the PM fiber includes a wave packet caused by multiple couplings. As shown in Fig. 11.4(b), four wave packets include the

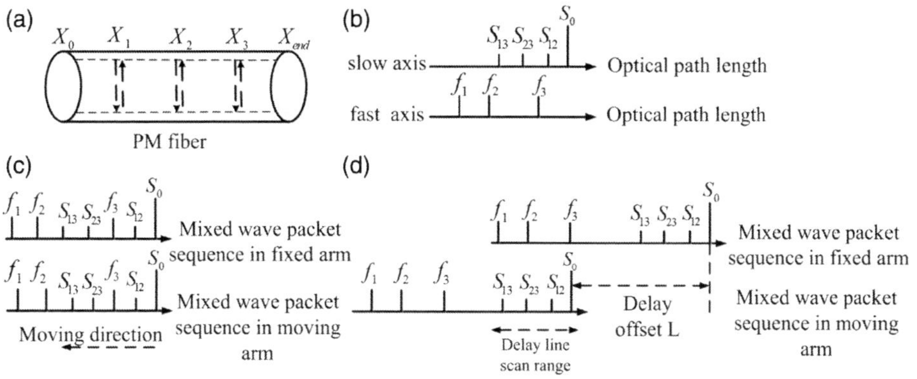

Figure 11.4 (a) Illustration of polarization coupling at locations X_1, X_2, and X_3 along the PM fiber. (b) The wave packet sequences polarized along the slow (denoted by S) and fast axes at output of the PM fiber (denoted by f). (c) Wave pockets in the two interferometer arms after light passing through the 45-deg-oriented analyzer, where the wave packets aligned to the slow and fast axes, are mixed together. When this mixed light is input to the interferometer in Fig. 11.3, a series of interference peaks, including multiple ghost peaks, will be observed as the delay in one arm of the interferometer is changed. (d) Wave pockets in the two interferometer arms after the delay device in Fig. 11.3 is inserted between the PM fiber's output and the 45-deg polarizer's input. This differential delay adds an additional delay between the slow axis and the fast axis of the PM fiber. Only the first-order interference peaks between S_0 and f_i and the third-order interference peaks between f_i and S_{ij} can be generated.

zeroth-order coupling S_0 (no coupling) and the second-order couplings S_{12}, S_{23}, and S_{13} emerging at the output, which are aligned to the slow axis of the PM fiber, and three main packets f_1, f_2, and f_3 emerging at the output, which are aligned to the fast axis of the PM fiber where they are generated by coupling from the slow axis to the fast axis (the first-order coupling) at points X_1, X_2, and X_3, respectively.

After passing through the 45-deg-oriented polarizer, the wave packets aligned to the slow and fast axes were projected onto a same direction of the linear polarizer axis, as shown in Fig. 11.4(c). When this mixed light is input to the interferometer, a series of interference peaks can be observed as the delay in one arm of the interferometer is changed, and these second-order couplings will cause ghost crosstalk peaks and result in confusion in a simple white-light interferometer.[20,22,23] As shown in Fig. 11.3, a differential group delay (delay device) can be used inside a DPXA to remove all ghost crosstalk peaks from the second-order couplings,[20] making it possible to accurately identify and measure a large number of polarization crosstalk along a PM fiber without ambiguity. In particular, as illustrated in Fig. 11.4(d), the delay device adds an additional delay L larger than $\Delta Z = \Delta n Z$ between polarization components in the slow and fast axes. Therefore, the two wave-packet sequences from the fast axis and slow axis are separated in time (or space) after the light passes through the analyzer. If the same delay offset between the fixed and moving arms in the interferometer were preset and the range of the variable delay line in the moving arm were restricted, then the undesired zeroth-order (S_0 with S_0, S_{ij} with S_{ij}, and f_i with f_i) and second-order (S_0 with S_{ij}, and f_i with f_j) interference signals (ghost interference peaks) will not be generated as the delay line scans. Only the desired first-order interference signals (S_0 with f_i) and the much weaker third-order interference signals (f_i with S_{nm}) will be present. Note that the third-order peaks are negligible (less than –75 dB) if the first-order coupling f_i is less than –25 dB. A more detailed description of such a ghost-peak-free DPXA can be found in Ref. 21.

11.2.2 Basics of polarization crosstalk in PM fibers

As an x-ray machine allows doctors to see inside of a human body, the DPXA helps engineers see inside a PM fiber or PM fiber coil. The x-ray machine requires doctors with a good understanding of the human body to interpret the x-ray images and identify the anomalies. Similarly, operators of the DPXA must have a good understanding of polarization crosstalk behaviors to be able to identify defects inside PM fiber or PM fiber coils. Therefore, the objective of this section is to give a basic background of polarization crosstalk or coupling in PM fibers and enable users to identify defects inside a fiber coil and their likely causes.

11.2.2.1 Classification of polarization crosstalk by causes

There are three main mechanisms causing polarization crosstalk in a PM fiber. The first mechanism is fiber-axis misalignment at fiber connection interfaces, such as connectors or fusion splice points, as shown in Fig. 11.5(a). Such crosstalk generally occur at a specific point in space and are large in amplitudes around –35 dB or more, determined by the misalignment angles. The second mechanism is from the imperfections of the PM fiber, such as local birefringence variations, internal shape variations, or internal stress. Such polarization crosstalk are generally small in amplitude, and the polarization coupling occurs gradually along a certain length of the PM fiber. The third mechanism is caused by external mechanical stresses on sections of the fiber, such as fiber bending, crossing, squeezing, or pressing. Such crosstalk are mostly complicated, with some occurring at sharp points in space and some occurring gradually along a length of fiber (with varied amplitudes that depend on the stress orientations with respect to the slow axis and the stress strengths, as shown in Fig. 11.5(a)).

11.2.2.2 Classification of polarization crosstalk by measurement results

In general, the DPXA can accurately measure the strengths of polarization crosstalk occurred at different locations with a spatial resolution of a few

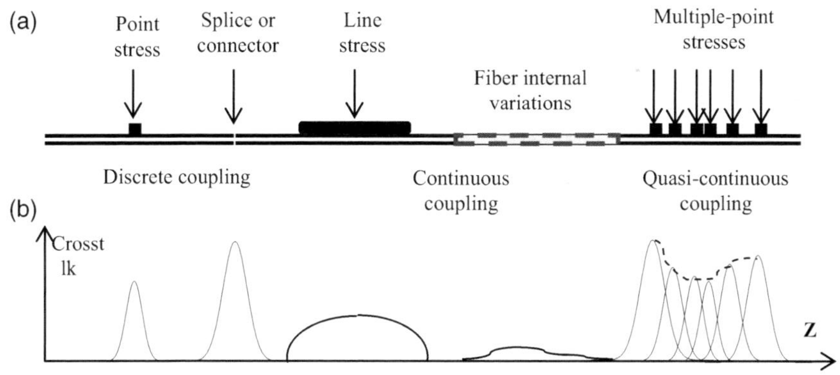

Figure 11.5 Illustration of different types of polarization crosstalk. On the left are discrete polarization crosstalk peaks induced by a point stress or a splice. The shape of the peak is Gaussian, which is the coherence function of the light source. The spatial resolution is also determined by the width of the coherence function. In the center are continuous polarization crosstalk induced by a line stress and internal fiber imperfection. On the right are quasi-continuous crosstalk induced by multiple densely packed stress points spaced on the order of or less than the resolution of the instrument.

centimeters. However, one may not be able to identify the causes of the crosstalk from the measurement results, although users can make educated guesses of what the causes might be based on the shape and strengths of measured crosstalk at these locations. Therefore it is also feasible to classify the crosstalk based on the shapes of measured polarization crosstalk, as in the following examples.

Crosstalk caused by discrete polarization coupling points refers to the polarization coupling induced by a sharp stress, a splice point, or multiple stress/splice points that are separately much larger than the resolution of the measurement instrument, as shown in Fig. 11.5(a). Such discrete couplings have distinctive peaks in the crosstalk measurements, with the width determined by the spatial resolution of the instrument, as shown in Fig. 11.5(b). For this type of coupling, it is meaningful to give a peak crosstalk value for each coupling point. Figure 11.6(a) shows the polarization crosstalk measurement result of a quadrupole fiber gyro coil using a ghost-free polarization crosstalk analyzer (PXA-1000). The crosstalk values listed in the table on the instrument's display screen are for such discrete coupling points, as shown in Fig. 11.6(b). The crosstalk table lists all crosstalk peaks above a threshold determined by a user, with both the location and the strength information of each crosstalk. Note that since the profile of a polarization crosstalk peak of a discrete coupling is the coherence function of the light source used, the PXA-1000 can also be used to measure the coherence function and the coherence length of a broadband light source.

Crosstalk caused by continuous polarization coupling occurs in a section of fiber induced by a line stress or by fiber internal imperfections, with the

(a)

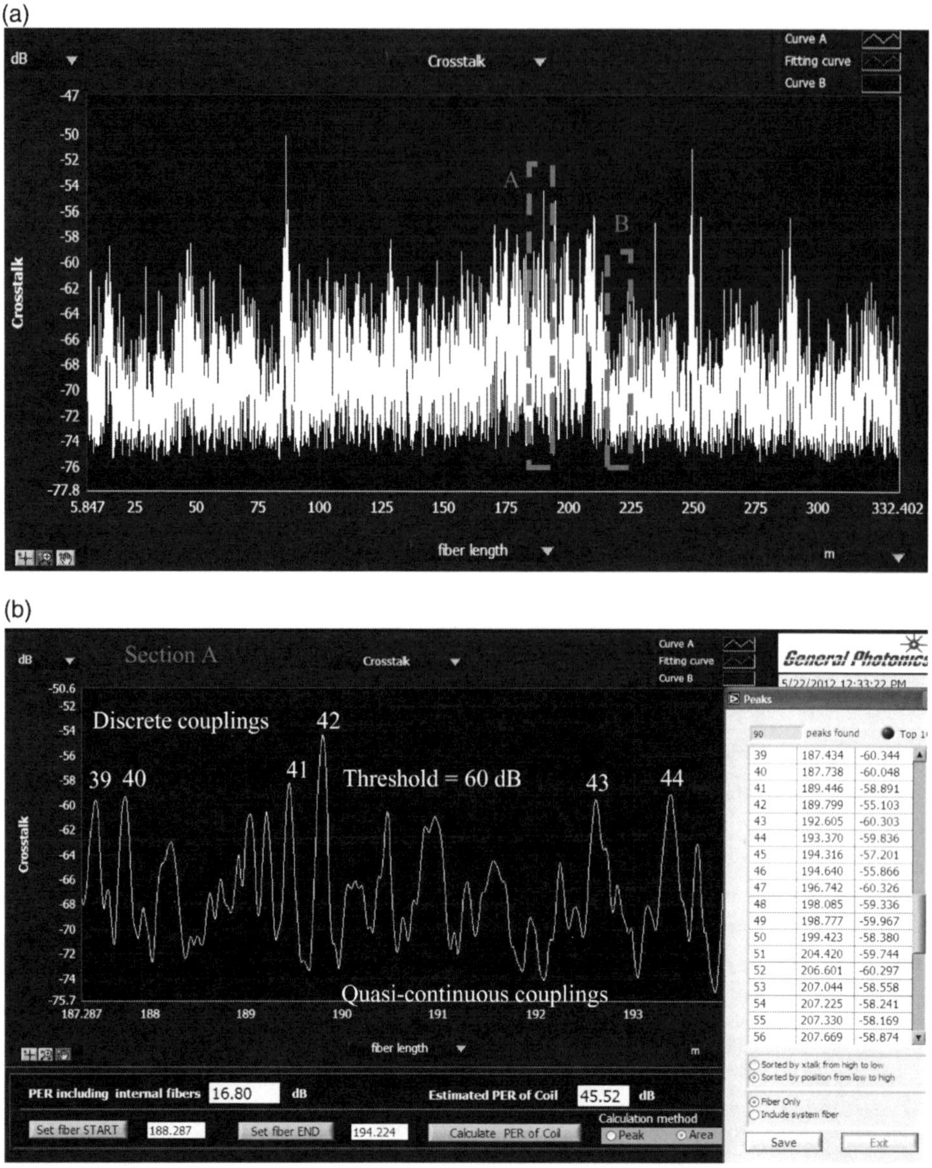

Figure 11.6 (a) Crosstalk measurement result of a 340-m-long fiber coil. (b) Enlarged view of section A showing discrete and quasi-continuous coupling peaks. The table on the right lists the crosstalk results of discrete peaks larger than –60 dB. Note that the shape and width of the discrete crosstalk peaks are determined by the coherence function of the light source.

section length comparable or larger than the resolution of the measurement instrument, as shown in Fig. 11.5(a). The crosstalk measurement result of such continuous coupling is a broad hump with a width and shape mainly determined by the length of the section stress, as shown in Fig. 11.5(b). In general, crosstalk caused by a section of fiber imperfection is very small in amplitude, on the order of –60 dB or lower. For continuous polarization coupling, it is not meaningful to give a peak crosstalk value. However, one is able to obtain the accumulative coupling occurred in a section of fiber by defining the starting and ending positions along the fiber using two cursors equipped in the PXA-1000 software, as shown in Figs. 11.7(a) and (b), with a result of –61.32 dB.

Crosstalk caused by quasi-continuous coupling induced by multiple stress points with spacing less than the resolution of the measurement instrument, as shown in Fig. 11.5(a). Such a polarization coupling will appear as a broad hump with height variations in polarization crosstalk measurements, with a width and shape determined by the number of stress points, their relative positions, and their relative strengths, as shown in Fig. 11.5(b). A quasi-continuous coupling cannot be distinguished from continuous coupling. Similar to the continuous coupling, for quasi-continuous coupling it is not meaningful to give a peak crosstalk value. However, similar to Fig. 11.7(b), one is able to obtain the cumulative coupling occurred in a section of fiber by defining the starting and ending positions along the fiber using two cursors in the PXA-1000 software.

Figure 11.8(a) shows the polarization crosstalk measurement of a low-quality fiber gyro coil with large polarization crosstalk peaks, whereas Fig. 11.8(b) is an enlarged view of the dashed box, having a large quasi-continuous coupling section with an accumulative polarization crosstalk of 26.46 dB between the two cursors.

The crosstalk events caused by quasi-continuous coupling fundamentally limit the bias stability of the FOG incorporating the coil because they indicate that (1) the distance between two such coupling points is less than the coherent length of the light source used; and (2) a portion of light coupled from the slow axis to the fast axis at one coupling point will couple back to the slow axis at a following coupling point and interfere with the light remaining in the slow axis. Such interference will induce a phase error and thus bias instability. Because the interfering light is entirely in the slow axis, it cannot be eliminated or reduced by the polarizing waveguide in the MIOC, no matter how high the extinction ratio is. Consequently, the coil with the crosstalk curve shown in Fig. 11.8(b) is particularly poor because it has not only high crosstalk but also a large number of quasi-continuous polarization couplings to cause bias instabilities.

11.2.3 Characterization of potting adhesive with a DPXA

DPXA measurement is effective in visualizing the stress induced polarization crosstalk degradation from the adhesive shrinkage at different temperatures, as shown in Fig. 11.9, and hence can be used to evaluate the performance of different coil bonding adhesives against temperature variations. Figures 11.9(a) and (b) show the polarization crosstalk along the fiber length

(a)

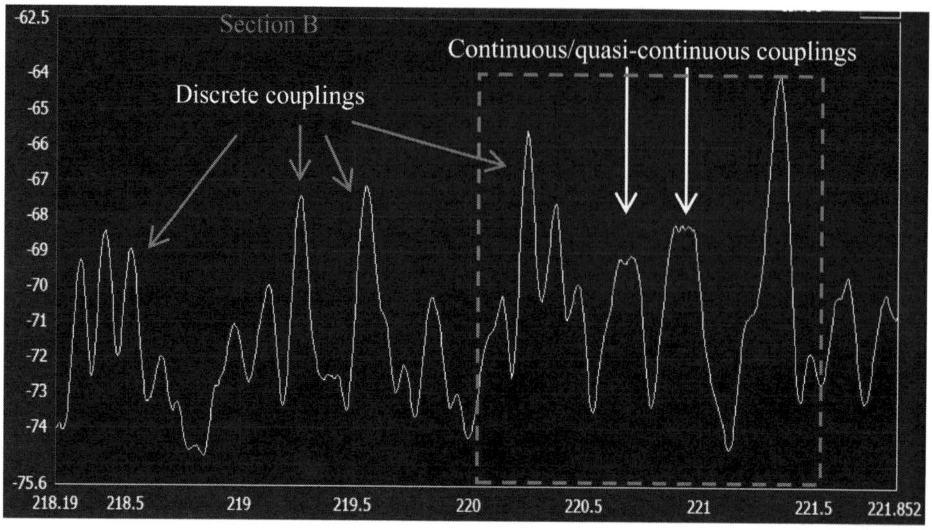

(b)

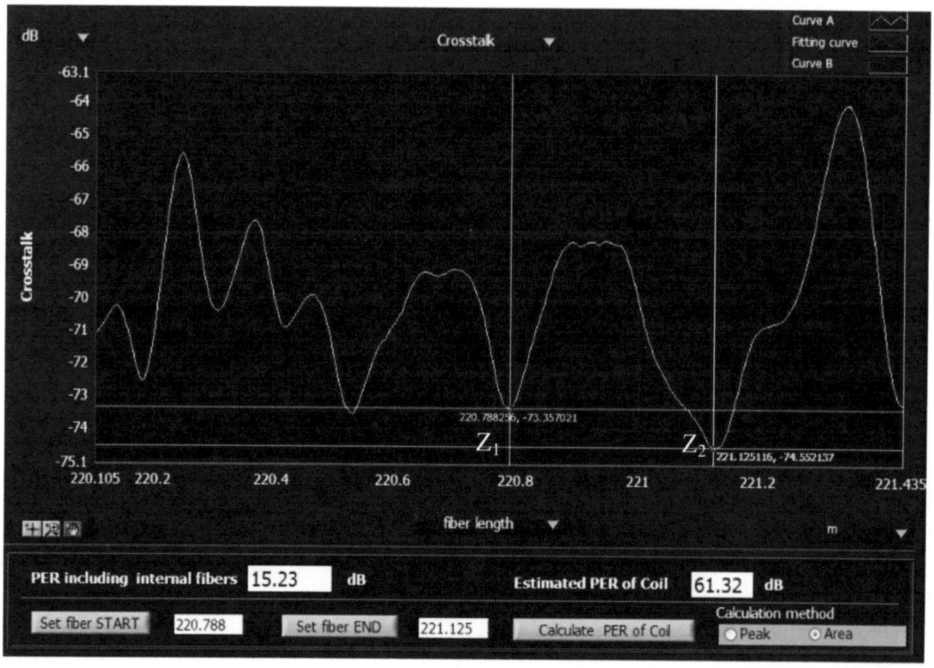

Figure 11.7 (a) Enlarged view of section B of the crosstalk measurement in Fig. 11.6(a), where two continuous or quasi-continuous crosstalk peaks are identified. (b) Enlarged view of the dotted box in (a), where the cumulative crosstalk value of a continuous/quasi-continuous coupling is obtained by setting Z_1 and Z_2 and calculating the integrated PER of the corresponding fiber section.

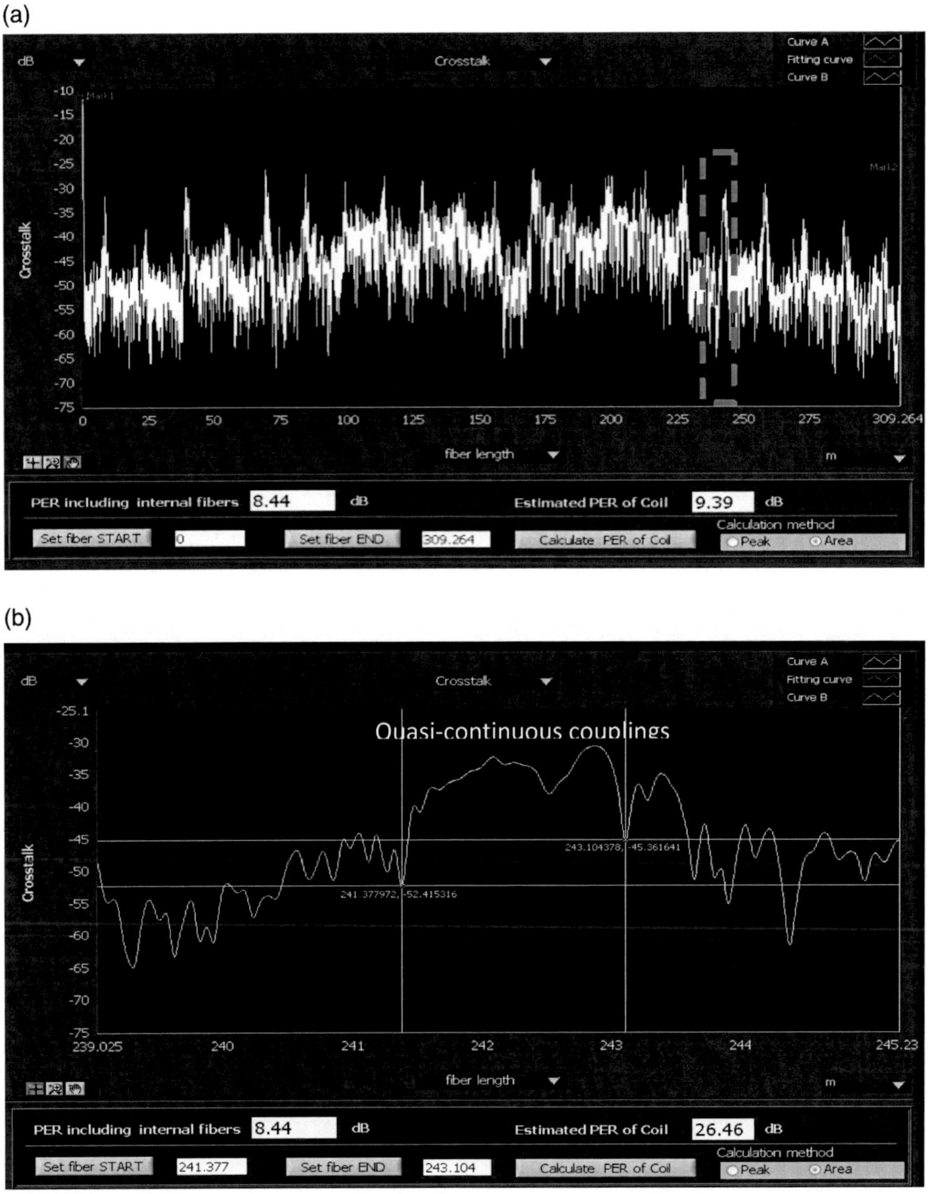

Figure 11.8 (a) The crosstalk curve of a low-quality PM coil of 309 m. (b) Enlarged view of the boxed section in (a), showing quasi-continuous couplings. The cumulative polarization crosstalk between two cursors is 26.46 dB.

for a 300-m coil potted with a non-ideal off-the-shelf adhesive available from a well-known adhesive manufacturer at temperatures of –40 and +80°C respectively, compared with the polarization crosstalk data at 25°C. A significant increase of polarization crosstalk across the entire length of

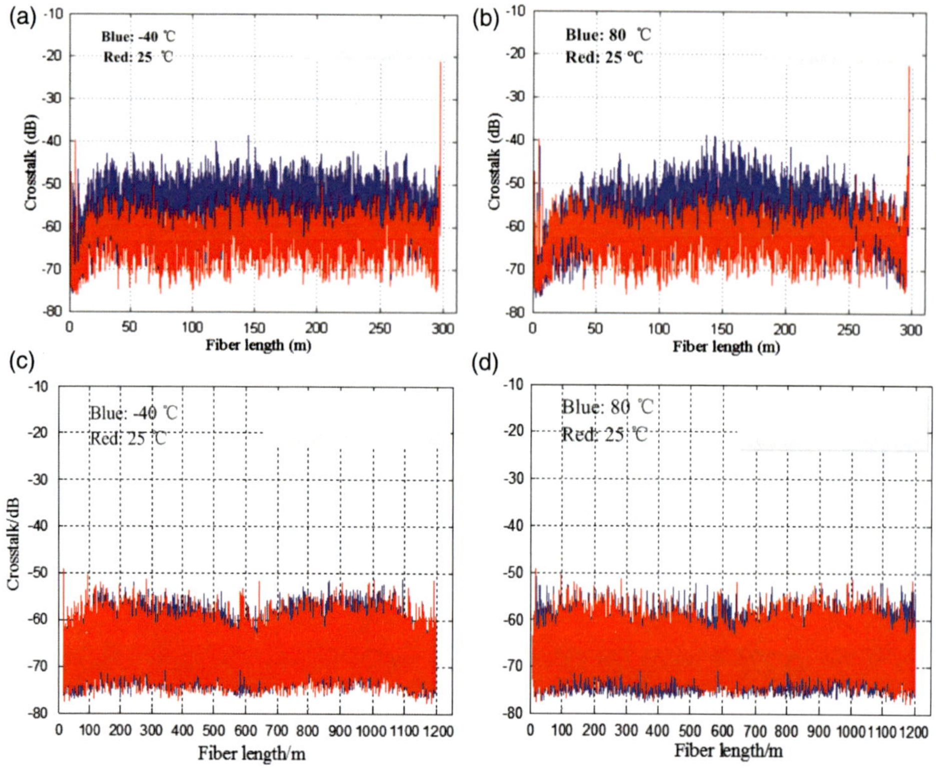

Figure 11.9 Comparison of coils made with different potting adhesives. a) and b) are polarization crosstalk curves at –40 and +80 degrees centigrade respectively made with an off-the-shelf coil potting adhesive, while c) and d) are the crosstalk curves of a coil made with optimized adhesive. The red curves are the reference polarization crosstalk curves at 25°C.

the coil is clearly observed at –40°C, indicting that excessive stress was exerted on the coil caused by the thermal shrinkage of the adhesive. At the high temperature of +80°C, the center portion of the coil experienced more stress than at the two ends, indicating that thermal expansion exerted more stress at the center portion.

Figures 11.9(c) and (d) show the distributed polarization crosstalk of a 1200-m coil potted with a General Photonics proprietary adhesive at –40 and +80°C respectively, compared with the crosstalk at 25°C. Negligible polarization crosstalk degradation is observed at the extreme temperatures, indicating that the adhesive induced extremely low stresses over the whole temperature range of interest.

Note that a large variation of polarization crosstalk induced by temperature is an indication of large local mechanical stress on a fiber coil and may contribute to temperature-dependent bias instabilities of a FOG incorporating the coil to enhance the Shupe effect.

11.2.4 Characterization of coil quality by polarization crosstalk analysis

In general, a good coil has a uniform polarization crosstalk along the fiber length, with a maximum crosstalk around –50 dB and an average crosstalk around –68 dB, with little crosstalk degradation at extreme temperatures. Figure 11.10 shows an example of such a coil manufactured by General Photonics, with polarization crosstalk versus distance curves at temperatures of 80, 60, –20, and –40°C, compared with the crosstalk data at 25°C. It is evident from Fig. 11.10(b) that even at 80°C, few polarization crosstalk peaks arise, indicating excellent performance stability at the extreme high temperature. Similarly at the extreme –40°C, no polarization crosstalk degradation is observed, indicating extraordinary performance at low temperatures.

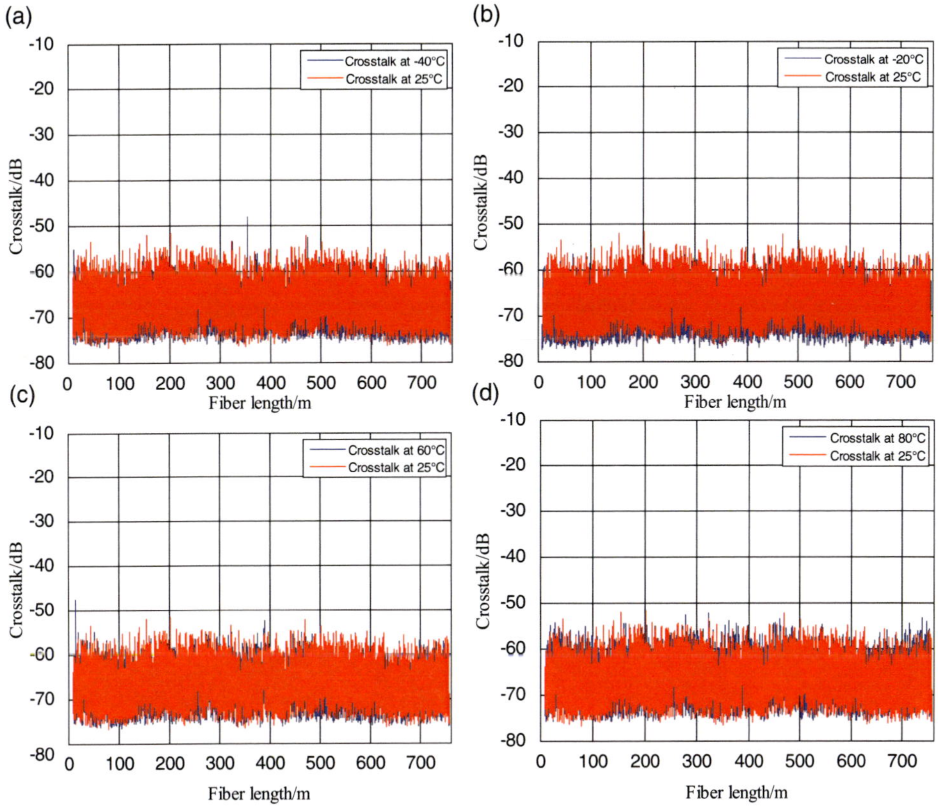

Figure 11.10 Example of a high-quality coil with polarization crosstalk curves measured at (a) –40°C, (b) –20°C, (c) 60°C, and (d) 80°C, showing excellent performance of the coil at extreme temperatures.

11.2.5 Polarization-maintaining fiber characterization and screening

In addition to the winding quality, the quality and characteristics of PM fiber itself will affect the final performance of the coil. Therefore, it is important to first screen the performance and fully understand the characteristics of the PM fiber purchased from a fiber vendor before using it to wind coils. Otherwise, it will be difficult to identify coil defects if problems arise, arguments with the fiber vendor may ensue if the coil manufacturer cannot isolate the fiber-quality issues, and it will be a big waste of both labor and material if low-quality fibers are used to manufacture coils.

The polarization-maintaining ability of a PM fiber is generally characterized by the PER or h-parameter (PER per unit length), whereas the fundamental parameter governing the performance of a PM fiber is actually characterized by its group birefringence. Therefore, it is important for the manufacturers and the users of a PM fiber to know not only the PER but also the group birefringence and all other group-birefringence-related parameters, including group birefringence variations with wavelength (group birefringence dispersion[24,25]), with temperature (group birefringence thermal coefficient[11,26]), and along the fiber (group birefringence uniformity[27]).

In order to accurately obtain these birefringence properties, a simple fixture to induce a series of polarization crosstalk peaks of equal spacing is introduced to assist the measurements. As will be shown in the following subsection, the fixture simplifies the measurement process, reduces many potential measurement errors, and therefore ensures accurate measurement results by least trained personnel. Furthermore, the DPXA can readily be used to identify polarization crosstalk induced by the connectors and splices during measurement setup and therefore eliminate their contributions to the total PER of a PM fiber, resulting in more accurate measurement results without the need to carefully align polarization at the input and output ends of the PM fiber. Finally, a set of measurement parameters obtained by a DPXA are proposed to quantify the performance of a PM fiber, with test examples for different PM fibers. The methods and processes described here can be widely applied throughout the industry for the complete characterization of PM fibers for FOG fiber coil applications.

11.2.5.1 Measurement fixture

To facilitate easy and accurate measurements of group birefringence related parameters, a spool-like fixture was designed to induce periodically spaced polarization crosstalk peaks along a PM fiber, as shown in Fig. 11.11(a). The experiment described here simply uses a standard fiber spool from the fiber manufacturer and affixes a piece of thin metal rod with a diameter of 2 mm across its width. A single layer of PM fiber with a length of 280 m is wound on the spool with a 10-gram winding tension, and each turn of the fiber is closely placed with the previous turn, as shown in Fig. 11.11(b).

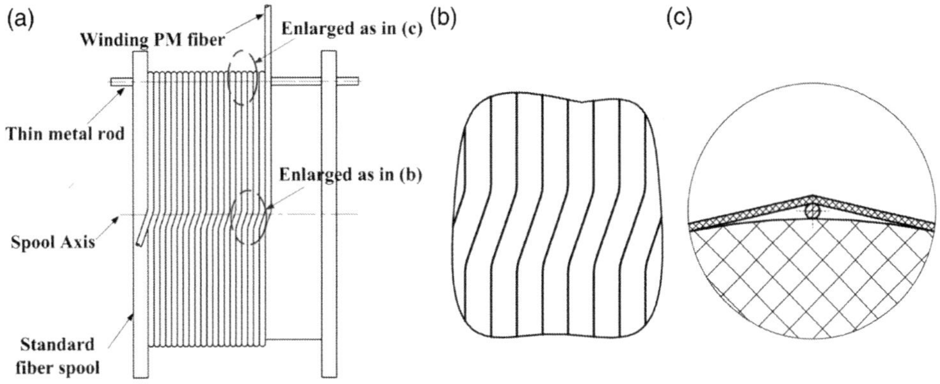

Figure 11.11 (a) Illustration of a length of PM fiber wound on a fiber spool with a thin metal rod to induce periodic transversal stresses on the PM fiber at locations the fiber crossing the rod. (b) Detailed view of the fiber when fiber is transitioning from one turn to the next. (c) Detailed view of the fiber when it crosses the metal rod.

The accuracy of the length of each fiber turn on the spool is critical to the measurement accuracies of the birefringence parameters of the PM fiber to be measured. As shown in Fig. 11.11, the length l of each fiber turn is the circumference l_c of the spool plus the additional length δ_{l1} caused by transitioning each turn of fiber to the next (Fig. 11(b)) and the additional length δ_{l2} by the metal rod (Fig. 11.11(c)):

$$\begin{aligned} l &= l_c + \delta_{l1} + \delta_{l2} \\ &= \pi d_c + d_c \cdot \left(\sqrt{d_c \cdot d_1}/(d_c - d_1) - \arccos\left((d_c - d_1)/(d_c + d_1) \right)/2 \right) \\ &\quad + 2 \cdot d_c \cdot \left(\sqrt{d_c \cdot d_2}/(d_c - d_2) - \arccos\left((d_c - d_2)/(d_c + d_2) \right)/2 \right). \end{aligned} \tag{11.2}$$

The value can be precisely determined when the diameters of the spool d_c, the fiber d_1, and the metal rod d_2 are known. In practice, one may use l_c to approximate l. For our experiment with $d_c = 0.17$ m, $d_1 = 1.65 \times 10^{-4}$ m and $d_2 = 2 \times 10^{-3}$ m, and the relative length error $(\delta_{l1}+\delta_{l2})/l_c$ for the approximation is about 0.1%. The relative length error can be reduced to less than 0.015% with the metal rod $d_2 = 5 \times 10^{-4}$ m. Note that the measurement accuracy of the circumference l_c is about 0.006% when a Vernier caliper is used to measure the diameter of the spool. In comparison, the fiber length measurement accuracy in Ref. 10 is limited by a OTDR (optical time domain reflectometer) or ruler on the order of 1%.

As expected, "point-like" stresses are automatically applied to the fiber at the points where the fiber crosses the metal rod (Fig. 11.11(c)) to produce multiple periodic polarization crosstalk peaks, with a periodicity defined precisely by Eq. (11.2). These periodic crosstalk peaks act like embedded ruler marks on the fiber, which automatically yield precise length information essential for group-birefringence-related measurements, as required by Eq. (11.1). Note that if the ghost peaks caused by second-order coupling are not eliminated, there would be many false peaks between the true periodic peaks, making it difficult to identify the positions of the true peaks.

In practice, such a spool-like fixture can be machined with a precise predetermined diameter (or circumference) and with a thin slot or bump across its width to induce periodic polarization crosstalk, making the embedded ruler more accurate. The reasons for making such a fixture are (1) to precisely define the lengths between crosstalk peaks because the accuracy of the group birefringence measurement is proportional to the accuracy of such lengths, as shown in Eq. (11.1), and (2) to create a distribution of crosstalk peaks to reflect the local group birefringence along the fiber.

Note that the spatial resolution of the local group birefringence is determined by the circumference of the fixture, e.g., on the order of the 0.5 m. It can be easily improved by reducing the diameter of the fixture, considering that the spatial resolution of the DPXA is around 6 cm (assuming the group birefringence of the PM fiber is 5×10^{-4}, although the 0.5-m resolution is sufficient for most PM fiber characterizations.

11.2.5.2 Group birefringence and group-birefringence-uniformity measurements

Figure 11.12(a) is the measured polarization crosstalk curve of a PM PANDA fiber with a core diameter of 6 μm, a cladding of 80 μm, and a buffer of

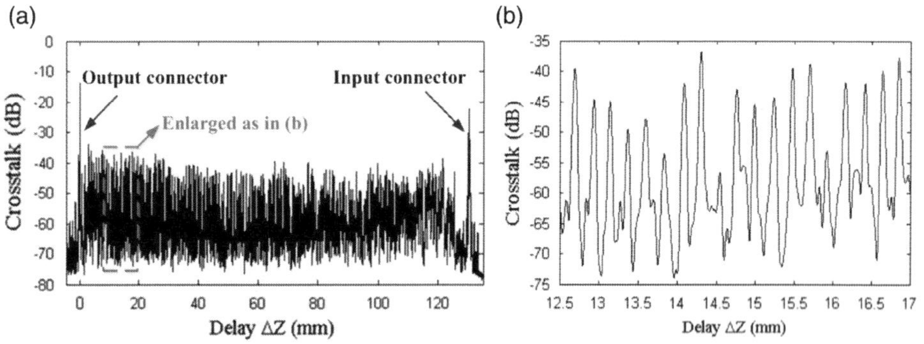

Figure 11.12 (a) Polarization crosstalk curve of 280-m PM fiber wound on the spool as a function of the interferometer delay ΔZ inside the DPXA. The peaks at the far right and far left correspond to the crosstalk induced at the input and output connectors, respectively, from slight birefringence axis misalignment between connecting PM fibers; (b) the equidistant periodic crosstalk peaks are induced by crossing the PM fiber over the metal rod with tension.

165 μm as a function of the interferometer delay ΔZ, showing the polarization crosstalk induced by the line pressure from the metal rod on the fiber. The peaks at the far left and far right correspond to polarization crosstalk induced at the output and input connectors, respectively, due to slight fiber birefringence axis misalignment. Figure 11.12(b) shows a detailed view of the equidistant periodic crosstalk peaks caused by transversal pressures induced whenever the fiber crosses the metal rod. These measured crosstalk amplitudes vary from peak to peak because of the angle variation between the direction of the transversal pressure and the fiber's principal axes during winding the fiber onto the spooling wheel, however such an amplitude variation does not affect the periodicity measurement which is important to the group birefringence measurement. One can readily obtain the spacing between any two stress crosstalk points by simply multiplying the circumference of spool with the number of stress-induced crosstalk peaks between two points. In addition, one can precisely obtain the relative delay ΔZ with the encoder of motorized delay line.

When Eq. (11.1) is used to obtain Δn, the total relative error $\delta_{\Delta n}/\Delta n$ can be expressed as[10]

$$\delta_{\Delta n}/\Delta n = \sqrt{(\delta_{\Delta Z}/\Delta Z)^2 + (\delta_Z/Z)^2} = \sqrt{(\delta_{\Delta Z}/\Delta n)^2 + (\delta_Z)^2}/Z, \qquad (11.3)$$

where $\delta_{\Delta n}$ is the group birefringence inaccuracy, $\delta_{\Delta z}$ is the reading error of the delay ΔZ of the variable delay line inside the DPXA, and δ_z is the measurement error of length Z. Note that in Refs. 10 and 11, the absolute length of the PM fiber under test must be accurately measured in order to obtain an accurate group birefringence Δn according to Eq. (11.3). Any length measurement error will proportionally contribute to the accuracy of Δn. In contrast, the relative length—defined by the circumference of the fiber spool—can be used to eliminate the need for an absolute length measurement and its associated error, and Eq. (11.1) can be rewritten as

$$\Delta n = \delta z/(Nl), \qquad (11.4)$$

where l is the fiber length corresponding to the period of the crosstalk peaks defined in Eq. (11.2), N is an integer to represent the number of periods chosen for the calculation, and δz is the corresponding delay in the interferometer for the N periods. The error sources for Δn are from both the relative location inaccuracy $\delta_{\Delta z}$ between the polarization crosstalk peaks measured with the variable delay line inside the DPXA and the error in the measurement of l. Note that the delay line generally has an error independent of the traveling distance, we therefore choose to use multiple periodicities ($N \gg 1$) in the experiment to reduce the effect of delay line error $\delta_{\Delta z}$, similar to the case of measuring the thickness of a stack of papers in order to accurately determine the thickness of a single paper. We found in our experiment that when $N \geq 5$, the measurement uncertainty is sufficiently small. The average Δn obtained is 4.65×10^{-4} when $N = 5$.

Figure 11.13(a) shows the variation of Δn as a function of distance Z along the fiber for the case of $N = 5$, where Z is obtained by dividing ΔZ by the average Δn obtained in Fig. 11.12, as defined in Eq. (11.1). The large data fluctuations at large distances are caused by the dispersion-induced peak broadening due to the group birefringence dispersion[12] (discussed in Section 11.2.5.3) because the broadening increases the uncertainty of $\delta_{\Delta z}$ in Eq. (11.4). The dispersion compensation procedures described in Ref. 12 can be used to further improve the measurement accuracy by multiplying the distributed crosstalk curve with a dispersion compensation function when group birefringence dispersion of the PM fiber is measured in Section 11.2.5.3. Figure 11.13(b) is the measured group birefringence as a function of distance along the fiber, showing that the measurement uncertainties are greatly reduced when the dispersion compensation procedure in Ref. 12 is applied. It is also evident that the mean Δn slightly varies along the fiber length for the FUT.

Note that the tension applied to the fiber when winding the fiber to the spool is not critical to the measurement accuracy. In a separate experiment, we found no noticeable differences in measured birefringence using the same method when three different tension settings (5, 10, and 15 grams) of the fiber winding machine were used to wind the spool.

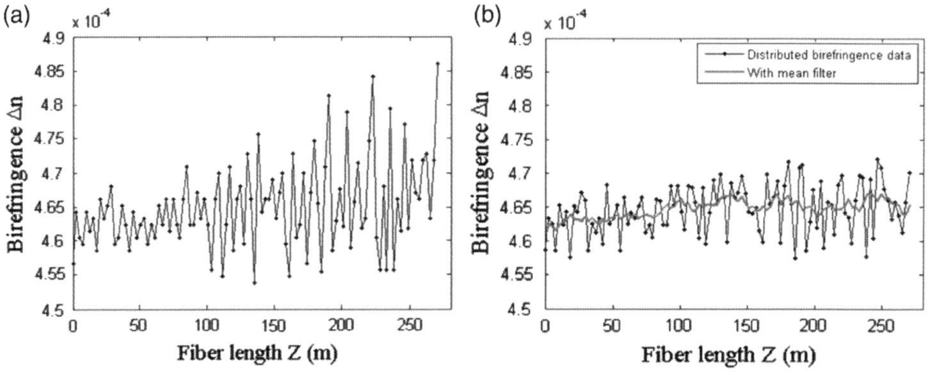

Figure 11.13 (a) Measured group birefringence as a function of distance along the 280-m fiber with $N = 5$ without applying dispersion compensation where the fiber length between any neighboring crosstalk peaks is defined by Eq. (11.2) (0.535 m in the experiment) of the fiber spool. The measurement uncertainty is shown to increase with the distance. (b) The same measured group birefringence as in (a), but the dispersion compensation is applied. The measurement uncertainty at large distances is significantly reduced. The six-point windowed sweeping average of the group birefringence as a function of distance is shown with the center line. In both (a) and (b), the distance zero is at the position of the output connector of the PM fiber under test.

11.2.5.3 Group birefringence dispersion measurement

As discussed in Ref. 22, the envelope of a measured crosstalk peak (i.e., the interference peak) is influenced by the SLED spectral distribution and group birefringence dispersion ΔD of the PM fiber. In fact, the envelope width increases quadratically with the distance Z due to the effect of the group birefringence dispersion, and a relationship between the envelope broadening W and group birefringence dispersion ΔD can be expressed as[12]

$$W/W_o = (1 + (\alpha\Delta D)^2 Z^2)^{1/2}, \tag{11.5}$$

where

$$\alpha = 2\pi c(\Delta\lambda/\lambda_0)^2. \tag{11.6}$$

In Eqs. (11.5) and (11.6), c is the speed of light in vacuum, $\Delta\lambda$ is the 3-dB spectral width of the light source with a Gaussian line shape, λ_0 is the center wavelength of the light source used for the measurement, and W_0 is the $1/e$ width of the interference envelope when the dispersion ΔD or Z equals zero. As shown in Ref. 12, one may simply measure the widths of any two polarization crosstalk peaks with a known spacing Z between them to obtain the dispersion ΔD using Eq. (11.5). However, in order to increase the measurement accuracy of ΔD, the widths of the crosstalk envelopes at multiple locations along the PM fiber are measured, and ΔD is then obtained by curve-fitting to Eq. (11.5).

Figure 11.14 shows the widths of the crosstalk peaks as a function of their locations along the fiber. The distance is measured from the first induced crosstalk peak and evaluated every 20 peaks (N=20 or 10.7 m). Unlike in Ref. 11.12 where the locations of crosstalk peaks were calculated using Eq. (11.1) where the total fiber length Z must be measured precisely, here only the relative locations are required and they can be precisely obtained by multiplying circumference of the fixture with the peak number, minimizing the error contribution of fiber distance in dispersion measurement. One may argue that the group birefringence dispersion may also be obtained using Eq. (11.5) by measuring the widths of the crosstalk peaks caused by the output and input connectors, however, the error of the fiber distance measurement also contributes to the dispersion measurement. In addition, such a two-point measurement is sensitive to measurement uncertainties of the widths. The method presented here can effectively avoid such error sources. As can be seen from Fig. 11.6, the widths of crosstalk peak start to show significant broadening at a distance large than 100 m. The group birefringence dispersion ΔD of the PM fiber is accurately obtained by a least-square fitting to Eq. (11.5) to be $\Delta D = 0.0079$ ps/(km $\cdot$ nm).

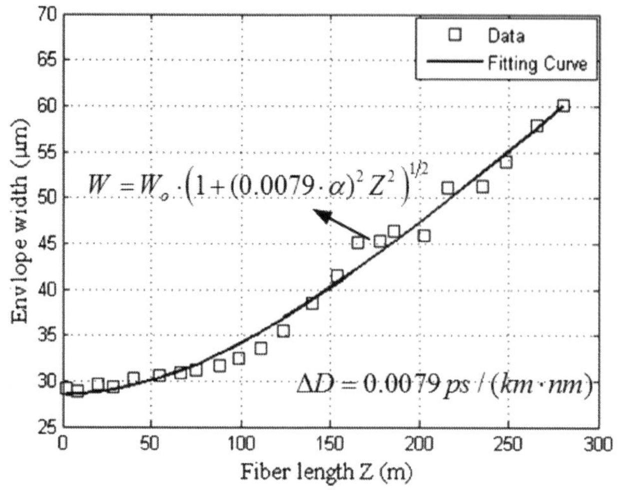

Figure 11.14 Envelope widths of crosstalk peaks induced by stress at various locations along the fiber. The envelope widths of polarization crosstalk peaks broaden as the fiber length increases due to group birefringence dispersion. Curve-fitting obtains the group birefringence dispersion of the PM fiber to be $\Delta D = 0.0079$ ps/(km · nm).

Note that a dispersion compensation function can be obtained once ΔD of the fiber is determined, as described in Ref. 12. This dispersion compensation function can be used to remove the broadening of the crosstalk peaks and hence reduce the measurement uncertainties of the group birefringence along the fiber, as described in Section 11.2.5.2.

11.2.5.4 Group birefringence thermal coefficient measurement

As discussed in Ref. 5, the birefringence of a PM fiber is expected to be sensitive to the temperature because it is the result of the anisotropic strain induced by the differential thermal expansion at different regions in the fiber cladding and varies linearly with temperature in the vicinity of room temperature. The group birefringence Δn can be written as[11]

$$\Delta n = \gamma(T_0 - T), \tag{11.7}$$

where T is the temperature of the PM fiber under test, T_0 is the softening temperature of the silica glass with dopants in the stress-inducing region of the cladding, and γ is the thermal coefficient of the group birefringence of the PM fiber to be measured.

As in Ref. 11, a fiber spool of Fig. 11.11(a) is placed inside a temperature chamber, with two fiber pigtails outside of the chamber. Figure 11.15(a) shows two typical polarization crosstalk curves of the PM as a function of relative delay for two different temperatures of 80°C and 40°C. The positions of all polarization crosstalk peaks are shifted with the temperature, as

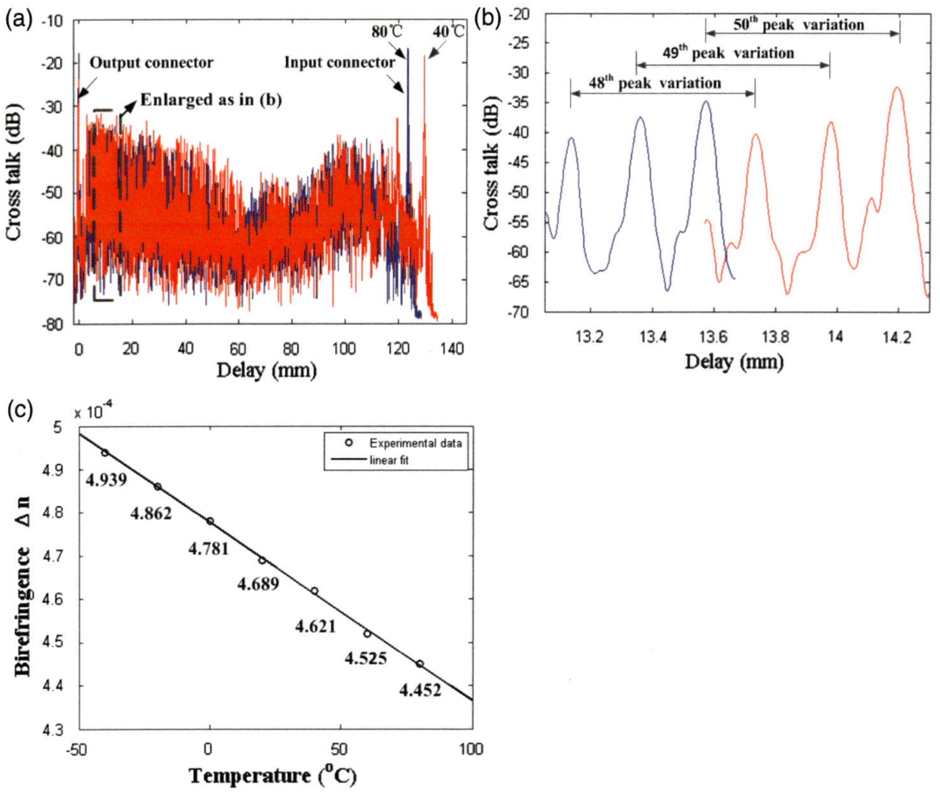

Figure 11.15 (a) Polarization crosstalk curves of a PM fiber as a function of the relative delay at 80°C (blue) and 40°C (red). (b) The expanded view of the positions of the 48st, 49th, and 50th peaks at 80°C (blue) and 40°C (red). (c) Δn obtained at seven different temperatures by measuring the spacing between the 1st and the 50th crosstalk peaks at different temperatures using Eq. (11.4).

predicted in Eq. (11.7). Note that in Ref. 11 the thermal coefficient of group birefringence was obtained by measuring the position of the crosstalk peak induced by the input connector as a function of temperature. However, as described in Ref. 10, there are two major error sources that affect the measurement accuracy: (1) the fiber length measurement error and (2) the fiber length outside the temperature chamber. In order to make an accurate measurement, the fiber inside the chamber must be sufficiently long and the fiber pigtails must be kept sufficiently short.

For more accurate measurement, both error sources can be effectively minimized for the following reasons: (1) because the periodic polarization crosstalk peaks induced by the measurement fixture acts as ruler marks along the fiber, the fiber length measurement between any two peaks can be easily obtained with a high precision; and (2) we only measure the relative position variations with temperature between any two periodic crosstalk peaks on the fiber section inside the chamber and therefore eliminate the error contribution of fiber sections outside of the chamber. In the experiment, the selected approach measures the spacing ΔZ between the 1st and 50th peaks, and then the group birefringence Δn as a function of temperature. As shown in Fig. 11.7(b), the peak positions of the 48th, 49th, and 50th peaks shifted to the left as the temperature increases, reducing the spacing ΔZ. The fact is that the spacing ΔZ decreases with the temperature indicates that Δn has a negative thermal coefficient. As mentioned previously, a thermal coefficient of the group birefringence γ can be obtained by a linear-fitting of Δn to Eq. (11.7) by using the least-square fitting method at different temperatures. Note that in order to reduce the effect of dispersion, crosstalk peaks close to the output end of the fiber under test are selected, although the dispersion compensation described in Section 11.2.5.2 may also be used to reduce the peak broadening and improve measurement accuracies to measure peaks close to the input connector ($N \gg 50$).

In our experiment we measured Δn for a PM fiber at seven different temperatures (i.e., –40°C, –20°C, 0°C, 20°C, 40°C, 60°C, and 80°C), and the results are plotted in Fig. 11.7(c). Linear-fitting Δn to Eq. (11.7) yields a group birefringence thermal coefficient $\gamma = -4.123 \times 10^{-7}$.

11.2.5.5 PER measurement

As mentioned in the introduction, traditional methods[28,29] or that using a PER meter to measure the PER of a PM fiber are susceptible to (1) polarization misalignment at the input end of the fiber under test, and (2) polarization misalignment between the light source and its fiber pigtail if a pigtailed light source is used. Using a DPXA, one can readily identify the crosstalk contributions from the polarization misalignments at the two fiber ends, as well as at the interface between the pigtail and the light source, and eliminate their contributions to the total PER because the corresponding polarization crosstalk peaks measured with a DPXA are spatially separated. Note that for the PER measurement described in this section, we do not induce the periodic polarization crosstalk peaks, as in the previous sections.

Figure 11.16 shows the measured polarization crosstalk curves of a PM fiber jumper with FC/PC connectors and a spool of PM fiber of 250 m directly from a PM fiber vendor, fusion spliced with two FC/PC connectors. When a PM fiber under test is connected to the DPXA, the polarization misalignment at the connection points induces significant crosstalk peaks. An auto-search program is implemented in the DPXA software to automatically identify those peaks, because the polarization crosstalk signatures of the fibers inside DPXA are known, as shown in Fig. 11.16. In addition, the polarization crosstalk peak resulting from the light source and its pigtail is located outside the region defined by the two connectors, and thus is not included for PER calculation. By definition, the PER of the fiber can be calculated as

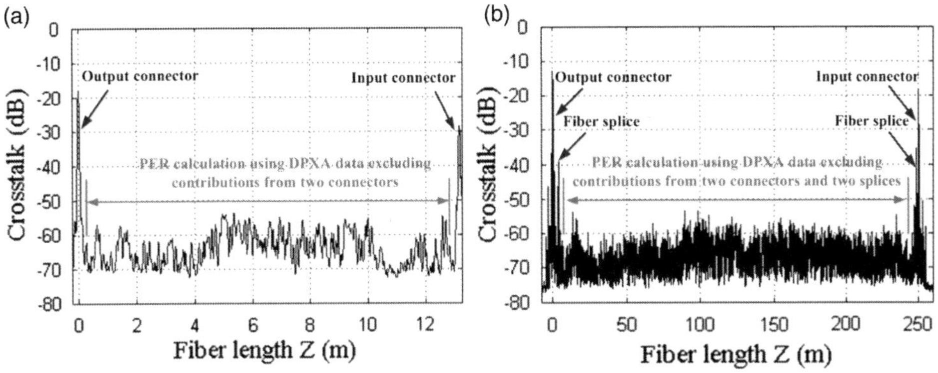

Figure 11.16 (a) Polarization crosstalk curves of a 13-m jumper with two FC/PC connectors and (b) a 250-m PM fiber coil spliced with two FC/PC connectors. PER measurement with a commercial PER always includes the contributions of the input connector and two splices, while the DPXA has the ability to identify and eliminate polarization crosstalk contributions of all connectors and splices in the measurement system. Note that fiber length in the horizontal axis is obtained by dividing the fiber delay line distance ΔZ with the average group birefringence obtained using the procedure described in Section 11.2.5.2.

$$PER = 10\log(P_f/P_s), \tag{11.8}$$

where P_f is the total power coupled to the fast axis from the slow axis and can be obtained by integrating of all polarization crosstalk between the two connectors, and P_s is the total power remain in the slow axis $P_s = P - P_f$, where P is the total received power at the fiber output.

We implement an algorithm in DPXA software to automatically calculate the PER excluding the contributions of the two end connectors from the crosstalk measurement curve, as shown in Fig. 16a. One may also use the DPXA software to calculate the total PER contribution between any two points along the fiber, and therefore to further exclude the contributions from the two fusion splicing points, as shown in Fig. 11.16(b). Table 11.1 compares multiple PER measurement results of a 13 meter fiber jumper and a 250 meter

Table 11.1 Comparison of PER measurements of a 13-m PM fiber jumper and a 250-m PM fiber coil obtained with a commercial PER meter and a DPXA.

	Fiber Jumper PER		Fiber Coil PER	
Measurement #	ERM (dB)	DPXA (dB)	ERM (dB)	DPXA (dB)
1	22.5	34.81	25.7	30.82
2	24.9	34.27	21.8	30.81
3	23.3	35.06	25.4	30.31
4	26.8	34.64	22.5	30.90
5	25.8	35.14	23.8	31.09
Uncertainty (max–min) (dB)	4.3	0.87	3.9	0.78

PM fiber coil obtained with a commercial PER meter and a DPXA. It is evidence that the PER value obtained by PER meter is several dB smaller than that obtained with a DPXA, due to the contributions of crosstalk from the polarization misalignment at the input connector. In addition, as anticipated, the measurement repeatability of a DPXA is much better that that of using a PER meter. Therefore, it is much easier to use a DPXA to obtain more accurate PER measurements than using a PER meter.

Note that we used a SLED source with a spectral width around 30 nm when using a PER meter for the PER measurement, in accordance with the requirement of test standard TIA-544-193 [29] to avoid measurement fluctuations caused by coherent effect resulting from the use of a narrow-linewidth laser.[30]

11.2.5.6 PM fiber-quality evaluation

Up to now, the only parameter that a user can use to characterize the polarization performance of a PM fiber from a vendor is the PER or h-parameter (PER per fiber length), which may not be able to reflect the true polarization performance of the fiber, especially considering that PER measurement using conventional methods may have significant fluctuations, as discussed previously. A set of parameters from a single DPXA scan are described here to fully describe the performance without ambiguity.

Figure 11.17 shows DPXA scans of three different PM fibers measured directly with the fiber on the fiber spool from the vendor. Four parameters characterize the quality of a PM fiber for the polarization-related performance:

1. the average polarization crosstalk,
2. the maximum crosstalk,
3. the number of crosstalk peaks above a certain threshold defined by the manufacturer or the user, and
4. PER.

The average crosstalk is the major contributor to the value of PER and closely relates to the h-parameter. The maximum crosstalk is an indication whether the PM fiber is degraded or damaged during manufacturing, spooling, or shipping of the PM fiber, although a single or few high-crosstalk peaks contribute insignificantly to the total PER of a long fiber. For some applications, such as fiber gyro coils, the high-crosstalk sections must be removed to assure high-quality fiber coil production. A large number of high-crosstalk peaks present in the fiber may indicate problems in the fiber manufacturing or spooling process, which also makes it impractical to sort out only good fiber sections for demanding applications.

Table 11.2 lists four parameters of the three different fibers under test, obtained from Fig. 11.17. Fiber I is a commercial PANDA fiber at 1310 nm

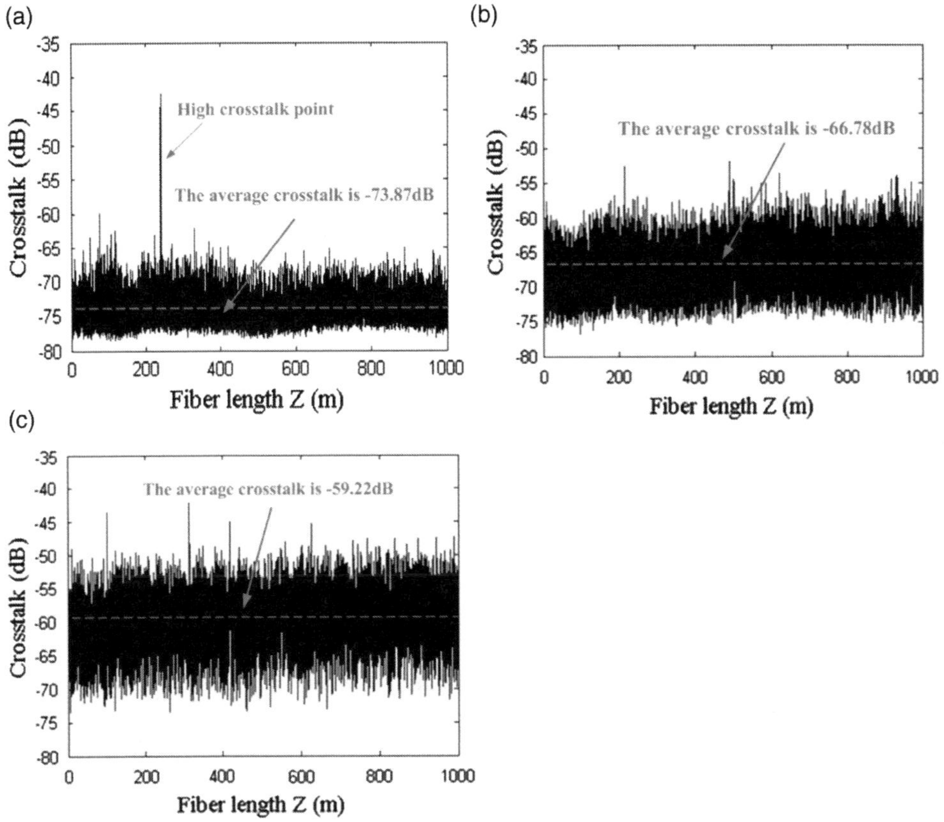

Figure 11.17 Polarization crosstalk curves of three different PM fibers: (a) a PANDA PM fiber at 1310 nm with a buffer diameter of 250 μm. A defect point is seen at around 220 m. (b) A PANDA fiber at 1310 nm of the same core/cladding diameters in (a) but with a reduced buffer diameter of 169 μm. (c) A third PANDA PM fiber at 1310 nm with the similar cladding diameters as in (a) but with an even more reduced buffer diameter of 136 μm. Two major defect points with crosstalk more than 45 dB are seen.

Table 11.2 Four parameters to fully characterize the quality of three different PM fibers.

	Average Crosstalk (dB)	Maximum Crosstalk (dB)	Number of Crosstalk Peaks above −55 dB	PER (dB)
Fiber I	−73.87	−42.36	1	28.8
Fiber II	−66.78	−51.88	23	23.6
Fiber III	−59.22	−42.15	1711	16.5

with a beat length of 2.57 mm, a core diameter of 6 um, a cladding diameter of 125 μm, and a buffer diameter of 250 μm, respectively. Fiber II is a different PANDA fiber at 1310 nm with the same core as fiber I but a different beat length of 2.13 mm, a cladding diameter of 80 μm, and buffer diameter of 169 μm, respectively. Finally, fiber III is a third type of PM fiber at 1310 nm with a beat length of 2.6 mm, a core diameter of 6.4 μm, a cladding diameter of 80 μm, and a buffer diameter of 136 μm. It is evident from Fig. 11.17(a) that fiber I has the lowest average crosstalk, resulting in a highest PER of 28.78 dB; however, it has a defect point about 220 m measured from the output connector with a high crosstalk peak of –42.36 dB, probably caused by mishandling when winding the fiber to the spool. Such a defect cannot be identified with a simple PER measurement. Experiments found that such a defect may be permanent, e.g., cannot be recovered even when the corresponding stress is released. On the other hand, fiber III has the highest average crosstalk of –59.22 dB, corresponding to a low PER of 16.25 dB. It also has a large number of high crosstalk peaks above –55 dB, probably because the thin buffer layer (136 μm) cannot effectively protect the fiber from external stresses. Therefore, all four parameters collectively give a full picture of the quality or performance of the PM fiber under test.

11.2.6 Single-mode fiber coil inspection

Depolarized FOGs that use low-cost single-mode (SM) fiber coils[31] are a viable alternative to PM-fiber–based FOGs. In addition to low fiber cost, another reason is the higher level of radiation tolerance that can be achieved with simpler single-mode-fiber designs. How to quantify the performance SM fiber coils, especially at different temperatures, has also been a challenge. This section begins with experimental data to use polarization-mode dispersion (PMD) and polarization-dependent loss (PDL)[32,33] to characterize SM fiber coils. Finally, polarization-analyzing optical frequency-domain reflectometry is used to obtain the distributed stress along the fiber coil to get space-resolved quality information about the fiber coils.

11.2.6.1 Lumped PMD and PDL measurements

PMD represents the accumulative retardation or birefringence caused by the stress on the fiber. A well engineered fiber coil is expected to have extremely low stress on the fiber, and hence low PMD. Excess stress on the various sections of fiber is expected to cause temperature instabilities and local asymmetries. Fig. 11.18(a) shows the measured PMD values of two SM coils made with two different potting adhesives as a function of temperature, using a highly accurate polarization-analyzing system (PSGA-001) based on binary polarization rotators.[33] Due to the superb measurement resolution and accuracy of the binary system, very small PMD variations can be clearly obtained. As can be seen, for the coil made with an off-the-shelf adhesive, a large PMD value increase is observed at low temperatures, indicating that high thermal stress was induced at low temperatures, probably caused by the shrinkage of the adhesive. In contrast, for a coil made with a specially formulated adhesive, low PMD values are observed at all temperatures, even at the low-temperature setting. Such experimental data indicates that this

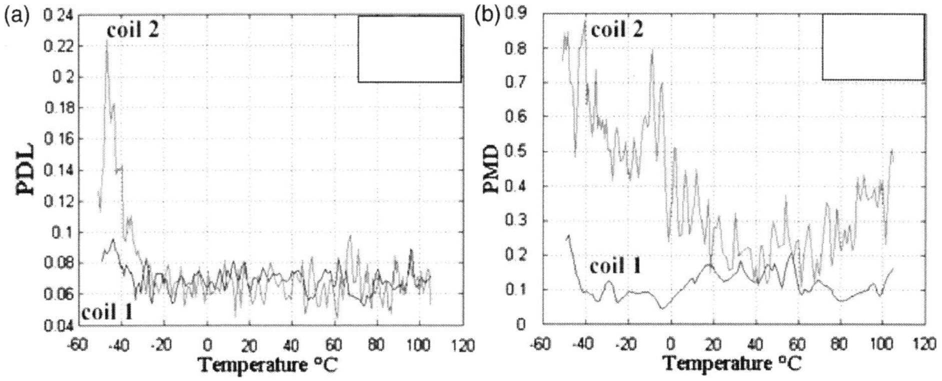

Figure 11.18 (a) Total accumulated PMD and (b) total PDL of coils with different potting adhesives as a function of temperatures. Coil 1 features a specially formulated adhesive developed by General Photonics for potting fiber gyro coils, whereas coil 2 uses an off-the-shelf adhesive.

adhesive is well suited for coil production because of the low shrinkage in the temperature range of interest.

Micro-bending of a fiber can generate PDL, and therefore micro-bending should be avoided during the fiber-winding process, especially when the fiber is transitioned from one layer to another, because PDL re-polarizes the depolarized light signal and causes instabilities. In addition, micro-bending–induced stress on the fiber also induces local asymmetries for the counter-propagating light signals. Figure 11.18(b) shows the experimental results of PDL measurement of the two different coils made with different potting adhesives, as in Fig. 11.18(a). For the coil made with the off-the-shelf adhesive, a large PDL increase is observed at low temperatures, consistent with the PMD measurement shown in Fig. 11.18(a), whereas the PDL of the coil made with the specially formulated adhesive shows almost no PDL increase at low temperatures, indicating the superb winding quality of the coil and great suitability of the adhesive for the coil fabrications.

11.2.6.2 Distributed transversal stress measurement

The PMD measurement described in Section 11.2.6.1 is actually the accumulative birefringence induced by the transversal stresses at different locations along the optical fiber coil during the winding process. Sometimes it is desirable to identify the exact locations at which the stresses are induced so that one can use the information to eliminate the stress points and improve the winding process.

Polarization sensitive optical time-domain reflectometers (OTDRs) were reported for measuring the distributed PMD in telecommunication fiber links.[35,36] However, the spatial resolution of the proposed method, on the order of 0.5 m, was not sufficient to detect the locations of high-stress points in fiber coils. In 2004, Froggatt et al.[37] used a technique of cross-correlating a current-measured optical frequency-domain reflectometry (OFDR) signal with a reference OFDR trace to obtain the space-resolved strain changes along the optical fiber. However, such a method also suffers from the deficiency that it cannot distinguish the strain from the temperature. In addition, no absolute strain or stress information can be obtained because such a measurement is only relative to the reference. It can be shown that polarimeteric OFDRs (P-OFDRs)[38–41] are capable of detecting distributed-stress-induced birefringence along the fiber, with sufficiently high spatial resolution, and thus such a P-OFDR was designed and built[42] to locate local stress points inside a fiber coil.

Figure 11.19 shows the setup for the P-OFDR measurement system, where a tunable laser with a linewidth around 100 kHz is tuned continuously up to 4 nm around 1550 nm. A small fraction of light (~5%) from coupler 1 is split and then sent to a Faraday rotation mirror (FRM) based Michelson interferometer with a relative time delay of 10 μs to generate clock signals at each "zero-crossing" position of the optical interference fringe signals. The generated real-time clock signals were then used for the P-OFDR raw data sampling. The light was further split by a 90/10 coupler (coupler 3), in which 10% of the light was used as a local oscillator light to coherently probe Rayleigh backscattering (RBS) light, and 90% of the light was sent to the fiber under test (FUT) via a circulator. The state of polarization (SOP) before entering and after exiting the FUT is randomly varied by an all-fiber polarization controller to uniformly cover the Poincaré sphere. The Rayleigh backscattered light from different locations of the fiber is brought to interfere with the portion of light directly from the tunable laser via a polarization

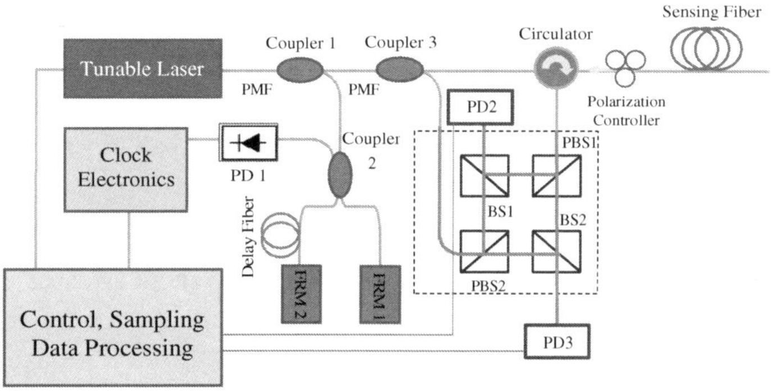

Figure 11.19 Experimental setup of distributed transversal stress measurement using a P-OFDR. PD 1-3: photo detectors. Note that the pigtails of couplers 1 and 3 are of PM material to avoid polarization fluctuations when light interferes. PBS1 and 2 are free-space polarization beamsplitters, whereas BS 1 and 2 are free-space polarization-insensitive beamsplitters. The PM fiber input to PBS2 is oriented such that the power is equally split by PBS2.

diversity interferometer (the dotted-line box in Fig. 11.19). The interference signals of the two polarization components are analyzed after detection by PD2 and PD3, and the local birefringence information is then calculated. The transverse stress related to the birefringence $W(z)$ is then obtained using the following equation:

$$W(z) = \frac{1}{2\Delta z} \cdot \sqrt{15\langle \Delta T^2(z) \rangle_{\text{I/O-SOP}}}, \tag{11.9}$$

where Δz is the distance between two locations at which the SOP evolution is analyzed, $\Delta T(z) = T(z + \Delta z/2) - T(z - \Delta z/2)$ is the normalized light power difference between two positions at $z + \Delta z/2$ and $z - \Delta z/2$, and $T(z) = P_i(z) / [P_2(z) + P_3(z)]$ is the normalized light power at the location z, where $P_i(z)$is the light power of one polarization component from PBS1 at z measured with one of the detectors Pdi ($i = 2{,}3$). In Eq. (11.9), $< >_{\text{I/O-SOP}}$ stands for averaging over the random input and output SOP variations. The fiber transverse stress

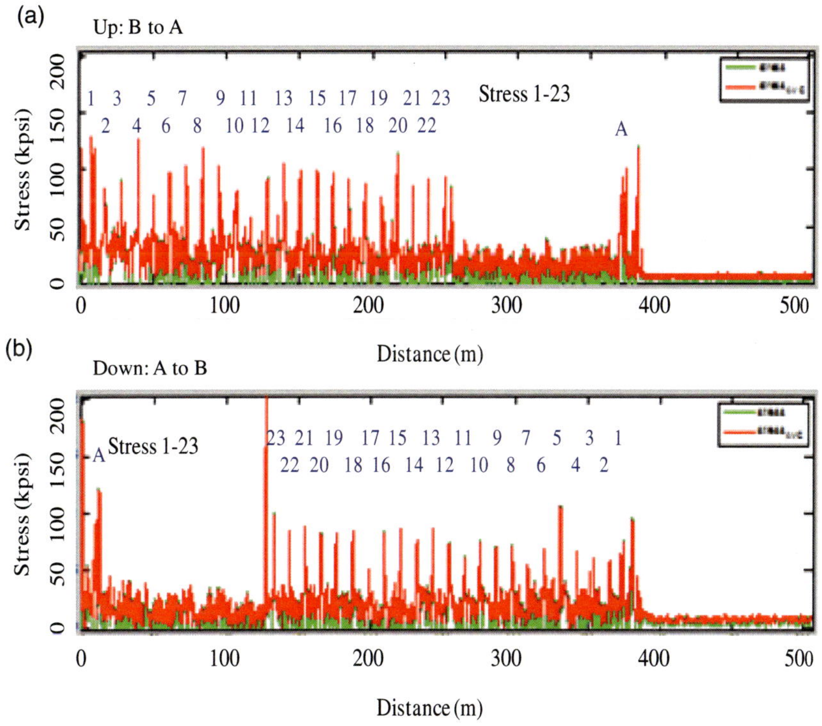

Figure 11.20 Distributed stress measurement of a 250-m fiber coil connected with a loose fiber spool ~130 m long. In the upper image, light enters the fiber from coil side B; in the lower image, light enters the fiber from spool side A. The fiber used is the bend-insensitive fiber Corning ClearCurve ZBL single-mode optical fiber, and the tension for winding the coil is ~10 grams.

as a function of the distance can be extracted from measured distributed fiber birefringence.[44]

To demonstrate the capabilities of the P-OFDR, the prototype system is used to measure a fiber coil with a diameter of 80 mm, which can clearly identify periodic stress peaks, as shown in Fig. 11.10. The periodicity of 23 peaks is the same as the length of each layer of the coil, consistent with the expectation that stresses are induced at fiber crossover points when the fiber transitions from one layer to the other during the winding process. There are a total of 23 peaks, corresponding to 23 layers of fiber in the fiber coil. To further prove that the periodic peaks are indeed from the coil winding, a length of fiber around 130 m loosely wound on a large spool of about 50 cm was spliced onto the coil and distributed stress tests were performed from both ends of the combined fiber A and B, with the A side being the large spool and the B side being the fiber coil. Periodic stress peaks were only observed in the fiber coil, not in the loosely wound fiber spool, no matter the direction of the light. This preliminary test result of transversal stresses on the fiber coil indicate that the example P-OFDR is sufficiently sensitive to detect such small winding-induced stresses and is promising for the production process control to improve the fiber-coil quality.

11.2.6.3 Degree-of-polarization tests

A depolarized FOG using SM fiber greatly reduces the cost of fiber coils because the cost of SM fiber is over 100 times less than that of PM fiber. An important parameter for the depolarized FOG is the degree of polarization (DOP), which must be sufficiently low for comparable performance with its PM fiber counterpart. In general, two depolarizers are required at each side of the coil, and the lengths of the depolarizers are on the order of 20 m, depending on the coherence length of the light source used in the FOG. It is desirable to make the depolarizer as part of the coil with the quadrupole winding pattern, otherwise the depolarizers may introduce asymmetry and cause bias instability of the gyro.

In practice, Lyot depolarizers are first made with two lengths of PM fibers with a 45-deg cross-splice, and the DOP of the depolarizers should be measured to ensure their quality. It is also important to measure the DOP of the final coil after the depolarizers are spliced onto the coil and are wound to the coil with the quadrupole winding pattern for the Shupe-effect reduction. The depolarized light may be re-polarized if the PDL and PMD of the coil are sufficiently high.

The DOP can be measured with polarimeters based on Muller matrix analysis. However, the accuracy and resolution for such measurement is on the order of 2%, which is insufficient because the DOP value in a typical depolarized FOG is less than 1%. In order to meet the demanding requirement for FOG applications, the maximum/minimum search method[45] has been adopted for DOP measurement with a DOP resolution of 0.1%.

11.3 Coil Transient Parameter Characterization

Previous discussions have mainly concentrated on the static parameters of fiber gyro coils. Another set of important parameters relates to the transient properties of the FOG, reflecting the performance of the coil under varying temperature (Shupe errors) and vibration. This section introduces a parameter called pointing-error thermal sensitivity (PETS) to quantify the performance of a quadrupole fiber coil subject to a time-dependent radial temperature gradient. The pointing error (asymptotic angular error) of a gyro system that incorporates the quadrupole fiber coil is linearly proportional to the final temperature gradient between the inner and outer coil surfaces in the radial direction, with PETS as the proportional constant. Furthermore, PETS is linearly proportional to a parameter called the effective asymmetric length of the fiber coil, caused by imperfections in fiber winding. In addition, by thermally inducing different time-varying radial temperature gradients on the fiber coil (as will be shown in inset (b) of Fig. 11.21) and measuring the corresponding pointing errors in a gyroscopic measurement setup, the PETS of the fiber coil and its associated effective asymmetric length can be determined with confidence. Such parameters can be directly used to trim the fiber coil [7, 46] for better temperature performance, feedback to coil production for improving winding process, or simply be used to quantitatively

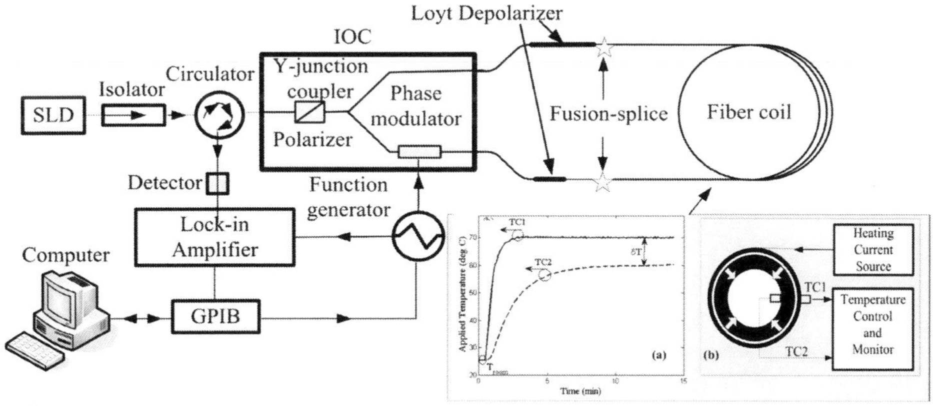

Figure 11.21 Gyroscopic setup for measuring thermal induced angular error signals. Inset (b) shows a ribbon heater with a width and a length equal to the width and the perimeter of the coil is wrapped around coil to apply a radial temperature gradient. TC1 and TC2 are the temperatures measured at the outer and inner surfaces of the fiber coil. Inset (a) shows TC1 and TC2 as a function of time when a radial temperature excitation is applied and are stabilized at 70.2°C and 54.3°C, respectively, resulting in a temperature gradient $\delta T = TC1 - TC2 = 15.9°C$.

determine the quality of fiber gyro coils. Finally, we trim multiple coils according to the obtained asymmetric lengths to achieve best thermal performance, and experimentally verify the usefulness and correctness of the PETS definition and the asymmetry length relations.

Three frameless SM fiber coils with conductive potting adhesives are used in this study and all of them were wound with the quadrupole winding scheme. The coils are approximately 1.25 cm tall with an inner diameter of 5.3 cm. The total length of fiber on each coil is 240 m, producing 24 fiber layers. Each fiber coil is placed in an open-loop gyroscopic setup shown in Fig. 11.21 to form a complete fiber gyro and to be evaluated for its thermal performance. As shown in insets (a) and (b), a temperature gradient is applied to a coil in the radial direction such that the fiber in each layer experiences the same temperature. The temperatures at the coil's inner and outer surfaces were monitored with two separate thermocouples. In the experiments, the lengths of fiber pigtails are kept constant at 2 m each.

Using the setup in Fig. 11.21, six different temperature excitation profiles were applied to the outer layer of each coil (TC_2 = 30°C, 40°C, 50°C, 60°C, 70°C, 80°C). Figure 11.22(a) shows the measured pointing error ψ_e as a function of δT of the three coils without trimming. The pointing error increases linearly with δT, agreeing well the theory.[15] The slope of each line is the PETS of the coil, and curve-fitting finds the PETS γ_T of three coils to be 0.086, 0.083, and 0.067 deg/°C, respectively. The corresponding asymmetry lengths of the three coils (coils 1, 2, and 3) can be obtained as 4.3 m, 4.1 m, and 3.33 m.[15] Figure 11.22(b) shows the effect of fiber trimming of coil 1. It is evident that the PETS coefficients of coil 1 become smaller and smaller, and eventually approaching zero when the trimming length approaches the asymmetry length. Finally, the PETS changes the sign when the coil is over-trimmed.

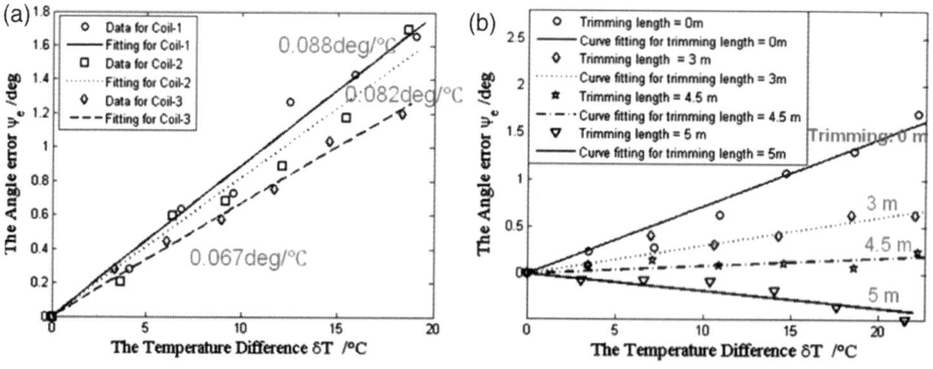

Figure 11.22 (a) Pointing error taken before coil trimming. The lines indicate the best linear fit through the data points, and the slope of each line is the PETS of the coil. (b) Pointing error of coil 2 with different trimming lengths on the "A" portion fiber.

To further demonstrate the effect of fiber trimming, the Shupe errors were measured by placing a coil of 1.2 km in a temperature chamber with different temperature profiles. The measurement was performed using a closed-loop fiber gyro system with the fiber coil detached from the main body of the gyro. Figure 11.23(a) shows the bias rate of the coil with a temperature profile shown with the green line with a peak-to-peak Shupe error of 0.4 deg/h. The peak-to-peak Shupe error is reduced to 0.08 deg/h after trimming, as shown in Fig. 11.23(b). Figure 11.23(c) shows the bias rate of the same coil under a stairway temperature profile, with a peak-to-peak Shupe error of 0.338 deg/h. The Shupe error is reduced to 0.11 deg/h after trimming, as shown in Fig. 11.23(d). The demonstration shows that coils can be produced using the fiber-trimming technique with extremely low temperature-induced rate errors.

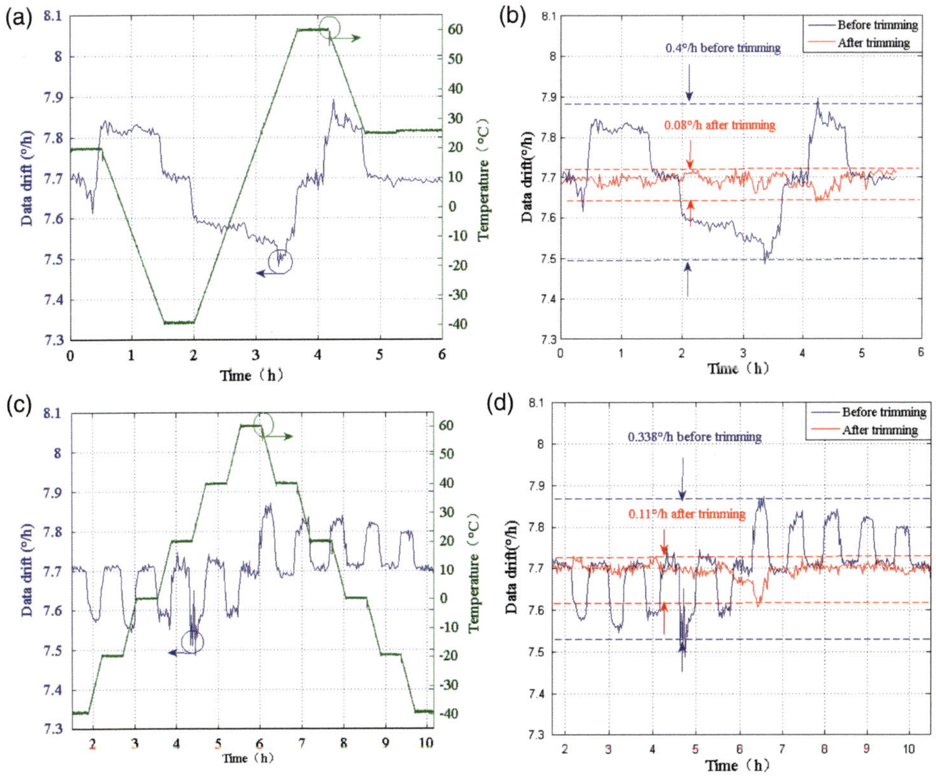

Figure 11.23 (a) Bias drift (blue) due to the Shupe effect when a coil of 1.2 km is placed in a temperature chamber with a temperature profile shown with the green line. (b) The Shupe error (red) after trimming with the same temperature profile of (a). (c) The bias drift due to the Shupe effect of the same coil in the temperature chamber with a stairway profile shown in green. (d) The Shupe error after trimming (red) with the same temperature profile as in (c). The temperature slope rate in both (a) and (c) is 1°C/minute. The bias drift data were obtained with 100-s averages.

11.4 Tomographic (3D) Inspection of Fiber Gyro Coils

As illustrated in Fig. 11.2(a), there are many winding defects more or less present in a final fiber coil product that compromise the performance of the resulting FOG, such as the bias stability and the rate or angular errors caused by thermal variation and mechanical vibrations. Therefore, it is important to evaluate the quality of fiber coils during the winding process and in the finished product to ensure high-quality production. Video inspection with CCD cameras is commonly used to monitor the winding of a fiber coil; however, it only provides surface images and is difficult to identify the defects and pinpoint the location of poorly wound sections beneath the surface of the fiber coil.

Optical coherence tomography (OCT) is a relatively new imaging technique for detecting the internal microstructures of a sample and obtaining its tomographic images with high spatial resolution. It is most widely used for medical applications, such as ophthalmology,[47,48] cardiovascular-disease diagnosis,[49] dentistry,[50] dermatology,[51] endoscopy,[52] blood-glucose monitoring,[53] and the diagnosis of bronchus[54] and macular problems.[55] Recently, OCT has been applied to industrial applications to exploit its noninvasive, high speed, and 3D qualities.[56–59]

OCT is inherently suitable to characterize fiber coils. First, a fiber coil is generally made by layers of optical fibers with diameters ranging from 100 to 250 microns. Such a layered structure is easily analyzed with tomographic images obtained by OCT. Each fiber layer can be clearly viewed and each fiber in each layer can be inspected individually. Other tomographic imaging techniques, such as x-ray computerized tomography (CT), may not have sufficient spatial resolution. Second, a fiber coil may also contain adhesives surrounding each turn of optical fiber, with an index refraction and a scattering coefficient different from those of the fiber. OCT images therefore can clearly identify the interfaces between the fiber turns and the adhesive, and detect any winding defects. Finally, all tomographic images can be reconstructed into 3D images to be viewed at different angles and different sections of viewer's choice, making the visual inspection more user friendly.

This section shows how to obtain different OCT images of a fiber coil in experiments and demonstrate the ability to use OCT images to visualize the conditions of fibers in different layers and identify winding anomalies. The following discusses how to implement the OCT inspection technique for fiber coil production and quality assurance. Finally, we discuss the limitations of current OCT system and area for improvement. We believe that OCT inspection can be a useful tool for the production of quality fiber coils.

Figure 11.24 shows the experimental setup for obtaining OCT images of a fiber coil. A commercial swept-source OCT (SSOCT) system is used to perform the experiments. A 3D reconstruction software was developed for the OCT system to allow the user to visualize data with different viewing angles and perspectives, with the flexibility of selecting proper experimental

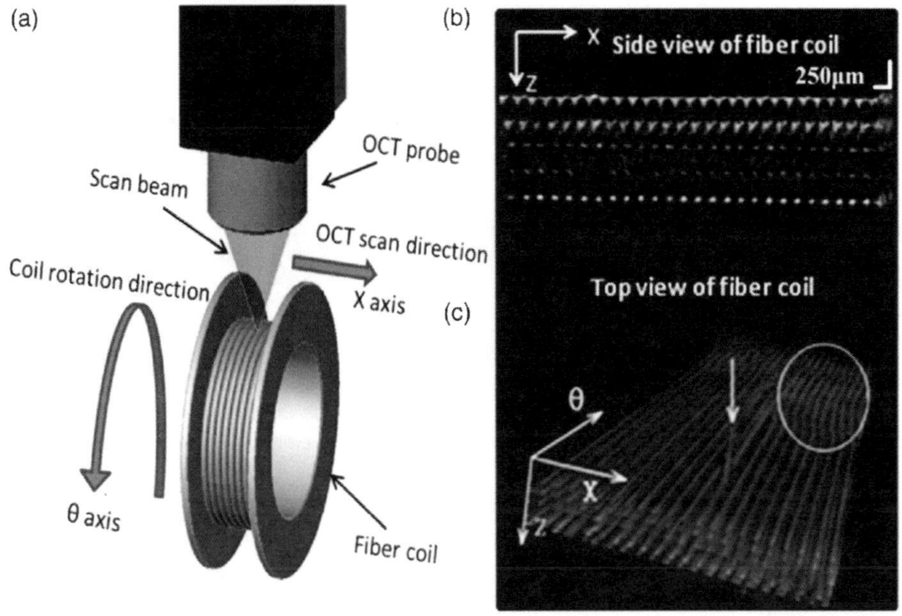

Figure 11.24 (a) Illustration of OCT image acquisition of a fiber coil. The fiber coil is affixed on a spindle of a lathe or a winding machine, and is rotated along its center axis at a fixed speed. The probe of the OCT system is above the fiber coil. The *x* axis is the OCT scan direction, and the θ axis is the rotating direction of the fiber coil. (b) The 2D tomographic image of the coil along the *x* axis. (c) The reconstructed 3D OCT image of a four-layer fiber coil without adhesive.

parameters, such as the lateral scanning range, the step width and the data sets to be accessed offline for further image processing and data analysis. In this experiment, a lateral scanning range of 8 mm and a step size of 25 μm were selected. As shown in Fig. 11.24, the center axis of the fiber coil is affixed to the spindle of a lathe or a fiber-winding machine, rotating slowly for a complete rotation at a constant rate of 1 rpm, whereas the OCT probe beam is scanned back and forth rapidly along the *x* axis at a high speed (~100 mm/s). The slow rotation rate, combined with the high scanning speed, ensures that (1) the whole coil surface can be scanned when the coil completes a 360-deg rotation; (2) each cross-section view of the coil is in the *xz* plane with a negligible tilting angle of 0.12 deg, as shown in Fig. 11.15; and (3) the resolution of the resulting OCT image is not compromised.

Figure 11.24(b) shows a typical 2D tomographic image obtained by the setup, and Fig. 11.24(c) shows the reconstructed 3D OCT image when the fiber coil makes one complete rotation (360 deg). By using the setup and process described previously, the circular fiber coil can be viewed as a rectangular volumetric image after 3D reconstruction is performed, making it more convenient for the viewer to identify fiber-winding defects inside the fiber coil.

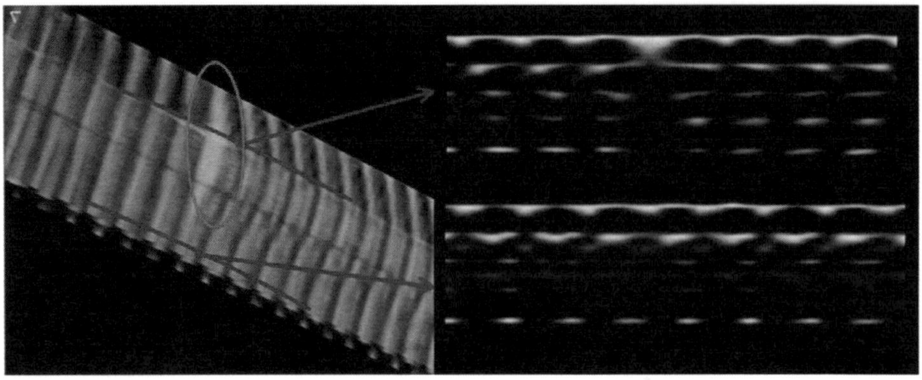

Figure 11.25 OCT image taken during the winding process of a four-layer fiber coil with potting adhesive, where the low-scattering fiber appears dark and the adhesive appears grey or white due to large scattering.

In Figs. 11.24(b) and (c), the dark areas are locations of optical fibers because of their weak scatting while the white areas are the interfaces of the fiber and air. In Fig. 11.24(c), a fiber climbing defect marked with an arrow and the nonuniform fiber turns marked with a circle are clearly identified.

Defects invisible to video inspection can be clearly seen by OCT tomographic images. For example, the nonuniform adhesive potting often occurs in a fiber coil, which may affect the thermal performance of the resulting FOG. Figure 11.25 shows an OCT image of a four-layer fiber coil with 200-μm coating fiber. At the location pointed with a white arrow, a defect of excessive adhesive can be clearly identified, because the fiber and the adhesive have different scattering coefficients and contrast sharply in OCT

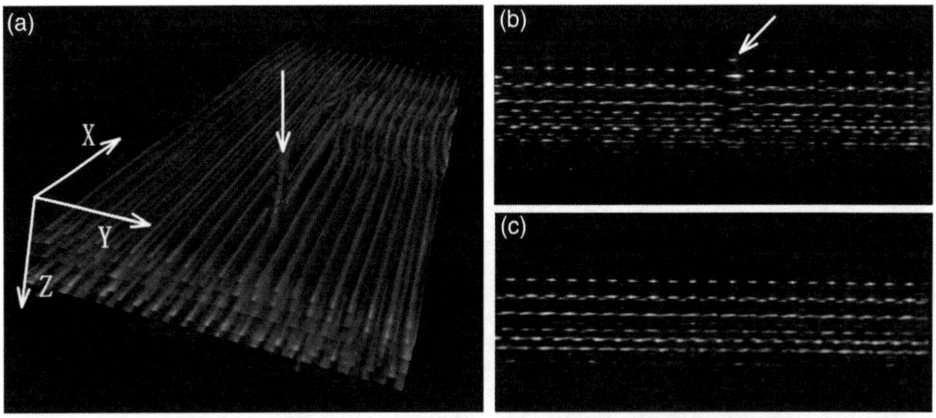

Figure 11.26 OCT tomographic image showing the fiber climbing defects. (a) The 3D tomographic view with the white arrow pointing at the climbed fiber. (b) The cross-section view showing the fiber climbing defect. C) The cross-section view of a good region without defects.

images. On the other hand, such a defect is not visible in a CCD image because the surfaces of both the fiber and the adhesive appear the same, and no information below the surface is available. Figure 11.26 shows the OCT images of a fiber coil with a fiber climbing defect. The image was taken while the coil was wound. The defects can be clearly seen, as indicated by the arrows.

Acknowledgement

The author would like to thank Suzhou Optoring Co., Ltd for providing some data used in this chapter and for productive collaborations. This work was supported by the internal research and development funding of General Photonics Corporation.

References

[1] H. C. Lefevre, "Fundamental of interferometeric fiber optic gyroscope," in *Fiber Optic Gyros: 20 Anniversary Conf., Proc. SPIE* **2837**, 46–60 (1996).
[2] H. C. Lefevre, The Fiber Optic Gyroscope, Artech House, Boston, 1993.
[3] F. Mohr, "Thermooptically induced bias drift in fiber optical Sagnac interferometers," *J. Lightwave Technol.* **14**(1), 27–41 (1996).
[4] G. A. Sanders, B. Szafraniec, R. Y. Liu, M. S. Bielas, and L. Strandjord, "Fiber-optic gyro development for a broad range of applications," *Proc.SPIE, Fbier Optic and Laser Sensors XIII* **2510**, 2–11 (1995).
[5] D. M. Shupe, "Thermally induced nonreciprocity in the fiber optic interferometer," *Applied Optics* **19**, 654–655 (1980).
[6] E. M. Troney, "Method of manufacture of quadrupole-wound fiber optic sensing coil," U. S. Patent 5351900 (1994).
[7] R. P. Goettsche and R. A. Bergh, "Trimming of fiber optic windng and method of achieving same," U.S. Patent 5528715 (1996).
[8] A. Kaliszek, B. Szafraniec, C. Lange, et al., "Low drift depolarizer for fiber optic gyroscope having legs wound in a winding pattern," US patent 6211963 (2001).
[9] T. De Fazio, K. Belsley, R. Smith, G. Shank, Jr., and W. Culver, "Development issues for automating quadrupole-pattern optical-fiber coil-winding for fiber-optic gyro manufacture," in *Proc. IEEE Int. Conf. Robot. Autom.* pp.202–207 (1994).
[10] M. Ivancevic, "Quadrupole-wound fiber optic sensing coil and method for manufacture thereof," U.S. Patent 4856900 (1989).
[11] Z. Ding, Z. Meng, X. Steve Yao, X. Chen, T. Liu, and M. Qin, "Accurate method for measuring the thermal coefficient of group birefringence of polarization-maintaining fibers," *Optics Letters*, **36**(11), 2173–2175 (2011).

[12] Z. Li, Z. Meng, X. Chen, T. Liu, and X. Steve Yao, "Method for improving the resolution and accuracy against birefringence dispersion in distributed polarization crosstalk measurements," *Opt. Lett.*, **37**, 2775–2777 (2012).

[13] X. Steve Yao, Z. Li, H. Chen, Z. Ding, Z. Meng, T. Liu, G. Li, Y. Su, and L. Wang, "Advances of distributed optical sensor systems for applications in industry and biomedicine" (Invited paper), SPIE Photonics Asia 2014.

[14] Z. Li, X. Steve Yao, X. Chen, H. Chen, Z. Meng, and T. Liu, "Complete characterization of polarization-maintaining fibers using distributed polarization analysis," *J. Lightwave Technol.* **33**(2), 372–379 (2015).

[15] Z. Li, Z. Meng, T. Liu, and X. Steve Yao, "A novel method for determining and improving the quality of a quadrupolar fiber gyro coil under temperature variations," *Optics Express*, **21**(2), 2521–2530 (2013).

[16] Z. Li, Z. Meng, L. Wang, T. Liu, and X. Steve Yao, "Tomographic inspection of fiber coils using optical coherence tomography," *IEEE Photonics Technology Letters* **27**(5), 549–552 (2015).

[17] B. Szafraniec and G.A. Sanders, "Theory of polarization evolution in interferometric Fiber-Optic Depolarized Gyros," *J. Lightwave Technol.* **17**(4), 579–590 (1999).

[18] X. Steve Yao, X. Chen, and T. Liu, "High accuracy polarization measurements using binary polarization rotators," *Optics Express*, **18**(7), 6667–6685 (2010).

[19] P. Martin, G. Le Boudec, and H. Lefevre, "Test apparatus of distributed polarization coupling in fiber gyro coils using white light interferometry," *SPIE Fiber optic gyros: 15th Anniversary Conference, Boston*, **1585**:173–179 (1991).

[20] M. Kemmler and H. Buschelberger, "White light interferometry for testing FOG components," *SPIE Fiber optic gyros: 15th Anniversary Conference, Boston*, **1585**:357–364 (1991).

[21] X. Chen and X. Steve Yao, "Measuring distributed polarization crosstalk in polarization maintaining fiber and optical birefringence material," U.S. Patent 8599385 (2013).

[22] F. Tang, X. Wang, Y. Zhang, and W. Jing, "Characterization of birefringence dispersion in polarization-maintaining fibers by use of white-light interferometry," *Applied Optics*, **46**(19), 4073–4080 (2007).

[23] F. Tang, X. Wang, Y. Zhang, and W. Jing, "Distributed measurement of birefringence dispersion in polarization-maintaining fibers," *Opt. Lett.*, **31**(23), 3411–3413 (2006).

[24] M. Tsubokawa, N. Shibata, T. Higashi, and S. Seikai, "Loss of longitudinal coherence as a result of the bierfringence effect," *J. Opt. Soc. Amer.*, **A4**, 1895–1901 (1987).

[25] T. Xu, W. Jing, H. Zhang, K. Liu, D. Jia, and Y. Zhang, "Influence of birefringence dispersion on a distributed stress sensor using birefringenct optical fiber," *Opt. Fiber Technol.*, vol. **15**, pp. 83–89 (2009).
[26] K. Mochizuki, Y. Namihira, and Y. Ejiri, "Birefringence variation with temperaure in elliptically caldded single-mode fibers," *Appl. Opt.*, **21**: 4223–4228 (1982).
[27] H. Zhang, G. Wen, Y. Ren, D. Jia, T. Liu, and Y. Zhang, "Measurement of beat length in polarization-maintaining fibers with external forces method," *Opt. Fiber Technol.*, **18**, 136–139 (2012).
[28] H-paremeter test method for Polarization-Maintaining optical fiber, TIA-455-192, 2005.
[29] Polarization crosstalk method for Polarization-Maintaining optical fier and components, TIA-455-193 (2005).
[30] A. Crant, P. Wysocki, and D. Holcomb, "Polarization effects in polarization-independent and polarization maintaining fiber amplifiers," presented at the IEEE/LEOS Summer Top. Metting, Quebec City, QC, Canada (2006).
[31] B. Szafraniec and G. Sanders, "Theory of polarization evolution in interferometeric fiber-optic depolarized gyros," *J. Lightwave Technol.*, **17**(4), 579–590 (1999).
[32] N. Gisin and B. Hutner, "Combined effects of polarization mode dispersion and polarization dependent losses in optical fibers," *Optics Communications*, **142**, 119–125 (1997).
[33] P. Lu, L. Chen, and X.Y. Bao, "Polarization mode dispersion and polarization dependent loss for a pulse in single-mode fibers," *J. Lightwave Technol.*, **19**(6), 856–860 (2001).
[34] X. Steve Yao, X. Chen, and T. Liu, "High accuracy polarization measurements using binary polarization rotators," *Optics Express*, **2010**, **18**(7), 6667–6685 (2001).
[35] A. Galtarossa, D. Grosso, and L. Schenato, "Distributed polarization-mode-dispersion measurement in fiber links by polarization-sensitive reflectometric techniques," *IEEE Photon. Technol Lett.*, **20**, 1944 (2008).
[36] N. Cyr, H. Chen, and G. Schinn, "Random-scrambling tunable pOTDR for distributed measurement of cumulative PMD," *J. Lightwave Technol.27*, 4164 (2009).
[37] M. Froggatt, B. Soller, D. Gifford, and M. Wolfe, in *Optical Fiber Communication Conference, Optical Society of Amerca, paper PDP17* (2004).
[38] X. Fan, Y. Koshikiya, and F. Ito, "Fully polametric phase-noise-compensated optical frequency domain reflectometry for distributed measurement of high-PMD fibers," *Opt. Lett.* **35** (1), 25–27 (2010).

[39] L. Palmieri, A. Galtarossa, and T. Geisler, "Distribued characterization of bending effects on the birefringence of single-mode optical fibers," *Opt. Lett.* **35**(14), 2481–2483 (2010).

[40] L Palmieri, "Distributed polarimetric measurements for optical fiber sensing" *Opt. Fiber Technol.* **19**, 720 (2013).

[41] A. Galtarossa, D. Grosso, and L. Palmieri, "Accurate characterization of twist-induced optical activity in single-mode fibers by means of polarization-sensitive reflectometry," *IEEE Photon. Technol. Lett.* **21**, 1713 (2009).

[42] C. Wei, H. Chen, X. Chen, D. Chen, Z. Li, and X. Steve Yao, "Distributed transverse stress measurement along an optic fiber using polarimetric OFDR," **41**(12), 2819–2822 (2016).

[43] R. Ulrich, S. C. Rashleigh, and W. Eickhoff, "Bending-induced birefringence in single-mode fibers," *Opt. Lett.* **5**, 273–275 (1980).

[44] A. Galtarossa, L. Palmieri, A. Pizzinat, M. Schiano, and T. Tambosso, "Measurement of Local Beat Length and Differential Group Delay in Installed Single-Mode Fibers," *J. Lightwave Technol.* **18**, 1389 (2000).

[45] Y. Shi, L. Yan, and X. Steve Yao, "Automatic maximum-minimum search method for accurate PMD and DOP characterization," *J. Lightwave Technol.* **24**(11), 4006–4012 (2006).

[46] C. Mao, Z. Gong, X. Mou, and G. Yang, "Finite difference method for temperature field in fiber optical gyroscope," *Piezoelectrics and Acoustooptics*, **25**, 98–101 (2003).

[47] D. Huang, E. Swanson, C. Lin, J. Schuman, W. Stinson, W. Chang, M. Hee, T. Flotte, K. Gregory, C. Puliafito, and J. Fujimoto, "Optical coherence tomography," *Science.* **254**, 1178–1181 (1991).

[48] W. Choi, B. Potsaid, V. Jayaraman, B. Baumann, I. Grulkowski, J. Liu, C. Lu, A. Cable, D. Huang, J. Duker, and J. Fujimoto, "Phase-sensitive swept-source optical coherence tomography imaging of the human retina with a vertical cavity surface-emitting laser light source," *Opt. Lett.* **38**, 338–340 (2013).

[49] G. Ughi, T. Adriaenssens, W. Desmet, and J. D'hooge, "Fully automatic three-dimensional visualization of intravascular optical coherence tomography images: methods and feasibility in vivo," *Biomed. Opt. Express.* **3**, 3291–3303 (2012).

[50] Z. Meng, X. Steve Yao, H. Yao, Y. Liang, T. Liu, Y. Li, G. Wang, and S. Lan, "Measurement of the refractive index of human teeth by optical coherence tomography", *J. Biomed. Opt.*, **14**, 304010 (2009).

[51] C. Blatter, J. Weingast, A. Alex, B. Grajciar, W. Wieser, W. Drexler, R. Huber, and R. A. Leitgeb, "In situ structural and microangiographic assessment of human skin lesions with high-speed OCT," *Biomed. Opt. Express.* **3**, 2636–2646 (2012).

[52] Y. Pan, H. Xie, and G.K. Fedder, "Endoscopic optical coherence tomography based on a microelectromechanical mirror," *Opt. Lett* **26**, 1966–1968 (2001).
[53] Y. Su, X. Steve Yao, Z. Meng, L. Wang, H. Yu, and T. Liu, "Effect of temperature on noninvasive blood glucose monitoring in vivo using optical coherence tomography," *Chinese Optics Letters* **12**, 111701 (2014).
[54] S. Lam, B. Standish, C. Baldwin, A. McWilliams, J. leRiche, A. Gazdar, A. Vitkin, V. Yang, N. Ikeda, and C. MacAulay, "In vivo optical coherence tomography imaging of preinvasive bronchial lesions," *Clinical Cancer Research* **14**, 2006–2011 (2008).
[55] S. Wolf and U. Wolf-Schnurrbusch, "Spectral-domain optical coherence tomography use in macular diseases: a review," *Ophthalmology* **224**, 333–340, (2010).
[56] R. Su, M. Kirillin, P. Ekberg, A. Roos, E. Sergeeva, and L. Mattsson, "Optical coherence tomography for quality assessment of embedded microchannels in alumina ceramic," *Opt. Express* **20**, 4603–4618 (2012).
[57] K. Fujiwara and O. Matoba, "High-speed cross-sectional imaging of valuable documents using common-path swept-source optical coherence tomography," *Appl. Opt.* **50**, H165–H170 (2011).
[58] D. Stifter, "Beyond biomedicine: a review of alternative applications and developments for optical coherence tomography," *Applied Physics B-Lasers and Optics* **88**, 337–357 (2007).
[59] P. Targowski and M. Iwanicka, "Optical coherence tomography: its role in the non-invasive structural examination and conservation of cultural heritage objects-a review," *Appl. Phys. A Mater. Sci. Process.* **106**, 265–277 (2012).
[60] T. Feng, Y. Shang, X. Wang, S. Wu, A. Khomenko, X. Chen, and X. Steve Yao, "Distributed polarization analysis with binary polarization rotators for the accurate measurement of distance resolved birefringence along a single mode fiber," *Opt. Express*, **26** (20), 25989–26002 (2018).

Chapter 12
A Personal History of the Fiber Optic Gyro

Eric Udd
Columbia Gorge Research, LLC

When I was a child, my father worked for the Naval Ordinance Plant in Forest Park, Illinois as an electrical engineer. He traveled frequently, flying out of Midway Airport or taking the train. For short trips during the winter, the train was almost as fast and more reliable. He traveled to Seattle, and these trips involved two or more stops to refuel. After O'Hare Airport opened, DC-6 and Lockheed Tri-Star planes began to support nonstop flights to Seattle with significant time savings. When 707s and DC-8s arrived, the long flights became much shorter.

My family would take our father to the airport to say goodbye and head to the observation deck to watch his plane take off. Plane delays because of "mechanical" problems were more often than not issues with mechanical gyros. These delays were substantial, and he would join us on the observation deck to pass the time. Because of weather conditions on these flights, safe, efficient travel depended heavily on the aircraft guidance system.

My first flight on an airplane was to Seattle in 1970 during a college Christmas break. It was a red eye on a 747 repositioning flight with about as many crew as passengers on board. Being a math and physics major, I thought at the time it would be wonderful to play a part in the design of this type of plane.

In 1977, I started working at McDonnell Douglas Astronautics Company (MDAC) in Huntington Beach, California after graduate work in the Aerospace and Mechanical Science Department at Princeton University. Having grown up during the 1960s, I was enthralled with anything to do with space flight and astronomy. I had two job offers out of grad school, one from Ford Aerospace and the other from MDAC.

While tempted by the slightly higher pay, discount car deals, and living in the Newport Beach, California location of Ford Aerospace, the visit to MDAC overwhelmed me. Stages of the Saturn V and Skylab artifacts were in the "bone yard" behind a test facility hanger. That hanger contained a shaker table capable of simulating severe launch conditions for the second stage of the Saturn V. Tests in process involved stages and fairings of the Delta and Titan rockets. A little later, we entered a long building. One section contained massive milling machines that were hogging out triangular patterns on huge sheets of aluminum. Further along the line, these were bent into cylinders and end caps for fuel tanks and outer shells for rocket stages. Another section of the same building contained Delta rockets in various stages of assembly. The rocket stages, fuel tanks, rocket engines, and inter-stages underwent final check out. The assembled rockets were then loaded by special overhead cranes onto oversize trailer trucks and shipped to either Cape Kennedy or Vandenberg AFB, depending on the customer-selected orbit. Adjacent buildings contained parts of the Space Shuttle solid rocket motors being refurbished, GaAs electronics being developed and assembled for space missions, and a MDAC payload assist module for launching satellites out of the Space Shuttle bay. Finally, we came to a 30-ft- (9-m) diameter man-rated vacuum chamber that could be pulled down to lower pressures than the atomic beam cell I spent months perfecting for my graduate research. It could reach 10^{-8} Torr in a few minutes while supporting sun, moon, and star simulators to test spacecraft. The small atomic beam cell I had built took several hours. The facilities and capabilities of the MDAC Electro-Optics Lab associated with the job included a variety of solar energy, launch vehicle, and defense projects. I accepted the position of associate engineer/scientist and started in early September 1977.

Partly because of excitement over my first "real" job and partly due to the work schedule I had become accustomed to at grad school, I made short work of my first assignment to build a handheld optical system to inspect cryogenic foam on the manufacturing floor. The first prototype was done after two weeks and transitioned the project to Al Pinkle, an optical technician, who was tasked to build ten more.

Richard Cahill, the manager of the Electro-Optics Lab, gave me a new assignment. The Guidance and Control Group had just completed a Delta redundant inertial measurement system (DRIMS) project to support Delta rocket launches. It used dry-tuned mechanical gyros and cost $500,000 per unit to build. The Hamilton Standard inertial measurement system it replaced cost MDAC $1,500,000. MDAC proposed that NASA fund the development, but they refused and instead suggested that MDAC fund development; the difference in cost would be split on subsequent launches. NASA included a provision in the final agreement that if another qualified inertial measurement unit became available at lower cost, it would have to be used. Since MDAC wanted to continue to enjoy the increased profit margin per launch, it was decided to increase awareness of next-generation rotation sensors. The Electro-Optics Lab was tasked with reviewing optical gyros. Cahill handed me this project. I looked at ring laser gyros (RLGs) and the recent work of Vali and Shorthill that demonstrated a FOG. Ring laser gyros had been under development for more than a decade, and while there had been significant progress, there were serious problems. Small amounts of backscatter off mirrors or impure gases caused severe lock-in, resulting in loss of signal. To overcome these and other problems, ultra-low-expansion ceramic blocks that required precision machining, high-level cleanrooms, and expensive subcomponents were necessary for assembly, driving up costs. The US Department of Defense (DoD) had invested hundreds of millions of dollars to enable a few companies to demonstrate production feasibility. Those companies had in turn invested heavily and were interested in supplying production inertial measurement units to as many applications as possible to recover their costs. MDAC decided not to make the investment necessary to build competitive RLGs. It seemed clear that the RLG manufacturers would not be interested in being suppliers for MDAC to build competitive inertial measurement units as this would increase their margins. The recent demonstration of the FOG had the potential to be implemented in solid state form. It could be scaled and configured into the shapes necessary for MDAC applications more easily than a RLG. After reporting to Cahill a week or so later, he remarked, "It is too bad the fiber gyro is nonlinear and sinusoidal. What's needed is a way to counterbalance rotationally induced phase shifts. But optics is reciprocal." He then walked away to discuss another project. I took this as a challenge to find a way to make the FOG "closed loop." A couple days later, I came up with the "dispersive laser gyro."

The idea was simple. The index of refraction of the optical fiber depends on the wavelength of light propagating through it. I knew I could change the wavelength of the light beam by introducing an acousto-optic modulator at one end of the fiber optic coil and using a first-order deflected light beam to cause a frequency shift. If I put the acousto-optic modulator at one end of the FOG coil, the light traversing in one direction, say clockwise, would have to be of a different optical path than the light beam traveling counter-clockwise. By having the two counter-propagating light beams traverse the same acousto-optic modulator in opposite directions, they arrive at the output detector at the same wavelength/frequency while the light beams are at different wavelengths in the fiber optic coil. Adjusting the frequency of the acousto-optic modulator in combination with the dispersion of the fiber coil allowed a nonreciprocal controlled phase shift. I had my notebook witnessed and at the first opportunity met with Cahill to go over my approach. He was excited and blocked out time for us to get together to work out the details. We soon discovered that the dispersion effect was much smaller than the phase shift due to the frequency difference of the light beams, and with our new disclosure the "phase-nulling laser gyro" was born.

Cahill and I continued to work on variations of implementing the closed-loop FOG and visited with George Finch, the chief McDonnell Douglas patent attorney based in Long Beach, California. Finch explained to us that McDonnell Douglas would not file for a patent until a working unit was built and a commercial market for the invention was defined. Significant commercial potential existed, so the main issue became the building of a working model.

The 1977 funds for FOGs were limited. With the help of a mechanical/optical technician, Bruce Turner, I began the design and assembly of a breadboard closed-loop fiber gyro. Fortunately, the Electro-Optics Lab and the abandoned former Douglas Research Lab located about 200 m away contained a significant selection of optical mounts and components that had supported prior projects. Turner and I searched what we needed, including a large aluminum plate. No one complained about the liberated aluminum plate, but procurement complained loudly about the Cadillac inner tube I bought on petty cash. I used it underneath the aluminum plate because early in the project we had limited air-table space. Cahill calmed the buyers for me, and I think later they were amused.

What I could not buy in 1977 I lined up so that orders could be placed as quickly as possible in 1978. Expediting these purchases meant walking it through the system and pleading with purchasing—the same people who objected to my inner tube purchase a few months before. This worked pretty well for some years, until major programs began to wonder why our purchase orders went through faster than theirs did.

By early 1978, we had solved a number of problems, including back reflection from the optical fiber ends. We built index-matching oil-filled reservoirs with angled windows on the ends. Microscope objectives were used to focus the light beams from the single-mode HeNe laser that served as a light source onto the fiber-coil ends. The fiber gyro coil used Valtec single-mode optical fiber with a 2-micron core. This made achieving good optical alignment a major challenge. The first configuration tested used an open-loop Sagnac interferometer configuration placed on a benchtop. This supported the characterization of noise sources. It became apparent early on that the Sagnac interferometer acted as an excellent acoustic sensor. After using a tuning fork as an acoustic source, it became clear that the ends of the fiber and the central beamsplitter had high sensitivity. Light feedback into the single-mode HeNe laser generated further acoustic signal amplification. These efforts resulted in our second patent filing on the Sagnac acoustic sensor.

After these initial tests, the closed-loop configuration of the first phase-nulling optical gyro used a flint-glass acousto-optic modulator as the frequency shifter. The breadboard mounted on the large aluminum plate and bolted to a guidance-and-control rate table served as the first optical demonstration model. In parallel, the electronics group built closed-loop circuitry to complement instruments already in place to support testing. The first test results were obtained in June 1978, and work on completing the utility patent for the phase-nulling optical gyro began in earnest. This filing contained design work on a 5-inch-diameter, 1-inch-thick all-solid-state phase-nulling optical gyro. (The guidance-and-control people wanted a 4-inch diameter, but we wanted 6, so the compromise resulted in the 5-inch design.) I worked on the optical design layout, Joe Kascin finished the final mechanical design, and Cahill contributed a number of design features, including sets of three in-line, angled wedges that helped with the final alignment of the optical beams into the ends of the fiber coil.

Two factors worked in favor of growing the MDAC fiber gyro group. The first involved a yearly review by the US Air Force (USAF) of all MDAC internal research and development programs. Grades were assigned, and the FOG project I presented received the highest score at the Huntington Beach facility. Overhead rates were set by the overall score received, and MDAC management was incentivized to see that the fiber gyro project received adequate funding. The second factor in favor of the fiber gyro group revolved around the USAF people reviewing internal research and development. They encouraged MDAC to visit Wright-Patterson AFB to explore the possibility of contracted research and development. As a result of these efforts, MDAC won the first competitive fiber gyro research and development contract from the USAF. The MDAC advantage of closed-loop operation enabled the prospect of linear output and wide dynamic range needed for long-range aircraft navigation.

In early 1979, the 5-inch-diameter fiber gyro using one of the first Hitachi single-mode laser diodes operated for the first time in an all-solid-state configuration. It quickly became apparent that achieving the 0.01 deg/h performance for inertial navigation would require significant improvements. The breadboard closed-loop fiber gyro had a noise floor of about 0.5 deg/s, and the 5-inch unit operated with a noise floor of about 10 deg/h with much higher bias drifts. Turner and I had spent about six weeks working out the processes necessary to align the solid state package and complete the final assembly. This included some modifications and configuration changes when one of the two acousto-optic modulators we planned to use failed; we resorted to a single modulator layout. With the unit fully assembled, tests began. To our horror, the $3000 Hitachi laser diode failed about 30 minutes into inertial testing, with the laser diode power dropping dramatically. Moments later, after some gain adjustments with our readout and data-recording equipment, we discovered that the SNR of the output remained about the same. It became clear that the performance of our 5-inch solid state fiber gyro with the Hitachi single-mode laser diode closely matched that of the version operating as a much-lower-power edge-emitting diode. It quickly became clear to us that coherent backscatter noise fell dramatically when the light source moved from a single-mode laser to LED operation. In our case, the net SNR came very close to not changing even though the power dropped from a few mW to tens of μW. We started to look for low-coherence light sources.

In parallel with MDAC efforts on early solid state fiber gyros, a series of important events were occurring. Reinhold Ulrich in Germany worked out the importance of polarization control and the need to place a high-quality polarizer between two beamsplitters. We closely followed his work, and when he asked to visit our laboratory, Cahill agreed and initiated the necessary paperwork. The visit lasted a few hours and included a tour of some of the early solid state fiber gyros and discussions on fiber gyro operations. Professor Herbert Shaw began his work at Stanford, and a number of other groups began looking into open-loop configurations, emphasizing noise limitations and performance issues.

During this time, the Naval Research Lab (NRL) focused heavily on fiber optic acoustic sensors based on the Mach–Zehnder interferometer. In 1979, Jack Donovan of NRL organized a fiber optic sensor workshop at Fort Eustis, inviting people working in the US. The workshop was dominated by discussions of fiber acoustic sensors led by Jack Donovan and Tom Giallorenzi of NRL and supported by Tony Daindridge, Jim Cole, and others. Victor Vali (University of Utah) and I (MDAC) presented our fiber gyro work. During one session on fiber sensor markets, I argued that the fiber gyro market would be larger than the fiber optic acoustic sensor market; Donovan was adamant that the fiber optic acoustic sensor market would be much larger. Looking back in 2018, he might have been partly right if I ignore that he was referring to Mach–Zehnder-based fiber optic acoustic sensors, which have evolved into Michelson interferometers that include fiber grating reflective elements. It is clear now that both markets are significant; which is larger and by how much is open to debate.

Of the 50 or so people attending the Fort Eustis workshop, Shaw and Vali are the only two I identified with who had interest in fiber gyros. There may have been others. Vali sought me out during breaks, and we spent quite a bit of time talking about our experiences building early open and closed loop fiber gyros as well as the theory behind their operation.

After the meeting, NRL awarded two contracts for development of fiber optic acoustic sensors to MDAC and Gould. Richard Cahill invented a superior read out system for the Mach-Zehnder interferometer that offered the Navy the wide dynamic range and linearity needed. Gould offered an alternative approach based on their unique capability associated with packaged biconical tapered fiber couplers. Vince Tekippe led the Gould effort. MDAC needed packaged fiber beamsplitters for both the fiber gyro projects and the fiber acoustic sensor projects. NRL had done some work but refused to provide any advice and Gould was a direct competitor. I decided to see what Bruce Turner and I could do with a butane torch and a series of simple pulling methods. After a couple days of effort produced only a few marginal beamsplitters and a wastebasket full of failed attempts, it became apparent that a serious effort would be needed to achieve consistent success.

A program was initiated led by Tim Clark and a senior MDAC mechanical engineer. About $200,000 and a year later, MDAC produced consistent, high-quality packaged fiber couplers to support both rotation and acoustic sensing programs.

After completion and demonstration of the 5-inch fiber gyro in early 1979, the Guidance and Control group wanted to have a much smaller package with a new target of 2.5 inches in diameter for 1980. Integrated optic chips and all-fiber components with the necessary size and performance were not available to MDAC at that time. Instead, I looked at using an acousto-optic modulator as a beamsplitter and a frequency shifter. Because of the small angle between the frequency-shifted and unshifted light beams, the placement of two fibers side by side in a single ferrule and with a reasonable-focal-length lens enabled this approach to work. We used precision stages for linear and rotational alignment that supported an ultra-small Mach–Zehnder readout system for the parallel fiber optic acoustic sensor project. Paul Theriault managed the final assembly with some alignment-system upgrades. Another innovative feature incorporated into the 2.5-inch package employed a small rutile crystal with its birefringent axes aligned at 45 deg to the central polarizer. By using a broadband light source, effective polarization scrambling avoided signal-level variations due to the changing birefringence in the fiber coil.

In parallel with the 2.5-inch fiber gyro effort, another internal customer wanted a fiber gyro that would fit into a 3-inch-long, 1-inch-square space. Wil Otaguro, a senior optical engineer, joined the group on a temporary basis and led its construction.

Several months after the 3-inch-long, 1-inch-square elliptical-coil fiber gyro demonstration ended, a group of engineers visited from Develco looking at the possibility of developing an oil-drilling tool using fiber gyros. They were shown a series of fiber gyros but stopped when shown the small elliptical-coil unit. Shortly thereafter, discussions began on the development of dual in-line fiber gyros that could be used to measure two axes of the earth rotation rate with 0.1-deg/h accuracy. The objective was to accurately measure true north without the limitations associated with measuring magnetic fields. The final qualification test included 30 minutes on a paint shaker.

The fiber sensor group I led continued to grow as more people joined it. Ray Wagoner turned out to be one of the most promising, with considerable skills at running a project. He managed the Develco oil-drilling project, which required several innovations, including a flipping mechanism that enabled consistent 0.1-deg/h accuracy. Thanks to expertise in fabricating electronics for space launches, MDAC knew how to produce hardened fiber gyro packages and electronic subassemblies.

The unit operated successfully at an oil show in Houston in 1982. It had performance far ahead of any comparable unit at that time. While other laboratories were still in the breadboard stage, the MDAC fiber gyro oil-drilling-tool prototype had withstood severe environmental testing and had been successfully demonstrated at a major oil show in Houston.

In parallel with the FOG oil-drilling tool in 1982, work continued with medium-grade FOGs for aerospace applications. A FOG usable on an attitude heading reference system (AHRS) was developed and evaluated by another commercial aerospace company. Eglin AFB funded work on a fiber gyro with performance characteristics associated with usage in a tactical missile. The Eglin AFB fiber gyros were completed in 1983.

Most universities and government laboratories in the early-to-mid-1980s focused on seeking the lowest possible bias drift and the highest sensitivity working with open loop configurations. In the US, Stanford worked closely with Litton on the development of an all-fiber configuration. The Naval Research Lab focused on a series of hero experiments pushing bias drift and sensitivity limits of open loop fiber gyros. Commercial companies realized that closed loop implementations would be necessary, and efforts in this area were made by Sperry, Litton, British Aerospace, Thompson CSF, Hitachi, Japan Aviation Electronics, and Mitsubishi. Hervé Arditty and Hervé LeFèvre at Thompson CSF championed the usage of integrated optic modulators and a minimum-configuration approach. In this case, the integrated optic chip contained all of the key central components of the closed-loop fiber gyro with an external light source, fiber coil, and detector.

In 1985, Glen Ramer and Emery Moore decided to turn over their Fiber Optic and Laser Sensor series of SPIE conferences they started in 1983 to Ramon DePaula and me. In conjunction with this effort, I decided to put together a Fiber Optic Gyros 10th Anniversary Conference[1] and proceeded to invite everyone I knew, including Vali, Bill Goss at NASA (who had sponsored Vali's work), Shaoul Ezekial at MIT, Arditty and LeFèvre, George Pavlath (who had assumed the lead at Litton Guidance and Control), and many more. To my surprise, they all accepted. The papers associated with that conference provide a good historical overview of progress made over the first decade of serious fiber gyro development. One of the events associated with that conference was a panel discussion I led consisting of Bill Burns of NRL, Pavlath, Ezekial, Arditty, and a few others. One of the issues that came up involved the relatively modest funding associated with fiber gyro development. For example, McDonnell Douglas had won all of the USAF contracts, but their total value was on the order of two million dollars. Draper Labs was awarded a large contract for twenty million by DARPA to develop a fiber optic rotation sensor and announced they were focusing their effort on passive ring resonators. I found this appalling as the passive ring resonator (at least as currently known) was inherently nonreciprocal and in my opinion would never have the required level of performance. I posed this to the panel, and many members expressed similar sentiments. We did not expect the Draper Labs effort on ring resonators to succeed (years later, after most of the funds were spent, they redirected their work to fiber gyros).

Significant progress had been made by Litton/Stanford and Thompson CSF moving FOG technology into a more reasonable prototype form.

McDonnell Douglas continued to win USAF contracts to develop tactical- and navigation-grade fiber gyros for Eglin AFB and Wright-Patterson AFB, respectively. Key guidance and control companies in the US started to lobby McDonnell Douglas management to license its closed-loop fiber gyro technology. They argued that as traditional suppliers to McDonnell Douglas they were best suited to carry on development. In 1986, McDonnell Douglas management asked for a budget to move the existing fiber gyro prototypes into a form necessary for production. The initial estimate was ten million dollars, which went up and down the management chain twice before a decision was made to license the technology. By 1987, the first licenses were completed to US companies, and worldwide licenses followed. McDonnell Douglas completed its efforts on fiber gyro prototypes with delivery of three tactical-grade fiber gyros with dimensions of $2 \times 2 \times 4$ inches to support a 4-inch-cube inertial measurement unit and the delivery of a navigation-grade FOG to Wright-Patterson AFB employing a novel configuration to support accurate wavelength measurements for scale-factor correction using frequency switching of four acousto-optic modulators.

Internal McDonnell Douglas resources supporting the fiber gyro effort were redirected to a Sagnac-interferometer-based secure fiber optic communication system and the emerging field of fiber optic smart structures, significantly enhancing the efforts of both of these programs. Efforts on Sagnac-interferometer-based sensors were largely directed at acoustic and distributed sensing in conjunction with the smart structure program. A notable exception was in an effort with the Jet Propulsion Lab to look at very-long-gauge-length strain sensors based on the Sagnac strain sensor to detect seismic signals using telecommunication cable. This early design effort ended with the adoption of GPS-based systems.

In 1991, the Fiber Optic Gyros: 15th Anniversary Conference[2] was held in Boston. I invited Ezekial to chair this conference with me and invited several companies that were making serious efforts to bring the FOG into production. Pavlath presented the effort of Litton Systems in the US; Litton had become the leader in the US in undertaking serious development efforts and had won a series of USAF contracts. Papers on production efforts in Japan were given by K. Sukuma of Japan Aviation Electronics, H. Kajioka of Hitachi, and Y. Hayakawa of Mitsubishi. Each of these groups had produced prototypes, with JAE showing units intended to support rockets, AHRSs for aircraft, and a high-speed-train position system. Hitachi showed high-, medium-, and low-grade navigation systems and a borehole prototype system with performance characteristics similar to the 1982 system with a number of design improvements. Mitsubishi reported on a series of both open- and closed-loop fiber gyros to support inertial navigation applications. Three fiber optic production efforts from Europe were presented: LeFèvre described the FOG 80 and FOG 50 prototypes; Alan Malvern of British Aerospace showed the Mark IIIa engineering demonstrator with electronics in a 76 × 76 × 80-mm package; and Wilfred Auch at Alcatel Standard Electric Lorenz overviewed early production of the PM1DE-FOG, with 40 units produced in 1989, followed by 80 units per year in 1990 and 1991. He announced plans for the production of 1000 units in 1992.

In 1993, I left McDonnell Douglas and started Blue Road Research, Inc. in Troutdale, Oregon in an effort to move closer to my wife's parents who both had serious medical problems. McDonnell Douglas provided licenses to Blue Road Research to all my fiber sensor patents, with the exception of those related to FOGs that were licensed worldwide and the Sagnac secure fiber optic communication system that was in the process of being certified under the NSA COMSEC Commercial Endorsement Program. In terms of the Sagnac interferometer, this left patents related to acoustic sensing, strain sensing, and distributed sensors. In the early years of Blue Road Research, I did some consulting work for US companies interested in entering the field, and in 1996 pulled together my last major conference devoted exclusively to fiber gyros—the Fiber Optic Gyros: 20th Anniversary Conference[3] with LeFèvre and Kazuo Hotate. By this time, the FOG had moved from early production to full-scale production at a number of companies, and the application areas had radically expanded. In the US, Litton and Honeywell had emerged as the primary contenders to supply fiber gyros for major aerospace platforms. Litton continued to win military contracts for a variety of platforms. Honeywell efforts were directed toward commercial applications with a major win of a FOG–based backup navigation system for the Boeing 777. Honeywell described plans for ship and satellite stabilization applications and a small tactical inertial measurement unit. Hitachi had built a factory able to support the manufacture of up to 5000 fiber gyros per month to support automobile navigation systems, commercial vehicle location, trucks, trains, robots, tunneling operations, ships, and autonomous vehicles. JAE described a series of aerospace applications and several unexpected efforts that included auto guidance for tractors working in fields, cleaning robots for malls, soccer grass-cutting systems, autonomous golf-ball-collecting carts, antenna stabilization, camera stabilization, and unmanned helicopters supporting applications such as crop spraying. Other companies describing production efforts included LITEF, Andrew Corporation, and Daimler-Benz Aerospace.

From 1997 to 1999, Blue Road Research supported a venture-funded startup COMTEL that had bought rights to the Sagnac secure fiber optic communication system from McDonnell Douglas. These efforts included expanding loop lengths from 10 km to 50 km and increasing bandwidth from 100 MBPS to 1 GBPS. Sagnac-interferometer sensor investigations were largely limited to acoustic and distributed sensing as Blue Road Research focused on fiber-grating-based applications.

In January 2006, I left Blue Road Research and started to work full time at my new startup, Columbia Gorge Research. My efforts on fiber gyros at Columbia Gorge Research have been limited to consulting efforts. Parallel efforts on the Sagnac interferometer were focused on intrusion detection and the potential to use Sagnac interferometers in oil and gas applications.

A special 30th Anniversary Fiber Optic Gyro session in 2006, organized by Paul Ruffin, was held in conjunction with the SPIE Photorefractive Fiber and Crystal Devices: Materials, Optical Properties and Applications XII Conference.[4] Pavlath, Ezekial, and I gave overview papers to support a series of papers on components and fiber gyro systems.

In 2016, I put together a 40th Anniversary Special Session in conjunction with the Fiber Optic Sensors and Applications XIII Conference[5] consisting of ten invited papers by the authors of the chapters in this book. Their efforts inspired this book, which has been expanded to update and more thoroughly

Figure 12.1 Informal photo courtesy of Herve LeFèvre at the 40th anniversary of the FOG session. From left: Glen Sanders (MIT, Honeywell), Eric Udd (McDonnell Douglas, Blue Road Research, Inc., Columbia Gorge Research LLC), Ralph Bergh (Stanford, Fibernetics LLC), Herve LeFèvre (Stanford, Thompson CSF, Photonetics, iXBlue), Michel Digonnet (Stanford-Ginzton Lab), and George Pavlath (Stanford, Litton, Northrup Grumman).

cover the field that has matured over more than 40 years. Fiber gyros based on Litton/Northrup Grumman designs were reported on all Mars rovers, including Opportunity, Spirit, and Curiosity. Hundreds of thousands of fiber gyros had been built and deployed by companies worldwide on aircraft, rockets, spacecraft, satellites, ground receivers, ships, submarines, boring tools, robots, and land vehicles. Figure 12.1 shows a group of former fiber gyro competitors at the 40th Anniversary Session in Baltimore in April 2016.

Manufacturing improvements continue at many companies as we move toward 2020 to reduce cost and size. Research efforts on new, higher-performance approaches to FOGs are ongoing at universities and companies. The FOG has found its place as one of the most successful fiber optic sensors. I believe it still has many years left before it begins to reach its full potential.

References

[1] E. Udd, Ed., *Fiber Optic Gyros: 10th Anniversary Conference, Proc. SPIE* **719**, SPIE Press, Bellingham, WA (1986).

[2] S. Ezekial and E. Udd, Eds., *Fiber Optic Gyros: 15th Anniversary Conference, Proc. SPIE* **1585**, SPIE Press, Bellingham, WA (1991).

[3] E. Udd, H. LeFèvre, and K. Hotate, *Fiber Optic Gyros: 20th Anniversary Conference, Proc. SPIE* **2837**, SPIE Press, Bellingham, WA (1996).

[4] F. Yu, R. Guo, and S. Yin, Eds., *Photorefractive Fiber and Crystal Devices: Materials, Optical Properties and Applications XII Conference, Proc. SPIE* **6314**, SPIE Press, Bellingham, WA (2006).

[5] E. Udd, G. Pickrell, and H. Du, *Fiber Optic Sensors and Applications XIII, Proc. SPIE* **9852**, SPIE Press, Bellingham, WA (2016).

Eric Udd, President of Columbia Gorge Research LLC, has been deeply involved with fiber optic sensors since 1977 and helped pioneer early work on fiber optic gyros, fiber optic smart structures for health monitoring, multi-axis strain sensors, high-speed fiber optic sensors systems, and fiber optic pressure and temperature sensors. He worked for McDonnell Douglas from 1977 to 1993, where he managed over 25 DOD, NASA, and internally funded fiber optic sensor programs. He held a series of positions, moving from Engineer/Scientist to Fiber Optics Manager, and in 1987 was appointed as one of 40 McDonnell Douglas Fellows. He invented, designed, and built the first closed-loop fiber gyros. Udd's work on fiber optic smart structures resulted in the instrumentation of the composite hydrogen tank of the Delta Clipper to measure strain and temperature to assess its condition after each flight. The success of the Delta Clipper inspired Spacex and Blue Origin to further develop vertical landings and reusable launch vehicles. In 1993, he started Blue Road Research, Inc. (BRR) and directed the growth of the company through its acquisition by Standard MEMS in January 2000. In January 2006, he left BRR to work full time at his new company, Columbia Gorge Research LLC (CGR). His efforts at BRR and CGR have involved applications in aerospace, defense, civil structures, composite manufacturing, robotic surgery, oil and gas field services, electrical utilities, and energetic material.

Udd has 54 issued US patents with many issued internationally, written over 200 papers, and chaired and helped organize 39 international conferences on fiber optic sensor technology, including the 10th, 15th and 20th Fiber Optic Gyro Anniversary Conferences. His books include *Fiber Optic Sensors: An Introduction for Engineers and Scientists, Second Edition* (Wiley, 2011), *Fiber Optic Smart Structures* (Wiley, 1995), and *Field Guide to Fiber Optic Sensors* (SPIE Press, 2014). He is a Fellow of SPIE and OSA, and a member of IEEE. He was awarded the Richardson Medal in 2009 by the Optical Society for his work on fiber optic sensors and the field of fiber optic smart structures.

Appendix
Additional Fiber Rotation Sensor Books, Papers, and Patents

A.1 Fiber Optic Rotation Sensor Contents in Books and Paper Collections

This section briefly reviews collections of papers associated with conferences and selected books with fiber optic rotation sensor content. The first major conference on fiber optic rotation sensors[1] took place in November 1981 in Cambridge, MA. Shaoul Ezekial and Hervé Ariditty acted as chairs. Coincidently, it took place about 5 years after Vali and Shorthill demonstrated the first open-loop fiber gyro. This conference emphasized the theory and operation of the fiber gyro, the components needed to support it, and early fiber gyro hardware. I organized and chaired the Fiber Optic Gyros: 10th Anniversary Conference,[2] which also took place in Cambridge in September 1986. I extended many invitations for papers and presentations and was pleased everyone accepted. Some papers were unsolicited. Robert Smith of Honeywell called and asked me if a bibliography of fiber gyro papers would be appropriate, and I told him it would be most welcome. After the conference, we met and talked about the possibility of doing a collection of important fiber gyro papers, and I encouraged him to do so. The result[3] is an SPIE Milestone volume published in 1988 that includes a bibliography of over 600 papers. The 1986 version had a bit over 400. The pace of publications picked up. I decided to organize a 15th anniversary conference,[4] held in September 1991 in Boston, and asked Shaoul Ezekial to chair it with me. A few companies had moved fiber gyros into early production prototypes. Also in 1991, my first book[5] came out with a chapter describing fiber rotation sensors, along with other chapters describing the components needed to realize a fiber gyro. A second edition of the book came out in 2011; other materials and chapters were updated, but the fiber gyro chapter remained unchanged as fundamentals still stood. Hervé Lefèvre came out with his book[6] in 1993 on the fiber gyro. This book, which now has a second edition, is a comprehensive overview of the theory and operation of the fiber gyro. It is the only one of its type by a single author. William Burns in 1994 edited a competitive book[7] asking leading experts worldwide to contribute chapters. This book goes into the state of the art at the time and includes a nice chapter on "industry reports." In many ways, the books[6,7] were complementary rather than competitive. By the 20th anniversary of the fiber gyro[8] (held in Denver in August 1996), production had begun to support autos, remotely piloted helicopters, aircraft, and robots. Overview production papers were presented from companies in the US, Japan, and Europe. Additional anniversary sessions were held at the 25th, 30th, 35th, and 40th anniversary years.[9–12] This book contains the content of the 40th year with, in many cases, greatly extended contributions and new chapters. Competition from MEMS-based gyros pushed the fiber gyro away from such mass markets as automobiles into higher-performance markets, which now include undersea vehicles, surface ships, military and commercial aircraft, satellites, ground tracking stations, launch vehicles, and rockets. Many of these applications are described in this book. To follow a particular development path associated with a specific company or technology, this collection of papers and books offers a good start.

References

[1] S. Ezekiel and H. J. Arditty, Eds., *Fiber Optic Rotation Sensors*, Springer-Verlag, New York (1982).

[2] E. Udd, Ed., *Fiber Optic Gyros: 10th Anniversary Conference, Proc. SPIE* 719, SPIE Press, Bellingham, WA (1986).

[3] R. B. Smith, Ed., *Selected Papers on Fiber Optic Gyros*, SPIE Milestone Series MS8, SPIE Press, Bellingham, WA (1989).

[4] S. Ezekiel and E. Udd, Eds., *Fiber Optic Gyros: 15th Anniversary Conference, Proc. SPIE* 1585, SPIE Press, Bellingham, WA (1991).

[5] E. Udd, "Fiber Optic Sensors Based on the Sagnac Interferometer and Passive Ring Resonator," in *Fiber Optic Sensors: An Introduction for Engineers and Scientists*, Wiley, New York (1991).

[6] H. Lefèvre, *The Fiber Optic Gyroscope*, Artech House, Boston (1993).

[7] W. K. Burns, Ed., *Optical Fiber Rotation Sensing*, Academic Press, Boston (1994).

[8] E. Udd, H. C. Lefèvre, and K. Hotate, Eds., *Fiber Optic Gyros 20th Anniversary Conference, Proc. SPIE* 2837, SPIE Press, Bellingham, WA (1996).

[9] Special 25th FOG Anniversary Session in OFS-15, 2002.

[10] Special 30th FOG Anniversary Session in OFS-18, 2006.

[11] Y. Liao, et al., Eds., *OFS2012 22nd International Conference on Optical Fiber Sensors, Proc. SPIE* 8421, SPIE Press, Bellingham, WA (2012).

[12] E. Udd, G. Pickrell, and H. H. Du, Eds., *Fiber Optic Sensors and Applications XIII, Proc. SPIE* 9852, SPIE Press, Bellingham, WA (2016).

A.2 Accessing the Fiber Optic Rotation Sensor Patent Literature

There is a second important avenue available to a FOG researcher or historian that offers a significant body of information that is often overlooked. Many business organizations do not allow their researchers to publish until patents are filed (and, in some cases, issued). As a result, the patent literature offers an important window into the technical history of the fiber gyro and the status of development at any given time. There are many places to start, and the number of patents on this topic is quite large; however, it can be interesting to research patents. They are free to download from www.uspto.gov or from Google patents. As an example, I have included a list of my patents on fiber gyros and derivative inventions[1–23] from work performed on fiber gyros. The new types of Sagnac interferometer sensors or, in one case, the Sagnac communication system often fell directly out of working on improving rotation sensing. It might be instructive to look at an early patent reference: the original closed-loop fiber gyro patent.[23] I collected and organized my thoughts and those of Richard Cahill in writing the body (specification) for this patent over about a three month period, spending more time than any other patent I have written (in part because it was my first patent application and also in an effort to cover a range of approaches). It collects our designs that were in place just prior to filing and the results of testing the original closed-loop fiber gyro breadboard (McDonnell Douglas at the time would not file without a working model). When the patent was filed, the US patent department had an unlimited amount of time to introduce new material that challenged the patent as it was not a "for profit" center (which happened during the Reagan administration later in the 1980s), so the examiner introduced a series of early patents. Victor Vali, Richard Shorthill, Raymond Goldstein, and Reuben Krogstad did have a patent[24] on a fiber gyro. John K. Corbin acted as the primary examiner and cited two patents by A. Wallace and J. B. Speller.[25,26] These patents have figures that have strong similarity to the open-loop fiber gyro demonstrated by Vali and Shorthill in 1976. The Wallace case was filed on July 9, 1958, and the Speller case on June 28, 1962. The same John K. Corbin acted as primary examiner for our closed-loop fiber gyro filing.[23] The list of prior US patents cited expanded from 3 in the Vali et al. filing[24] on June 11, 1975 to 14 US patents and 2 German patents with our case filed on December 7, 1978. Many of the new patents filed preceded the Vali et al. filing. By the time our patent was allowed and I had looked at all of the prior art cited by Corbin and his assisting examiner, Bruce Y. Arnold, it became clear that there were a number of candidates for the inventor of the most basic fiber optic gyro. Using the first hardware demonstration of the fiber optic gyro by Vali and Shorthill seems to me to be a good starting point for marking anniversaries. Making hardware work properly is quite different from a paper design.

One can go backward or forward from any patent. Looking at patents cited as prior art on the patent allows one to go back to more fundamental inventions. Looking for patents that have cited the patent allows one to move forward in time as inventions intended to improve performance, reduce cost, or enable other advantages. While I have included my patents here as an example, patents by any developer of fiber gyro technology could serve as well. The reader would have to look no further than the authors of the chapters in this book to find them.

References

[1] E. Udd, et al, "Single fiber Sagnac com system," US patent 6690890 (Feb 10, 2004).
[2] E. Udd and Mike Morrel, "Single fiber Sagnac system," US patent 6459486 (Oct 1, 2002).
[3] E. Udd, "Coherent Alarm System," US patent 5694114 (Dec 2, 1997).
[4] E. Udd, "Sagnac Michelson sensing system," US patent 5636021 (June 3, 1997).
[5] E. Udd, "Secure fiber optic secure com alarm system," US patent 5455698 (Oct 3, 1995).
[6] E. Udd and P. Theriault, "Fiber optic measuring apparatus," US patent 5446533 (Aug 29, 1995).
[7] E. Udd and S. Higley, "Secure for networks," US patent 5422772 (Jun 6, 1995).
[8] E. Udd, "Distributed sagnac sensor systems," US patent 5402231, (Mar 28, 1995).
[9] E. Udd, "Sagnac interferometer based secure communication system," US patent 5311592 (May 10, 1994).
[10] E. Udd, "Secure fiber optic secure com system," US patent 5274488 (Dec 28, 1993).
[11] E. Udd, "Secure fiber optic secure com system," US patent 5223967 (Jun 29, 1993).
[12] E. Udd, "Optic detection system using a Sagnac interferometer," US patent 5046848 (Sep 10, 1991).
[13] E. Udd, "Sagnac distributed sensor," US patent 4976507 (Dec 11, 1990).
[14] E. Udd, "Sagnac distributed sensor," US patent 4898468 (Feb 6, 1990).
[15] R. Michal, E. Udd, and R. F. Cahill, "Reciprocally switched four modulator system," US patent 4875775 (Oct 24, 1989).
[16] R. Michal, E. Udd, and R. F. Cahill, "Optical fiber sensing systems having acouto optical modulation and deflection devices," US patent 4789241 (Dec 6, 1988).
[17] E. Udd, et al., "Fiber optic sensor," US patent 4787741 (Nov 29, 1988).
[18] R. Cahill and E. Udd, "Vibrating optic fiber accelerometer and gyro," US patent 4755051 (Jul 5, 1988).

[19] E. Udd, et al, "Micro-bending accelerometer and gyro," US patent 4743116 (May 10, 1988).
[20] R. Cahill, E. Udd, et al., "Fiber optic earth rotation sensor," US patent, 4712306 (Dec 15, 1987).
[21] R. Cahill and E. Udd, "Compact optical gyro," US patent 4588296 (May 13, 1986).
[22] R. Cahill and E. Udd, "Optical acoustic sensor," US patent 4375680 (Mar 1, 1983).
[23] R. Cahill and E. Udd, "Phase nulling optical gyro," US patent 4299490 (Nov 10, 1981).
[24] V. Vali, R. W. Shorthill, R. Goldstein, and R. Krogstad, "Laser Gyroscope," US Patent 4,013,365 (Mar 22, 1977).
[25] A. Wallace, "Electromagnetic wave gyroscopes or angular velocity measuring systems," US Patent 3,102,953 (Sep 3, 1963).
[26] J. B. Speller, "Relativistic inertial reference device," US Patent 3,395,270 (Jul 30, 1968).

Index

O

P

Q

R

S